The Fallen Sky

The Fallen Sky

An Intimate History *of* Shooting Stars

Christopher Cokinos

Jeremy P. Tarcher/Penguin
a member of Penguin Group (USA) Inc.
New York

JEREMY P. TARCHER/PENGUIN
Published by the Penguin Group
Penguin Group (USA) Inc., 375 Hudson Street, New York, New York 10014, USA •
Penguin Group (Canada), 90 Eglinton Avenue East, Suite 700, Toronto, Ontario M4P 2Y3, Canada
(a division of Pearson Canada Inc.) • Penguin Books Ltd, 80 Strand,
London WC2R 0RL, England • Penguin Ireland, 25 St Stephen's Green, Dublin 2,
Ireland (a division of Penguin Books Ltd) • Penguin Group (Australia), 250 Camberwell Road,
Camberwell, Victoria 3124, Australia (a division of Pearson Australia Group Pty Ltd) •
Penguin Books India Pvt Ltd, 11 Community Centre, Panchsheel Park, New Delhi–110 017, India •
Penguin Group (NZ), 67 Apollo Drive, Rosedale, North Shore 0632, New Zealand
(a division of Pearson New Zealand Ltd) • Penguin Books (South Africa)
(Pty) Ltd, 24 Sturdee Avenue, Rosebank, Johannesburg 2196, South Africa

Penguin Books Ltd, Registered Offices: 80 Strand, London WC2R 0RL, England

Published simultaneously in Canada

Page 503 constitutes an extension of this copyright page.

Library of Congress Cataloging-in-Publication Data

Cokinos, Christopher.
The fallen sky : an intimate history of shooting stars / Christopher Cokinos.
p. cm.
ISBN 978-1-58542-720-8
1. Meteorites—Miscellanea. I. Title.
QB755.C65 2009 2009017493
523.5'1—dc22

Printed in the United States of America
1 3 5 7 9 10 8 6 4 2

Book design by Amanda Dewey

Then Ninsun, who is well-beloved and wise, said to Gilgamesh, "This star of heaven which descended like a meteor from the sky; which you tried to lift, but found too heavy, when you tried to move it it would not budge, and so you brought it to my feet; I made it for you, a goad and spur, and you were drawn as though to a woman."

—*The Epic of Gilgamesh,* translated by N. K. Sandars

Catch a falling star an' put it in your pocket,
Never let it fade away!
Catch a falling star an' put it in your pocket,
Save it for a rainy day!

For love may come an' tap you on the shoulder,
Some star-less night!
Just in case you feel you wanna' hold her,
You'll have a pocketful of starlight!

— Paul Vance and Lee Pockriss, "Catch a Falling Star"

But a lifetime burning in every moment
And not the lifetime of one man only
But of old stones that cannot be deciphered.

—T. S. Eliot, "East Coker," *Four Quartets*

For Kathe,
my starlight

Contents

Introduction

On any clear night, under a dark enough sky, we can see shooting stars. We wish upon them, even if we don't quite know what they are—of course they're not really stars—and even if we don't know where they come from or what they might tell us about the universe. It's as if we're eager to pin our chances on something strange and sudden, something beautiful beyond our ken. Across cultures and time, we have written ourselves into the sky. We create constellations, transforming the random spatter of stars into shapes and stories. We name planets after gods. And we associate meteors and meteorites—the light of dust or rocks burning passage through the air, and the stones, after such fire, that sometimes fall to Earth—with the most elemental aspects of our lives: good luck, ill fortune, and even death.

Meteorites are, in fact, implicated in the seeding of life's ingredients on Earth. And even the most indifferent know that these bits of former asteroids have rained devastation in the past and threaten to do so in the future. Meteorites are the alpha and omega of geology. These rocks—mere rocks—encompass the origins of life and the reality of death on our planet.

Not surprisingly, we try to tame the wildness of meteorites by incorporating them into popular culture. Movies show humanity outgunning asteroids or comets headed for Earth or, at least, surviving the effects of massive impacts.

Accounts of actual space rocks whizzing by are now relegated to the inside pages of newspapers; it seems we've had too many close calls to get excited about them. Then again, the front page of the July 5, 2004, *Weekly World News* announced, "Another Meteor Fells Pope!" Amazingly, the 900-pound stone did not kill him, though a photo shows the pontiff unhappily pinned under all that weight. An episode of *Gilligan's Island* called "Meet the Meteor" featured a meteorite that accelerated the aging of the hapless castaways, but they survived, and the Professor even made a Geiger counter out of bamboo. In the otherwise forgettable movie *My Super Ex-Girlfriend,* Uma Thurman got her hottie superpowers from a meteorite. So humor dilutes threats—as does possession, literal or metaphoric. We can set meteorites on our mantels, displaying them as "specimens," including some iron meteorites that have been ablated into fantastic shapes, like Henry Moore sculptures and sometimes just as pricey. An English breakfast cereal, Shreddies, once included little packets of meteorite dust as a promotional giveaway, and I've wondered how many kids sprinkled their new treats over cereal and milk. Grown-ups might be interested to know that a sex-toy manufacturer offers a line of shapely "Meteor Plugs" in small, medium, and large, while a lingerie company sells stockings in a color dubbed "meteorite gray." More poignantly, we christen asteroids to make them serve our needs: Astronomers have bestowed the names Solidarity, Magnanimity, and Compassion upon three asteroids as tributes to the victims of the September 11 attacks. Eros and Thanatos indeed, Freud might have said.

Some of the meteors—shooting stars—that flash across the sky contain rocks that are large enough to fall more or less intact onto the Earth's surface: meteorites. This has been happening for billions of years. Covertly dark or blandly gray, often woefully misshapen—that's what many meteorites look like, lumps so ordinary-seeming that most people never notice these rocks have landed on farm fields, deserts, shorelines, or backyards. The untrained eye can even mistake some meteorites for chunks of concrete, as if a cosmic road crew had jackhammered a solar highway and sent its discarded congeries spilling down, briefly lit, to land among soybeans or ferns. Those who quest for meteorites, however, recognize them amidst the average rubble of the Earth the way a birder hears rare song untangling itself from a forest full of sound. Passion does that: It sharpens one's senses, it changes the world. Amateur collectors, professional dealers, and planetary scientists are eager to obtain and understand these stones. They have gone and still go to great lengths to gain such exotica as a water-trapping H5 chondrite, a "snick" from Mars, or an igneous, calcium-poor ureilite.

MANY PEOPLE, myself among them, discount the notions of heavenly jurisdiction over a person's life, whether it's thinking your wish-upon-a-falling-star has come true or simply believing in a horoscope. Yet I have found that in actual and often moving ways the fallen sky can reveal secrets not only of the solar system but of our hearts. That is why this is an intimate history of shooting stars. We go out hunting meteorites, and some of us find ourselves as well.

Years ago, when I lived in eastern Kansas, I was trying to deepen my connection to the tallgrass prairie, to find stories that might enrich those austere horizons. Thumbing through a volume by the nature writer Edwin Way Teale, I came across the history of a homesteading farmwife who harvested not only wheat but one of the rarest meteorites known to science. I was then also writing a book about extinct birds and I couldn't yet fathom that the grief I felt about their fates was also, in part, an expression of many inarticulate griefs I carried in my life. Prompted by those inner darknesses and by the words of Teale and others, I went outside at night and looked up. I saw and learned the stars. I used them as a balm. And I saw meteors—sudden, thin streaks on any given night; showers of them in the summer, watching with my wife as we camped beside a marsh and sand dunes in Colorado; and in our front yard, the one I'd walk away from, long after fireballs had exploded one November, golden shocks in the sky.

All this led me to learn more about meteorites than I ever thought I could, and so this book has also become a chronicle of some of the most important meteorites we know of—stones that have altered our knowledge of the solar system and our place in it—and as such serves as a kind of informal record of scientific recognitions over the past 200 years. The science—so fast-moving it's been difficult to keep up—ranges from studying the primordial delivery of amino acids in meteorites to assessing the dangers of asteroids classified by the epic names "Atens" and "Apollos." I've especially tried to convey how individuals and cultures have valued meteorites, how they have been venerated as objects of power and continue to be objects of profit. But also, and this is crucial, how meteorites have become, through the rigors of science and the marvels of story, objective correlatives for our desire to live with both explanation and mystery.

Whether someone wishes to possess a meteorite to sell it or to crack one open in a laboratory for discovery, the meteorite must first be found or hunted. Which often means you have to be willing to go where the meteorites are (rather than have them mailed to you by suppliers). I've traveled to meteorite sites around the world, from the nearby to the far-flung, from a small-town street corner in Kansas to the iceberg-clotted coast of northwest Greenland, from a stranger's driveway in Portland, Oregon, to a German church made of rock born in the fire of a meteorite impact. Such journeys have impressed on me that wonder—whether from discovering a geological rarity or tracking down a hidden history or finding a lover—is not as pristine a feeling as some would think. I found that mine was a journey into wonder and its costs. Along the way, I bore changes in my life and realized that I was hunting the lives of the meteorite hunters—not just the stones themselves—and I began to understand these strangers' lives better when I accepted my own. Quests, after all, can come at a very high price. This was never more clear than when I suffered a breakdown in a wind-racked tent in Antarctica; outside were boxes of meteorites.

As to the meteorite clan, they're a complicated, colorful lot. There's an early twentieth-century backwoodsman who stole a 15-*ton* meteorite, then lost it and his marriage; an engineer and businessman whose heart burst after receiving calculations concerning the vaporization of high-velocity iron; a little-known and single-minded naturalist who changed the course of meteoritics (the science of meteorite study) while pushing himself, his wife, and their children to the brink of disaster during the Depression; a man who once sold "space passports" at a mall before becoming an affluent and controversial meteorite dealer; and a researcher nicknamed "Mongo," whose annual expeditions retrieve thousands of meteorites from the hostile polar plateau at the bottom of the planet.

If a meteoroid (as small rocks are called when they're in space) becomes a meteor (as the flashes of light are called when grit and rocks burn through the atmosphere), then if the meteor survives to become a meteorite (the rock that lands), and if someone happens along to find it . . . what an unlikely chain of events! Yet this happens, and to be the first human to touch that bit of space, to catch that fallen star, is exhilarating, is sweet and beautiful, and, for some, feels like a brief mastering of the universe and its cold indifference. The rock and iron that cross the Earth's orbit can seem less daunting if, in some wild country, you can place a sliver of former asteroid into a bag that elsewhere might hold your lunch-hour sandwich.

And what feelings fill, not the hunter, but the shopper, who might be standing in an air-conditioned room before a tray of meteorites, contemplating whether to buy a nubbin of Mars? Should he fork over $3,500 for a Martian meteorite the size of a pencil eraser?

The first time I held meteorites was in a natural-history store, and I half-expected them to be warm with creation. One even looked like a bakery roll. They were disappointingly cool and metallic to the touch. Still, I hefted them in my palm, felt their ridges and pits, imagined that mineral skin on fire. Then I set the meteorites down, thinking I could not buy the sky, though I wanted, somehow, to find my own piece of it.

So of course this is a study of obsession, and obsession is a characteristic with the potential to cause harm as well as nurture calm. As one veteran space-rock aficionado told me years ago, "Meteorite people . . . are a little weird." We all carry our compulsions, and soon after I began this project I began to wonder if compulsions not only lead us to extremities of experience but also, at least sometimes, to places of connection and peace. Recall Gilgamesh, who, raging over his kingdom of Uruk, dreamed of a meteorite he could not move, then found his fast friend and lover Enkidu, with whom he went on adventures and misadventures. The heroes of this book have had their own versions of the same. Some of my protagonists have even triumphed. They have found rare meteorites that are the envy of others and that convey discoveries about the deep past, they have found meteorites that have brought wealth and fame, they have found meteorites that infused purpose into lives that otherwise might have been too commonplace. They have found, in a word, joy.

It's these stories and how I came to discover them that have helped me to organize the book the way I have. I begin at the beginning—with both the origin of the solar system and the origin of my quest to understand meteorites and those who are so taken by them. My first research trip happened to fall on an important scientific anniversary, one that took me back to the first discovered asteroid. From then on, the chapters unfold from around the beginning of the twentieth century to the present day (with a chapter-length flashback about the timeless folklore of meteorites and our first scientific discoveries about them in the late eighteenth and early nineteenth centuries). At the end of the book, readers can find both a glossary and a chart to help keep track of terms and types of meteorites. Rather than take the approach of a textbook, introducing technical information and historical facts in dry sequence, I found myself learning about meteorites and our relationships to them as the events of my life dictated. The book mirrors that personal process of discovery. The advantage to this approach was that I learned about my characters and their insights regarding meteorites at the same time I learned some things about myself. In a strict sense, this isn't a book about meteorites. It's a book about my pursuit to understand the passions of meteorite hunters. This pursuit concludes not with my finding meteorites in the wild, which I did, but at my home in Utah, the book having also become, unexpectedly, an extended meditation on departure and arrival, on distance and place.

WHEN CHICKEN LITTLE FELT AN ACORN (or rose petal or pea) land on her head, she thought the sky was falling and sounded her insistent alarm. In some versions of this story, many of Chicken Little's animal friends are eaten by the duplicitous fox. A mistaken perception about some wider accident is enough to trigger a different disaster closer to home. And we're meant to understand, therefore, that skepticism and courage are laudable traits—and many who seek shooting stars have them—but it's important to recall that in other versions of the story, the sky does fall. Perhaps it matters little whether that sky is the big one we walk under or the one we carry inside ourselves.

When I began this book, I believed the story of Chicken Little was just a children's tale I could deploy for poetic effect or maybe as a parable for the threats to civilization posed by rogue asteroids. I couldn't yet see the story as a parable for my own life or even the lives of meteorite hunters, lives I hoped would be so ascendant, so engaging, so full of simple awe that I could skim across difficulties like a stone thrown across water. I could write an adventure. About this, as with other things, I was mistaken.

I did recover a sense of wonder and I did have adventures, and it turns out that in spite of—or perhaps because of—the price that can be paid for astonishment, I can better understand many of the ancient metaphors: meteors as blood, as tears, as burning hearts in the sky, and meteorites as gods, gods being, always, of course, versions of ourselves.

We each have found ourselves lost in the dark wood, whatever we thought the true way had been or can be, but for me, in no small measure, the path out was lit at times with the passage of shooting stars. This book is an exploration of lives, including my own, caught in such light.

· Prologue ·

Dust: A Brief Memoir of Overlooked Things

Dust gets a bad rap. It's an adjective for products whose nouns are "mop," "remover," or "buster." In Queen's song—so catchy it can be irritatingly impossible to forget—"Another one bites the dust." The rock group Kansas whimpered that "all we are is dust in the wind." The slow shall be left in the dust. The feckless shall eat my dust. History, when forgotten, dies in a dustbin. Dust to dust, we're told, emphasis on the dry conclusion. Despite the welcome richness of gold dust, which is preferable, a rainstorm or a dust storm? Given its rarity, gold dust doesn't kick up in high winds to smother homesteads and towns. Other dust—from prosaic to suffocating—dominates our lives. Twenty years ago, when I first moved to Kansas from Indiana, I kept brushing what seemed like eraser shavings from my desk, then looked up to see an orange sky: a dust storm. I thought I had moved to the apocalypse. There are no dams that will stop a flood of dust. Most of us prefer rain.

The dictionary offers these disagreeable definitions of "dust": "(1) powdery earth or other matter in bits fine enough to be easily suspended in air; (2) a cloud of such matter; (3) confusion, turmoil; (4) (*a*) earth, esp. as the place of burial; (*b*) mortal remains disintegrated or thought of as disintegrating to earth or dust; (5) a humble or abject condition; (6) anything worthless."

When I lived in Kansas with my then wife, nearly every morning I'd wet old rags, then tie them to a broom. I'd push, turn, and drag the jury-rigged mop across oak floors, retrieving especially egregious aggregations from beneath the bed. Then I would step to the front stoop, lift the yellow dust mop to the sky as if it were a banner in my war on entropy, and shake it like mad. Hair, grit, mites, skin, floss, lint, insect bits, pollen, seeds, puff, all manner of heinous detritus—with cheer, I launched them all. The cloud usually traveled east toward a neighbor's yard, but he never complained. Had he, I might have told him that dust is life, even if it makes us sneeze. Dust, it turns out, is genesis.

THE STORY OF THE SOLAR SYSTEM—and all that follows, from meteorites and those who cherish them—begins with dust. Stars died. They exploded. There were, researchers have conjectured, perhaps up to ten supernovae whose traces we can read in the solar system today. When such stars explode, the remains gather together at a far distance to start again as dust, entrained with capes of gas.

Supernovae occur when large stars—those about 25 times heftier than our sun—can no longer burn. They've run out of fuel, one element after another having been gobbled up through hunger and change. After it has burned hydrogen, helium, carbon, neon, and oxygen in successively shorter ages, comes the last day in the life of such a star, a few hours fueled by silicon. Within the star, iron is about to be shredded like a puffball because iron makes less energy than it needs in order to be a fusion source. One ordinary second ends it all: the star collapses, sending shock waves slamming through it, and the star just rips apart.

Previously, the star had expelled carbon gas, which eventually condenses into diamonds. "Countless trillions of diamonds," as scientist Harry Y. McSween, Jr. puts it, "but not a single one will ever grace a pendant or seal a marriage vow." The supernova itself casts out, among other things, a rare form of xenon that infuses into those interstellar diamonds. And some of these micro-diamonds make their way into meteorites that land on Earth. It took two decades of painstaking work by chemist Edward Anders and other scientists to coax from meteorites the traces of all that interstellar material. "The . . . grains . . . " McSween writes, "are the grime of a dozen other stars and the smoke of at least one supernova." Oxygen, plutonium, and iodine are forged in the hearts of dying stars. Gold too. All that sent forth in a flash of iron demise.

Stardust soared from explosions brighter than galaxies. Gas billowed. Shock waves with their linty freight hit other particles and gases already lingering in one small pocket of the Milky Way. Gravity rescued them.

About 4.6 billion years ago this cloud of dust and gas began slowly to contract, rotate, and flatten. The heart of the presolar nebula began to quicken, and within a few thousand years a protostar—not yet fusing hydrogen—went through phases of contraction, expansion, and varied brightness. It took some 10,000 years for outward gas pressure to balance against gravitational contraction, then another 10 million years for the protostar to condense just enough to create what became our sun, whose interior pressures forced a sustained hydrogen fusion reaction at 10 million degrees.

When I first began reading in detail about the development of the solar system, news arrived that researcher Eric Feigelson and others had used the orbiting Chandra X-ray Observatory to stare deep into the star womb of Orion, where each winter the green wing of that nebula—some 1,500 light-years away—flies into the view through my backyard telescope. In northern Utah, where I've lived since 2002, I'll stand outside in my parka hunched over the scope, just staring into the nebula as it rises over the Bear River Range, my face cold, my nose running. I love seeing the Orion Nebula. I love knowing that I'm looking back at a version of what preceded our own solar system. Feigelson found that Orion's young stars—stars like our own in its childhood—flare up in the X-ray portion of the spectrum, producing isotopes of beryllium-10, calcium-41, and aluminum-26, all substances found in meteorites and once thought to be made only by supernovae. I've also learned that at least one huge star in its old age shed some mass in an interstellar wind rich in aluminum-26, which also could have precipitated the contraction of materials that became our solar system. Not long after shedding its aluminum-26, the star went supernova. So the presolar nebula contained stuff cast off by massive stars as well as the remains of a supernova—or more than one—and after our young sun formed, it too gifted all manner of energy and isotopes.

The presolar nebula's contraction, the outward pressure of the protostar, then of the sun, and the rotation of the entire ensemble led to a complicated dance of creation. (I use the word "creation"—it's a good word—in a non-deistic sense.) Like a figure skater's scratch spin—legs crossed, arms pulling in, beautiful dervish, quicker and quicker on ice, the greater the contraction, the faster the spin—this not-quite-a-solar-system drew in more particles and gas molecules into faster orbits. Working against the contraction was, some say, a magnetic field generated by the shrinking cloud itself, which may have slowed the solar system's formation.

What material didn't flow inward to the warming center accreted in "small, loosely bound dustballs," researcher David Kring says. Particles clumped, bumped, stuck, and flew. Grains and flecks became pebbles. Electrostatic attraction overcame the shredding violence of impacts, swirling gases, and solar winds. Data from the Stardust space probe, which returned actual samples of a comet, suggest that in addition to stuff falling toward the center, gas pressure and turbulence pushed heated materials outward along the plane of the developing solar system to the chillier reaches of this messy, mixed-up cloud. Such stuff can get out beyond the so-called snow line, where it's cold enough for ice to form in space. Scientists are finding evidence that isotopes varied in different parts of the proto–solar system, which means conditions varied not only by temperature but by composition of materials.

Grains, flecks, and pebbles accreted into rocks. They became careering piles of rocks, then boulders upon boulders. Once these objects were more than a half-mile wide, gravity superseded chance accretion, electrostatic attraction, and drag, thereby speeding up formation of the planetesimals. These minor planets, the asteroids, were born. From them would come meteorites. Easily told in a few sentences, this transformation from dusty disk to solar system actually took millions of years.

On days so cold that water vapor freezes directly in the air—ice crystals appearing as if by magic in a clear sky, catching sunlight in a dazzling curtain of glitter—I can almost believe I'm at the beginning, floating through the outer reaches of the solar system. I think of what Kring said about hot dust that cooled rapidly: that those grains "condensed like snow from the air." On winter afternoons in Utah, on a hike, say, with my partner Kathe, I sometimes try to imagine that I'm standing on a minor planet billions of years ago instead of in a snowy draw flanked by juniper and Douglas fir. I try to imagine that I'm looking toward Neptune, toward Pluto, toward the distant Oort cloud with its encircling moat of billions of comets.

Snow falling from a cold, cloudless sky has a name. It's called "diamond dust."

LONG BEFORE MOVING TO UTAH WITH KATHE—I didn't even know her then—I camped one summer night in Idaho with the woman I'd been married to for more than a decade. We had driven there from Kansas on one of our annual trips to the American West, where we'd drive remote roads, hike, backpack, and pitch a tent. We were high up in the mountains, and the air was cold and clear. These adventures allowed us to escape the hot prairie and gave us something physical to share in a relationship that was primarily, and often richly, emotional, intellectual, and creative. Whatever passions tied other couples together—and I wondered about those passions—we had this, and it strengthened muscles for vistas, words for poems, stories to share at dinner parties. We loved these trips. They were also deeply lonely.

That night while she was asleep in the tent, I stayed up late looking at the sky, which seemed in one wide blotch to be strangely white, vaguely so, and I wondered if I was seeing a faint swath of northern lights. By that point, I'd been stargazing for a few seasons, but I couldn't quite tell what was up there, if anything at all. I rubbed my eyes. I stood up and changed the tilt of my head. No, it was still there, a chalky scrim, a pallid aurora if one at all.

Very recently it occurred to me that I probably had witnessed something called the "gegenschein," which is sunlight hitting interplanetary dust that orbits in the same plane as the planets. Dust lit by the sun, long since set, like the moon reflecting sunlight, only far more subtle and diffuse. When I saw that pale light in Idaho, I was a long way from writing about asteroids and meteorites, about watchers of the former and seekers of the latter, but now it seems apt that before I knew what journey I was on, I saw a glow coming from the direction of the asteroid belt. I saw something beautiful in detritus. And I didn't know it. That too is apt. I was well practiced in not knowing.

Whether on that trip or another, as we drove one time from the Flint Hills of Kansas to summer mountains, I leaned my head on the glass of the passenger-side window. I remember that I thought without cessation about someone else, a woman I'd been obsessed by, befriended, even wrote a poem about, a poem whose deep yearnings my wife never questioned. Who did she recognize in it? Anyone at all? Was that my stuttered, deceptive attempt at talking? Was her silence not a closure but an invitation for me to speak more plainly instead of in literary code? Or was such code safer for both of us? In many ways, in many places, I wrote of the sky, of stars, of nature, when really I was writing about the body.

I didn't know that dust which seems to obscure also relays light, has a name, has origins. I didn't know that, in the years ahead, just before and just as I was falling in love with Kathe, I would hitch some of my passion to those origins and to meteorites and to the people who care about both.

So I HAD SEEN THE REMAINS. I had witnessed, it seems, the dusty denouement of the solar system's creation. What had for a time been a small puzzle, then a memory, then a mostly forgotten moment, and then a realization, straddled, I see now, crucial aspects of my life—skin/sadness (my own, others) and sky/solace (birds, stars). I was ashamed of who I was or could be in rooms of touch and whisper, so falling in love outward—with the world that isn't human—seemed a good remedy, until it wasn't. In the company of shame and realization, and without the honest dignity of the wider world I had run to for years, I would end my marriage and begin anew. Then I would come to care about the beginnings of things and their lineages, in the cosmos and in my psyche, and it would not seem at all strange to yoke them together. I would feel currents—passion for Kathe, passion for new stories and facts—and these would take me, like energy jetted from a star, to many unexpected places.

That's why I find myself on snowy days in Utah thinking about luck, about failure, about love, about diamond dust. That's why I find that places in the sky and places on the ground need not be, as they had been, an escape. The sky and the history of the sky can be connection, another place, like the bedroom, where one can be whole. That's why I find myself imagining planets so huge they'd swallow many earths but so light that they could float on our collective sweat, no Atlas needed.

Just what kind of universe is this? Gas-giant worlds that could bob like beach toys on a sea big enough to hold them? Saturn in the bath? Diamond dust? Exploding stars? And people, happy or otherwise, to contemplate the same? Occasionally, when we see something especially wonderful, a western tanager against the blue sky or snow streaked on red rock in the desert or lightning and virga from a gray bank of clouds, Kathe and I will mutter, "Crazy. Just crazy."

Chance and facts are good enough for me, and so when I return to the pre-heartbreak era of random gatherings, the desire-free eons of searing sands in space and epic snow-line blizzards, to the start of the solar system, I learn that the first full-fledged planets to congeal were the beach-ball gas giants, such as Jupiter and Saturn, worlds of psychedelic clouds and storms so big they'd make a hurricane look like a raindrop.

Then followed the rocky inner planets, ours included. Earth took 10^8 years to form.

Crazy, just crazy.

The gas giants formed, most scientists think, by accreting rocky-icy cores that then attracted their huge sheaths of gas. A few believe the gas giants formed lightning-fast, in just about 1,000 years, as gas balls that then attracted ice, which migrated down toward the planets' centers. A kind of blizzard raining through something like sky.

As to the asteroids, they grew as wide as states, as thick as mountains I know, while their roiling innards cooled and, on some worlds, lava erupted and flowed just as it does here on Earth. Inside many asteroids, the heat produced by the decay of aluminum-26 and other elements kept the cores molten, along with, some say, the heat created by countless collisions. The cores of asteroids began to cool 20 to 30 million years after the originating supernovae had exploded.

Jupiter's massive gravitational field prevented the formation of a full-fledged planet between Mars and itself, so only asteroids, solid or otherwise, were able to form in a zone between 2.2 and 3.3 astronomical units from the sun. (One a.u. is the distance from the Earth to the sun, or about 93 million miles.) It was once thought that a planet had formed there and that the asteroid belt was the remains of its destruction, but there never was such a world.

Collisions between the minor planets led to "families" of asteroid types. A parent body shatters from collision, and the consequent piles become, in the words of one writer, "chips off the same block." After a parent body breaks up, the many smaller offspring essentially hang around together, bound by gravity. This helps to explain why many asteroids have tiny satellites orbiting them, why some asteroids are binary—two rocks twirling around each other—and why a lot of asteroids are loosely bound rubble piles. I like to think of such asteroids as floating cumulus, only a lot harder.

The planets that formed 4 billion years ago—Earth, Mars, Venus, and the rest—were bombarded by meteorites, asteroids, and comets. Researchers in 2001 suggested that a "final depletion of [the] leftover population" of asteroids that were in orbits "highly inclined" to the rest of the solar system helps to account for what some refer to as "the late heavy bombardment." This is a conjectured period of intense impacts on the Moon and the terrestrial planets about 3.8 to 4.1 billion years ago. No one explanation for the event holds sway—and some scientists doubt it even happened—but changes in the orbits of Jupiter and Saturn may have swept impactors toward the Earth. A highly inclined orbit means a steeper angle of impact and hence a more violent encounter. In sum, it probably was a rocking rough time to be a planet.

Then again, creation sometimes necessitates rancor.

LIKE MEMORY AND ART, like trauma and argument, physics and chemistry circumscribe things—what is included, what is excluded, what is transformed, what is left alone. From countless grains of humble star grit gathered and melted and smashed came not only entire worlds but just three broad types of meteorites: iron, stony, and stony-iron. And within some of those meteorites may be the key, the very reason that the planets were able to evolve in the first place.

Meteorites are essentially organized by amounts of metal, in particular nickel-iron, and various kinds of silicates, which are compounds of silicon and oxygen. Irons are 98 percent nickel-iron. (It's the nickel-iron that makes many meteorites heavier than Earth rocks of similar size.) Stony meteorites are mostly made of silicates, but can contain up to 25 percent nickel-iron; further, the stony meteorites are divided into those that usually have little glassy globes called chondrules (the "chondrites") and those meteorites that derive from the melted interiors of asteroids (the "achondrites"). The chondrites, never having been fully melted, can be extremely valuable to scientists searching for the original material of the presolar nebula. As meteoriticist Robert Dodd puts it, "When a body melts . . . it effectively burns its birth certificate and shreds its baby pictures." Such is the case with the achondrites. Finally, the stony-iron meteorites are composed of both stone and metal, as that hyphen suggests.

Authors David Coleman and Sarah Kennedy offer an ingenious visualization. They ask readers to imagine placing a dark iron meteorite on a table, then stacking on top of it a stony-iron meteorite, then one of the stony meteorite types, an achondrite, say, which can look like a piece of a driveway. The iron meteorite represents the heavier core of an asteroid. The stony-iron meteorite represents a mixing zone between core and crust. Finally, as they and others explain, the achondrite represents the igneous interior, brought up to the broken surface by impacts, a surface that can also contain unmelted rock and rubble—chondrites. This generally accepted and long-standing view of an asteroid's internal layers is called the "onion model"—or, more precisely, such a layered world is a "differentiated body." Not all asteroids have been large enough, however, to have contained molten cores or other layers. Within each broad category of meteorite—iron, stony, stony-iron—are numerous divisions and subcategories.

Perhaps the central mystery concerning the composition of meteorites has to do with those little chondrules, those wee glassy spheres—about a millimeter in size, sometimes less, sometimes more—that are suspended inside chondrites. (Think of millet seeds or BBs to imagine their size and shape.) Their formation continues to puzzle researchers. To understand chondrules is to know more intimately how our solar system came to be, and chondrules are one reason meteorites are considered, in an oft-repeated phrase, "the poor man's space probe." No one knows how, exactly, they formed, but their formation probably helped planets build past a certain bottleneck of accreting dust. Chondrules seem to be the bridge between small bits of stuff in the early solar system and the planets themselves.

It was nineteenth-century scientist Henry Sorby who first really studied chondrules and who was right in deducing that these globes of melt matter must have solidified prior to being caught up in chondrites themselves. "Melted globules with well-defined outlines could not have formed in a mass of rock pressing against them on all sides," Sorby wrote, "and I therefore argue that some at least of the constituent particles of meteorites were originally detached glassy globules, like drops of a fiery rain."

But it's particularly vexing to account for this, because chondrules clearly were once very hot and then ended up inside meteorites that have *not* melted. Cosmochemist Derek Sears reports 72 theories to account for their formation—72! Within the presolar nebula and the early solar system, any number of events may have brought on a storm of fiery rain. Lightning might have heated dust to melting points. Dust might have simply heated up closer to the sun. Colliding dust particles might have melted. Frictional heating from differing rates of rotation in pockets of gas and dust could have formed chondrules. Now that we know material near the center of the forming solar system was transported to its outer regions, we can see that chondrules could have formed in the hot zone near the sun then been zipped far away whenever the sun threw out more mass and energy in fits and starts. The x-wind theory posits an x-shaped pattern spraying hot matter above and below the disk of the solar system; the chondrules formed in the solar winds then were mixed into the unmelted chondritic material farther away. Many researchers have pointed to shock waves in the nebula that would have heated dust into chondrules; the shock waves might have been internal to the nebula or have come from an interstellar explosion, producing the needed chondrule-formation temperature of some 2,000 degrees Kelvin. But it's not clear what can keep the shock waves going for the three-million-year time span apparently needed to make chondrules. And while the shock-wave model is popular, no one has actually observed one passing through a presolar nebula. That may change with better telescopes, of course. As of now, every model of chondrule formation has its problems.

Despite Sorby's observation that chondrules have "well-defined outlines," some chondrules show evidence of abrasion and fracturing; that is, they may also have formed from impact events. A study in *Nature* suggested that some chondrules were created when two planet-or-moon-sized objects smashed into each other and melted. Another research article suggested that impacts between "molten planetesimals" released "impact plumes" whose "droplets conceivably cooled to form chondrules." The droplets, once solid, got taken into asteroids made of the stuff that chondrite meteorites are made of.

Just how weird are chondrules? Harold Connolly and S. J. Desch write that "if chondrules were not already known to exist, they would not be predicted to exist." Scientists hate—and love—things like that. It keeps them well occupied. Furthermore, as the two point out, astrophysicists "have long recognized a bottleneck in the accretion process at the scale of centimeters." In other words, it's not entirely clear how you grow dust from "a few centimeters" to objects about a yard across. Small bits can bunch up nicely with other small bits, and objects about a half-mile wide are massive enough for gravitational attraction to work against fragmentation. In between, collisions tend to break up aggregated material. Connolly and Desch argue that chondrules were the glue that allowed objects to grow. Without these tiny beads, ignored by all but specialists in this esoteric field, there probably wouldn't have been a solar system.

And there's more. Before the chondrules formed, the first solids in the presolar nebula were things called "CAIs"—calcium-aluminum inclusions. These are blebby, formerly melted bits of minerals that condensed out of the early solar system at an age approaching 5 billion years ago. Some meteorites have these inclusions; if you touch an open slice, you're touching the oldest thing you ever can. Calcium-aluminum inclusions formed a million or more years before chondrules. Their formation is also not yet nailed down, but the inclusions may also have been a kind of glue. Like chondrules, they seem to have drifted into chondritic matter and been swallowed up to become part of these meteorites' asteroid parents.

I like the fact that something so elemental, so vital to our lives, so small, remains a mystery—and that our curiosity impels us toward clarity and explanation. When the mystery's no more, what we know about chondrule formation will lead to other questions, such as the formation of solar systems around other stars, where meteorites with chondrules might be landing on other earths where other hands pick them up. Though we've not yet found Earth-like worlds around different stars—we will soon enough—we've seen disks of dust around other suns, we've counted hundreds of "exoplanets," and we've discovered at least three extrasolar asteroid belts. And where there are asteroids there should be rocky worlds—terrestrial planets like Venus, Earth, and Mars.

On this world, pausing in their work, scientists, like the rest of us, sometimes admire the beauty of what they reach toward. In multicolored photos, meteorites cut into thin, translucent slices—"thin sections"—can reveal chondrules that look like a pointillist's stained-glass mosaic. I think part of their beauty comes not only from their age and importance but from their size and shape as well. They helped make planets and they look like them too. At meteorite shows, collectors with lenses will *ooh* and *ahh* over how chondrules bespeckle a meteorite specimen cupped in one's palm. Chondrules not only give goose bumps, they are goose bumps; they are stipples on creation's body.

YOU NEED NOT BE A METEORITE SCIENTIST or a meteorite collector to touch this body, for it reaches us every day, though in forms—dust and micrometeorites—too small to notice without fancy equipment. A scientist who studies space dust, Donald Brownlee, once said that "if you had lettuce for lunch, you probably ate a few . . . [cosmic] dust particles." I love the image of the cosmos begetting us and our hungers, and our hunger eating up the cosmos in a salad at Denny's. If you've stuck your tongue out on a winter's day, tasting the snow, you may have swallowed meteor dust. We're all eating stars! Every day! What a marvelous annulus of accident and need.

Best, though, is to catch the stuff on high instead of at the buffet, so high-altitude collecting of cosmic dust or micrometeorites—technically called "interplanetary dust particles"—now routinely takes place. (Interest in cosmic dust actually dates back to the nineteenth century.) Researchers such as Scott Messenger have found that such dust contains material from other star systems. Without mastering what science-fiction writers term "FTL"—faster than light travel—humans have sampled actual grains of sand from interstellar shores, and the dust of other suns has sifted together with our own homegrown solar-system dust. Think of Blake's world-in-a-grain-of-sand; now it's many worlds in a grain-within-a-grain-within-a-grain. Scientist Monika Kress, who worked with Brownlee, writes that "most micrometeorites are only a few hundred micrometers across—about the diameter of a human hair. They are utterly dwarfed by the grains of sand you shake out of your shoe after a walk on the beach." But mixed into the beach sand is star sand too.

This stuff is everywhere. In the 1990s, the water well at the South Pole was recognized as "the largest source of micrometeorites yet discovered," and sophisticated suction devices gathered them up. They're in the seas. Kress notes that many ocean-dwellers "gobble up just about anything they run into." Interplanetary dust particles encased in fish droppings thus sink faster to the bottom of the sea. So we're not the only star-eaters. I could begin a bad poem by invoking the cosmic gar, the astral sturgeon.

Researchers who wish to study cosmic particles—those free and those expelled by fish—must sit "in a darkened room at a console . . . staring at ghostly images . . . on a TV . . . turning knobs and dials expectantly," writes Kress. Workers must "focus the beam of [a] scanning electron microscope" and spot micrometeorites. Finding ET, or extraterrestrial, elements in proper quantities, such as a nonearthly isotope of helium, nails a grain as alien. Like some meteorites, interplanetary dust particles can contain organic matter—not life itself but some of its building blocks.

This microscopic dust and little grains of space sand you could measure on a ruler (oh, this one's about $\frac{1}{16}$ inch) and detached chondrules and small pebbles and larger stones and meteoroids the size of carriages and those the size of houses—without fail, they plow into our air every minute, every night and day. Creation's body sheds its skin like the rest of us.

Meteoroids hit the upper reaches of the atmosphere traveling from between about 7 miles per second to some 44 miles per second; they start burning up at altitudes of about 70 miles. An object no wider than a man's leg is long will produce a fireball that's 200 times larger. Even a rock the size of a raisin can produce a huge fireball. For a time, the object's surface is heated to about 2,000 degrees Fahrenheit, only a few hundred degrees cooler than a sunspot. You can tell what the meteor's made of by the color of its tail—red for silicon, yellow for iron, orange-yellow for sodium, bluish-green for magnesium, violet for calcium. The glow comes from heating up the meteoroid and from the ionization of the air as it passes.

Eventually all but the very biggest objects are slowed by the thickening air until they're traveling more or less as fast as commercial airplanes might, about 300 miles per hour give or take 100. Some meteors explode in the air from the pressure of the event, and rocks fall to the ground at far slower speeds. Perhaps counterintuitively, the smallest particles are the ones with the best chances of survival, because their mass is so little relative to surface area. Dust sifts instead of plummets.

About 40,000 tons of space dust reach us each year. Or it's more like 100,000 tons of both dust and rocks, another source says, which works out to some 274 tons each day. Robert Dodd calculates that if the annual infall of space material is somewhere between 100 and 1,000 tons daily, then after 4.5 to 4.6 billion years this infall would amount to a layer 5 feet deep. Meteorite collectors would only need a shovel. But weather and geology—rain, deposition, lava, shifting land forms—all conspire to hide what falls.

A French meteorite dealer, Alain Carion, claims that only some 500 "large meteorites" land annually, and of those, perhaps two dozen or so are ever recovered. Dodd says that there should be eight new meteorites of about a gram ($\frac{1}{28}$ ounce) or heavier per square mile each year; a stony meteorite that weighs 10 grams would be about the size of a "hazelnut," he notes, but meteorites need to be about 100 times bigger to have a decent chance of being found. Only a handful of all meteorites are witnessed to land, and these are known as "falls." Meteorites whose fall was not witnessed but are recovered are called "finds." Meteorites are named for the places where they're found.

A high number of meteorites in the two-pound range, about 1 million, hit the Earth each year—falling without preference for any given place or any given time of day or year—but as Dodd remarks, "This number translates to a very low probability for a particular spot." Objects weighing about 13 pounds fall to the Earth five times a day, according to a sign in the Arthur Ross Hall of Meteorites at the American Museum of Natural History. Meteoriticist Ralph Harvey tells me that "one in 10 falls results in 10 or more recovered pieces," and that if you take 20 random meteorites, you'll be likely to find they represent 10 or 11 different falls. David Kring notes that in his state of Arizona, by way of example, just 31 of the nearly 800 meteorites in the two-pound range that have fallen since the late seventeenth century have been located. Sometimes a meteorite can drop dozens, hundreds, or thousands of fragments—seen or unseen.

When we happen onto meteorites, we happen onto irregular dark clumps, heavy and often magnetic. Iron meteorites show little depressions called "regmaglypts" or "thumbprints," while stonies display a dark skin—sometimes also pitted—called the fusion crust, black or brown depending on age. The thumbprints result from uneven heating and pressure. Fusion crusts, which can be prominent on stony meteorites, form when the thin melting exterior of a space rock passing through the air cools and solidifies. Iron meteorites, though relatively rare, are the easiest to spot because they look so dark, heavy, and often weirdly shaped.

Or else we're fooled by iron oxide concretions, fooled by slag, or fooled by a chunk of lava. "Meteorwrongs," they're called. And I've lost track over the years of the number of people who've told me they have a meteorite in their yard, just out front, by Uncle Ned's silver maple, how the rock fell "just over there," how Uncle Ned or So-and-so, his friend—very reliable—saw it himself and man how it's heavy! And you won't believe what good luck after it landed and do you wanna see it? Once I had lunch with a treasure hunter who'd claimed in the newspaper that he'd found a meteorite from a fireball. He showed me—the thing was black but hardly magnetic. Many meteorites are strongly magnetic. It had a sheared-off face, as though a saw had sliced the whatever-it-was open. Worse, it had little holes where gas had escaped (meteorites rarely have these and some experts say they never do). Plus, the rock didn't seem to weigh enough. Was this his test to see if I noticed what was wrong with it? Did he keep a real iron in his pocket? I didn't have the gumption to question the thing he showed me and I told him to call once he'd heard from a real geologist.

We're so hungry for rarity and connection that we hitch our hopes to the lights in the sky and think they come all the way to the ground just for *moi.* Many people, when told their "meteorite" isn't one at all, refuse to believe the verdict. I don't expect the treasure hunter will let me know what a geologist tells him.

WE'RE ALL TREASURE HUNTERS OF A KIND. Clear pilgrims, inchoate seekers, both.

When I lived in Kansas, I often drove from our house in the college-town suburbs out to a dirt road that cuts through ranch-land prairie. At that time, I had just a small telescope, but I cherished it. From Wildflower Road, I saw faint blurs of galaxies, star clusters, and nebulae. I've kept lists of them. I've collected them by sight. They were beautiful, and I could not touch them. No one could. Once I saw the staggering sight of a comet hanging in the sky, its tail flung out heedlessly, and below, in the stepped, grassy hills, lines of prairie fire curved, their orange beads and tongues of flame giving rise to thin smoke.

Usually I went there alone. A few times I did not. Before Kathe, there had been one other. Some dust from that road still coats me, a reminder of how I'd wronged and how I'd foolishly thought I could turn heartsickness into pure quests for the things of the world—birds, stars, words, maybe meteorites. How did I not see that for the cruelty and cowardice it was?

Sometime in those years, I learned the story of Gilgamesh, with its loves, its mistakes, its dream of a meteorite—a *meteorite?*—that beckoned a man to change. In that ancient tale, Gilgamesh and Enkidu experience battles and journeys, but none are as compelling as the passion the two feel for each other and the grief that lashes Gilgamesh when his companion dies at the hands of a vengeful god. Love and loss is the susurrus behind the plot. Was knowing that story the start of my changing, the start of a journey with shooting stars? Was loving that story a nascent hope, a goad and spur I didn't fully recognize?

IF WE COULD SEE THEM during the day—meteors, not hopes—we would have more chances to wish than wishes themselves. Meteors washed out by sunlight are beauty invisible. With ideal conditions, including a clear sky across the entire planet, we could see with our naked eyes more than 25 million meteors every 24 hours. Our sky is trellised with hidden fire—vaporizing tracers, mundane dramas of destruction and creation. They're as real as the "slender fire" of which Sappho wrote, the fire, she said, that was "quick under my skin."

In one respect this isn't just metaphor. Bright meteors or fireballs can produce plasma trails that alter electrical fields in the air. When these excitations encounter some objects, including one's own body, electricity is transduced into sound. Centuries of reports that meteors were accompanied by a variety of sounds—whooshing, breaking twigs, and humming bees—had been dismissed as illusory until physicist and astronomer Colin Keay verified the cause as transduction. In one of his experiments, test subjects "were hearing their own hair vibrating in time with the varying electric field." As Ralph Harvey explains, "When you ionize gas, it turns out that that produces mechanical energy in the ultrasound frequency . . . we can feel it or we can hear it as a hum." There is a music of the spheres after all, and it's nothing like what Kepler had supposed.

The Swedish say that "dust is always dust, however near to heaven it may be blown," but, we know now, dust made the heavens. Dust dies as fire in the sky or else floats like dandelion seeds. The sky is more alive with friction than most of us know. Sometimes we hear it hum and zing, a crackling glissando out of the blue. The sky can make our bodies sing. I like this version of "dust to dust." It's more lively.

The Indo-European base for "dust" derives from **dhus-no,* "dust-colored, mist-gray." **Dhus-* is the base also for "fury," so that **dhus-* means "to rage, storm." That seems right. After all, it was a dizzying dust storm of creation some 5 billion years ago that gave rise to all this, for, as another folk saying goes, "Even dust, if amassed enough, will form a great mountain." Heated, impacted, melted, cooled, changed, changed again and again, left alone, gathered and regathered, stardust—like some morphing comic-book superhero—became lawyers, apples, Jupiter, sidewalks, Olympus Mons, water, the Museé des Beaux-Arts, lunar regolith, a lover's shoulder blades, a mouth open in anger, the lips you've kissed, burning meteors, the two-kilometer diameter asteroid 1986 DA, along with its 10,000 tons of gold, enough, an engineer once told me, to make 2 billion wedding bands. It became the card that Kathe gave me early in our affair, the card with this quote from Nietzsche: "One must still have chaos in oneself to be able to give birth to a dancing star."

When I dust, I fetch and dispense whole worlds. When I dust, I know that this stuff of the universe, which, in quantity, obscures vision, can also make me see. Dust is both what the Taoists say it is—a symbol of the tiresome bustle of the quotidian—and something more. It is a deep expression of the way of the cosmos. I need only step outside on a summer night and stare toward Sagittarius, then up higher toward Cygnus, to see this, for the great black rifts in the Milky Way are vast clouds of gas and dust.

The tellurian dust bunnies we hunt without mercy or gratitude should be reminders of realms and ancestry. Historian Jeffrey Burton Russell tells us "a grain of house dust is roughly halfway in size between a subatomic particle and the planet Earth." I am a giant and I am a speck. And a few ancient grains of dust in your house, my house—right now—are probably unaltered from the stars; this is the dust rich in iron and nickel, dust rich in an isotope of helium created only in space, dust rich in exotic amino acids not found on Earth. The cosmos isn't just up there; it's underfoot, the cosmos is on your feather duster, it's on your plate. It floats as sunlit motes, dancing stars.

An itchy totem of a vast past stretched behind us, dust is something we can brush off old documents long-sealed in libraries, organized in cabinets, or kept in boxes in the closets of strangers just met, hands opening file folders so we can learn of the dead, who, in my case, also loved meteorites and had their own misfortunes, their own pinnacles. We can wipe our hands on our jeans, we can wash, but it never leaves us, never, this grit on our skin and the sky that burns, the one that drops without warning into our lives.

Book I

Distances Measured in Various Units

· *Chapter I.1* ·

The First Asteroid

January 1, 2001

By chance my travels with shooting stars begin on this, the first day of the twenty-first century, on a bicentennial mostly forgotten. Though I do not know it yet, this is also the last New Year's Day I'll spend with my wife. We sit quietly together on a flight from Kansas City to New York City on my way to see my first meteorite ever—a huge one, a storied one, probably the most contested meteorite in history—but right now I'm mostly thinking about my accident of timing and a priest who stood on a rooftop two centuries ago to this very day.

On January 1, 1801, on a Thursday night in Palermo, a priest and astronomer named Giuseppe Piazzi saw something new swim into the glass of his small refractor. He had been moving his eye across the stars of Taurus the Bull, whose own bright eye, Aldebaran, a glossy orange star, is easily learned and, in our lifetimes at least, never absent but for seasons and clouds. At a palace observatory, Piazzi puzzled out one small light from all the rest. What was this new thing? Days later he wrote, "I have announced this star as a comet." But it had no coma, no tail. In fact, it didn't look like a comet at all. He wondered if "it might be something better."

In a painting I've seen only in reproduction, the bald-headed, large-nosed, and taciturn Piazzi points upward to the mysterious light with his right index finger. He wears a coat nearly as dark as the sky behind him. His head and hand seem almost to float in space, as if the universe were putting him together or taking him apart.

The new light was where theory had predicted a planet should be—a missing globe between the orbits of Mars and Jupiter. For a time it seemed as though Piazzi had discovered that long-sought world, but soon enough it became clear that what he had found was too small. Neither comet nor planet, it was the first discovered asteroid.

He named the body Ceres, after the Roman goddess of grain, who was also Sicily's genius loci. For a time, the asteroid had various other names, including Cupido, Roman god of sensual love, a figure that reduced the complex Greek god Eros—who represented love as a unifying, creative force—to an absurd, chubby boy flinging arrows of random desire.

Later, Ceres would be found to be quite small by planetary standards, but at some 590 miles in diameter, it's still the largest, most massive body in the asteroid belt. The diameter of Ceres is almost exactly the distance between the college town of Bloomington, Indiana, where my wife and I first met and courted, to the college town of Manhattan, Kansas, where we moved for her teaching job and where our marriage will end. I will realize this many years later, how, driving a highway flanked by corn, I had several times traversed the distance of a world I didn't know.

Ceres is just large enough, scientists have found, to have a differentiated interior, one that probably includes a layer of water ice in its mantle. In 1995, astronomers spotted a large dark blotch about 150 miles across on the asteroid's surface; no one knows if it's a plain or a crater, but they proposed calling the feature Piazzi. In 2011, the robotic probe Dawn will arrive at Ceres, after having spent a year studying another asteroid, Vesta.

After the Ceres discovery, more asteroid finds followed: Pallas in 1802, Juno in 1804, Vesta in 1807. By the end of the nineteenth century, some 400 asteroids had been catalogued. By 1923, there would be 1,000, and the 1,000th would be named "Piazzia."

Now, using sophisticated computers and telescopes, astronomers find asteroids almost every day. In fact, the number of known asteroids began to double annually at the start of the twenty-first century. At the beginning of the new millennium, approximately 20,000 asteroids were known from their orbital motions. Many more exist, but for them we lack precise orbital data. So they await further scrutiny at the Minor Planet Center in Cambridge, Massachusetts, the world's repository for such information.

A few decades ago one astronomer dubbed asteroids "the vermin of the skies." The skies seethe with silent stones that hurtle and spin and crash. They stay there or fall toward us or tear toward the void. Piazzi might have suspected but couldn't have known for certain that asteroids are the progenitors of most meteorites, but only a year after Ceres was discovered, the astronomer Heinrich Olbers—who famously asked why the sky is dark at night—wondered publicly if meteorites might be chips of asteroids. He was the first to do so.

I know little else of Piazzi's life, apart from this fortunate find of his. But my traveling on this day, this anniversary of Piazzi's discovery, gives me an unexpected trace of inheritance, a movement backward and forward, and perhaps I like this feeling because I'm well acquainted with duality. Piazzi's hand still pauses in the sky, and the plane lances distance.

Some autumns before, at the beginning of my adult obsession with the night sky, I walked to our Kansas front yard with my wife, and we nestled into sleeping bags. Leafless hackberries spindled the dark. I explained a few things as I understood them—meteor, bolide, fireball, radiant. No meteorites reach the ground during meteor showers, I reasoned, because the showers are made up of such small objects, the spindrift tails of former comets, so that the grains burn up in the air or maybe some particles just drift about rather than land. I wasn't sure. We fell silent and watched with our backs to the earth, beside the prickly branches of a quince bush. It was a night of the Leonids, an annual meteor shower that every thirty-three years or so increases dramatically in its rate of fall—when a former comet's debris and the Earth cross paths. It was a night that promised spectacular visions.

Every few minutes, bright meteors surged by, all seeming to blast forth from the constellation Leo, the shower's radiant, the place where the trajectories appeared to converge. Three separate ones glowed as bright as Venus. Fireballs! At about 1:45 A.M. one streaked toward Procyon in Canis Minor. This fireball blazed a blue-white brighter than a full moon, illuminating the surrounding sky, and for five seconds of its journey, the ground itself, including the cozy, astonished watchers. It was as if a lightning bolt had struck but just stayed on and on. Suddenly, a fragment arced away from the meteor, like a spark from a campfire, and the remainder exploded in golds and reds. I'd never seen anything like it before.

I drew charts of the meteors, my penciled lines showing end points of trajectories, indicating where some of the fireballs turned into bolides—explosions—while my words noted colors. The charts are scientifically useless but I've kept them, in part because they represented an attempt at some methodology. Mostly, they've allowed me to relive that night.

After a while, we dragged our sleeping bags back inside, and despite the sweetness of our watching, I fell into a fitful sleep, not unusual, dreaming of vast ziggurats and *Metropolis*-like cities. I had to flee some planetary ending, I had to scuttle beneath a floor, light shafting between the slats. Lower down, lower, I had to find places where there might be safety. I had to hide, despite a dream-sky that had, at first, presented a welcome aspect, that soft, darkening blue.

AS I SIT IN THE AIRPLANE'S SEAT, my mind wanders to the fireballs of the past and to Piazzi's day, to today and to the half-glimpsed facts of the middle of a life. I shuffle some files on my lap and try to refocus. I'm going to visit the American Museum of Natural History, to root around in its archives for anything I can find about meteorites and meteorite hunters and, especially, to gawk at displays, to see for the first time one of the sky vermin—an ex-asteroid, a meteorite, a famous one, the first I wish to chronicle. I pull out photos of the Willamette, an all-iron, 15.5-ton meteorite that looks like a goitrous Frisbee, flat on one side, cone-shaped on the other.

Found in Oregon's Willamette Valley, the meteorite is the largest ever discovered in the United States. The meteorite's flat side is about 10 feet long and about 7 feet wide; from flat top to cone tip the iron is 4 feet thick—impressively big. Another measure of the meteorite is its history. It's been tussled over for the better part of the twentieth century. The first white claimant was the meteorite's "discoverer," a backwoodsman named Ellis Hughes, the only person, it can safely be said, who ever stole anything weighing 15 tons. A 1922 article said of him, "The pluck, energy and mechanical resourcefulness of Hughes furnished a human interest . . . that not even the dry-as-dust scientists were able to ignore." I'd say.

The story is bigger even than Ellis Hughes's heist, encompassing, I'll come to learn, three dramatic and sometimes comic trials, an obscure history of secret tribal ceremonies, high-stakes negotiations, a scheming naturalist, then, much later, earnest grade school lobbyists, *The Tonight Show*, and a challenge to the museum's ownership of the behemoth that threatened to wrest the meteorite from the planetarium, which had been literally built around it. The scientific facts are also compelling, for the meteorite gives evidence of a violent past. When the Willamette was sliced open and polished, its interior revealed striking patterns that show both crystalline growth and hot collision. The stories of the Willamette—or, as the Clackamas Indians called it, Tomanowos—embody all the forms of veneration ever accorded meteorites: spiritual, financial, scientific, educational.

I gape at the photos, so hefty is the meteorite yet so jaunty, like a sumo wrestler en pointe. Then there are all those holes, deep cavities, shallow bowls, apparent canyons, a kind of a cave system unearthed to earthbound eyes. What to make of the absences? The holes—and how did they get there?—give this serious stone an air that is at once grave and playful. This meteorite is rock-hard Swiss cheese with an attitude, a morsel from Medusa's dinner-party trays.

I think again of Piazzi with a fondness I want to invest with camaraderie. He was mapping the sky when he noticed something new. I am mapping too. Not lights in the sky but stories of its yield, the factual poetry of our desires. ("Desire," from the Latin *desiderare*, meaning "to await from the stars.") And I admit that when I'm not thinking of Piazzi or the Leonids or the Willamette or Ellis Hughes, I daydream about strangers I might meet at the museum. At this time, I don't yet know what befell Hughes and his wife.

Discovery and disaster, I will find, are sometimes kin.

Chapter I.2

Ellis Hughes's 15-Ton Caper

Today I leave my wife to her own sightseeing and cross the street from the hotel to the American Museum of Natural History so I can work in the museum's archives, searching for a trail that might lead to Ellis Hughes. Before I get to the library, I pass the cubic, glass-walled Rose Center for Earth and Space, which looks a bit like an oil refinery trapped inside a shopping-mall atrium. Its acre of windows curtain a giant sphere that looks light enough to float, like a spaceship-dirigible bound for the stars.

I spend hours in the library note-taking and copying. And I think on and off about what I save for last. This long walk past the dinosaur displays, then into the Rose Center, where I descend a sweeping staircase and finally see—for I've been delaying the pleasure all day—the Willamette meteorite, mounted at a rakish angle, suddenly actual, its flat top exposed outward, dramatically lit, while the bell-shaped side that was stuck in the ground points down toward a screen in the floor that plays movies of craters and alien surfaces.

The meteorite is its own world, with edges sharp and dull, with diminutive mesas and iceberg-like shapes protruding above valleys large enough to hold a child. Holes open to vistas, like arches in canyon country. You can grab it, and its heft is arrogant. Some spots gleam, made silvery by chains used the few times humans have managed to move it. Honestly, I'm giddy. The Willamette's as big as a small car, and I slap it in glee.

Near the waist-high railing that contains video panels and signs and that rims the entire meteorite, I overhear someone mention "the Indian reservation controversy." Someone else says, "This looks like modern art, almost, you know?" A man walks up, brushes it with a finger, shrugs, and walks off. Others, like this blonde schoolgirl, remain entranced. A woman with a deep Brooklyn accent asks, "Did they make this?" Her husband responds, "Noooo! It's a reeel meteorite—it fell from da sky to da Earth." Again and again, people caress the stone.

Just as the crowd clears away, I'm alone with the Willamette—with Tomanowos—and I touch a clean, flat surface where someone years ago had sliced off a specimen.

A woman, alone, in a red cashmere sweater and khaki skirt puts her hand on the rock, her bright red fingernails gleaming. Pensive, she stands nearby.

It's closing time, and outside I see that above bare trees, above the glass cube, there is a line in the sky formed by—and, really, this is too perfect—Aldebaran, the eye of Taurus, where Piazzi found the first asteroid, then Jupiter, then Saturn, then the first-quarter Moon and Venus. Serene, they hang there like lights on a dark arch, a place to walk beneath.

The following day, at the end of several hours of dusty files and photocopy requests, I'm back with the meteorite when a twenty-something man asks me, "Is it the real thing?" Yes, I tell him, it is. He nods, smiling, and I want to crawl into one of the cavities while he reaches to touch the iron.

Then I'm touching the meteorite—again—when my left hand grazes the side of a large depression on the apex; the texture is rough, a place of oxidation, and sliding my hand away I feel how my forefinger—*holy shit*—dislodges a thin shaving of meteorite, a slice of sky no wider than a thumbtack. I hold the gritty slice tight between thumb and forefinger as if it were a gem and I go to sit on a bench, where I pull out a small bag—it had held a bagel—and think that was amazingly, weirdly, freakishly too easy. Nervous, I look around, then slide the tiny chip into the baggie, go back to the meteorite, close my eyes, and thank it, though later I'll be uncertain as to why I engaged in such personification. The meteorite couldn't hear me speak my gratitude.

The next night, the evening before my wife and I leave New York, we're back inside the Rose Center, this blue ice cube of a building in which models of planets and galaxies are placed such that we see them as a little cosmos surrounded by glass walls; beyond them loom the buildings of the city and the falling night. We're at a jazz performance. At white-cloth tables with votive candles, listeners eat fancy snacks and sip beer, while I lean against the green-lit letters "Planetary Impacts" on the rim around the Willamette, at least until a guard asks me to not sit on an exhibit. Beside a metal post with the red-lit letters "Ammonia molecule," the Marc Carey Trio kicks into a Brazilian piece. The organ echoes a spacey sustain, Buck Rogers semi-salsa, pretty cool. Children dance and couples kiss.

"To learn more about the search for life visit the PLANET WALL."

To learn more about Ellis Hughes—well, good luck. There's no wall to visit, no sign to explain his discovery. Despite otherwise superlative exhibits, the museum labels don't tell the story of the man who stole this meteorite. The story of the Clackamas tribe's worship of Tomanowos, however, is conveyed—and there's a reason for that—and, of course, the science of impacts gets serious attention. Yet the very person who ripped the iron from the ground? All evening I've circled the meteorite looking at each sign, thinking, *Surely, this time, I'll find his name.*

ELLIS HUGHES HAD NOTICED the rock a day earlier on his way back from cutting firewood for a school, but paid it no mind.

The next day, though, a rusted saw near the big rock caught his eye, and he sat down on the flat boulder for a spell. What he did between that moment and when Bill Dale happened by we can't say. Ellis probably looked it over, noticed the cavities, watched out for sharp edges. He may have cupped his hands into one of the bowls to splash rainwater on his face. The rock stuck out of the ground by about 18 inches.

When Dale, an acquaintance, arrived, he asked Hughes if he'd seen the rock before.

"Yes," Ellis answered. "I saw it yesterday." Perhaps knowing what it was he'd found, Ellis Hughes chose that moment to pound the rock with a white stone.

"It rang," he recalled, "like a bell."

Some meteorites are iron bards. When struck, irons reverberate, and Bill Dale declared, "Hughes, I'll bet that is a meteor."

That fall day in 1902 Ellis Hughes—forty-two-years old, farmer, father of four children, owner of nearly 29 acres—made music with a moss-covered meteorite. Did he startle chickadees and jays and wild dogs? Did waves ripple in the water-filled depressions?

The sounds died. Ellis got up and pushed aside ferns and hazel brush tangled close to the rock. He noticed that the soil all around was the deep, dark red of blood or the last burnt umber of sunset. Rust. The meteorite had been there for a good while. The land, unfenced, had once been heavily timbered, but now fallen logs and tree stumps—some as wide as 7 feet—littered the scrubby hillsides. In a few years the meteorite would have been completely hidden.

The dark-haired, strapping Ellis Hughes wondered at his prospect. Born in Wales, he had arrived in Oregon about 1890 after having worked in mines in Australia and Canada. His wife, Phebe, a lumber-camp cook, had grown up in Oregon. They lived a hardscrabble life, one more of a piece with the nineteenth century than the twentieth, a century that in 1902 saw the invention of the automobile, the spread of telephone lines, and the first transmission by radio.

The men camouflaged the meteorite with hazel brush. And though Ellis later denied it, they may have placed the head of a dead calf on the rock—a strange thing to do if the goal was concealment. The land was littered with dead cows.

Believing the meteorite valuable, the two men wanted to purchase the land on which it sat, land owned by, of all entities, the Oregon Iron and Steel Company. According to author Douglas Preston, Dale headed "to eastern Oregon to sell some property to raise the necessary capital." Reportedly, the company was asking $17.50 per acre, and it's not clear whether the two prospectors ever made a bid; if they did, they were rebuffed. In any case, Dale vanished, and, Preston says, Phebe Hughes "began nagging" her husband.

"It would probably be there yet, but my wife . . . had ideas," Ellis Hughes said years afterward. "She was afraid somebody would go up and get it the next day."

Get it the next day? "Get" is too small a verb for this caper and "the next day" too short by about 365.

. . .

IF HE COULD GET IT to his land, Ellis Hughes could charge visitors to see the meteorite, then maybe sell it to a museum or scientist. That the meteorite was on—or rather *in*—someone else's property didn't matter. So throughout spring and summer 1903, he cut swaths through logs, saplings, and brush; no standing trees were more than a foot thick in trunk. Bizarrely, Ellis even cut an 800-foot clearing from the meteorite in the direction leading *away* from his house before cutting the path toward his home, nearly a mile away. This he intended as deception, but it's not clear why he thought a wide path in the woods wouldn't draw interest.

He wove wire to make a long rope. He borrowed block and tackle and more rope from a paper mill. He built a wagon, with tree-trunk wheels and a Spanish windlass or capstan. With that equipment, along with his fifteen-year-old stepson Edward, a horse, and some pulleys, Ellis Hughes was ready that summer to lasso a visitor from space. Clearing brush and moss from the meteorite, he set to, first attaching his tackle to the root of a fir stump. Using levers, he slowly ripped the meteorite from the ground—dirt dropping, roots dangling from its cone-shaped bottom. It loomed, poised between where it had been and where it was about to go.

"Excellent luck followed the laborious jacking-up process," astronomer and historian J. H. Pruett noted, "for when [the meteorite] was sufficiently raised . . . [it] flopped over and tumbled onto the truck flat side down." Ellis would later say, in satisfied tones, "It couldn't have been done better if you'd laid it there with your own hands."

Using some of the holes that hollowed through the meteorite, Ellis chained the rock to his creaking wagon. He had just lifted and lowered the equivalent of a half-dozen blocks from an Egyptian pyramid. He thought the meteorite weighed perhaps 5 tons.

Day after day, week after week, when other chores allowed, Ellis and young Edward Hughes and their horse—later Ellis would say he used two horses, not one—worked at moving the meteorite. In a 1960s newspaper photo, Edward displays a squarish chunk of the meteorite and the roundish stone with which

Ellis cleaned the meteorite. Edward, with receding hairline, big ears, glasses, and a smile, holds one in each hand. Asked how he helped, he said, "You know what shovels are for." The men attached one end of the cable to the wagon and secured the other end to the capstan, which was chained to trees in sequence, as he moved the whole assemblage. Sometimes he first had to fill in holes on the rough downslope.

"Slowly the wagon was edged down hill. Then the horse was driven in an endless path around the capstan, applying a tremendous . . . pull on the cable," wrote Portland science professor Erwin Lange in his account of the saga. "The wagon moved slowly. Some days there was no forward progress, and the greatest distance traversed in any one day was 150 feet. Each time the cable was completely wound up, the capstan had to be moved and reanchored."

Rains that autumn caused the wagon to settle into mud, making the capstan even more of a necessity. Ellis also laid out planks over the muck and moved them as the wagon reached the end of the rough road.

After weeks of labor, Ellis Hughes brought it home. He had moved 15.5 tons of iron three-quarters of a mile from its longtime resting place to his property. There he built a shed to protect the meteorite. But if Phebe exclaimed over the prize, Ellis never said.

SOON ENOUGH, from the end of a streetcar line two miles away, visitors from Oregon City and Portland trudged up the hill to see a bell-shaped rock from space. Many of them had seen or knew of a fireball that streaked above the area in 1883 and mistakenly thought the meteorite had fallen then. Visitors could gawk at the meteorite for a quarter, the price of a vaudeville show at the Marquam Grand Theatre back in the city.

According to the *Oregonian*, a reporter had found Ellis Hughes while he was still moving the rock—over which he'd thrown burlap because "the sun might warp it."

A San Francisco paper announced "Enormous and Valuable Meteor Is Found by Chance in Oregon." The article reported incorrectly that Bill Dale had found the rock, bought the land from Oregon Iron, and moved the meteorite ("several trucks were wrecked") to his property. "Trespass notices" were discouraging "kodak enthusiasts and curio hunters, as well as . . . reporters and . . . artists. The meteor is now under canvas and closely guarded."

In fact, anyone who paid could see the space rock, including excited naturalists such as Henry A. Ward. One Oregon researcher "secured several small pieces" while a local photographer took pictures of "the strange monster."

The authenticity of the iron was unquestioned. What became less clear was to whom it belonged. Thus the *Oregon City Enterprise* on November 6, 1903, put in print the not-so-quiet rumors: that the "heavenly monster" had been stolen. "With the official announcement as to the genuineness of the meteor," the paper opined, "the strife to determine its lawful ownership will likely prove interesting." That would be a serious understatement. The paper reported that the meteorite belonged at present to—Bill Dale, its discoverer.

Perhaps Dale had not disappeared. Perhaps he and Ellis Hughes were caught in a rivalry for the possession of the meteorite. But Ellis's main rival would be a different entity.

Among the guests to the one-man meteorite show were individuals representing Oregon Iron and Steel, men who offered Ellis Hughes $50 for the rock. Hughes said no. They offered $100. He declined. Then company official A. S. Pattullo reportedly said, "I'm not going to pay to see my own property."

OF THAT PROPERTY much more can be said now than could be said in Pattullo's day. For one thing, we know the Willamette Meteorite didn't fall in 1883. It didn't even fall in Oregon. Several thousand years ago, during the last ice age, an ice dam half a mile high periodically blocked the waters of ancient Lake Missoula. From time to time, the water's pressure burst the dam open, sending catastrophic floods, hundreds of feet high, that scraped across vast areas of the Northwest. It's believed that the Willamette Meteorite arrived embedded in an iceberg during such a flood. Though it resided in the Oregon woods for thousands of years, the Willamette probably fell somewhere in what is now Canada.

Henry Ward wrote the first scientific description of the meteorite and its discovery, noting, with the breathlessness of a salesperson—Ward had been a purveyor of natural-history objects—that "the region is a wild one, covered by a primeval forest of pines and birch, little visited and largely inaccessible." In fact, the area was cut-over, with that streetcar line only an hour's walk from the Ellis Hughes house. Ward had taken a cross-country train trip to get to Oregon, where he found Ellis to be courteous, modest, and no one's fool. Rooting around in mud, chilled by the winter rains, Ward was mesmerized by the Willamette meteorite, still the most visually striking meteorite in the world. Of the cavities on the flat side, Ward exulted, "Nothing can exceed the labyrinthine and chaotic outspread of these. . . . They make a confusion of kettle-holes; of wash-bowls; of small bath-tubs!" Unless I've missed something in Tom Wolfe, this may be the only instance of the word "bath-tubs" being followed by an exclamation point. But I understand his excitement.

With less poetry, others have since deduced even more about the Willamette. Researchers writing in *Oregon Geology* said that in the 13,000 years since the meteorite landed, rust had whittled the rock down from its original mass of more than 20 tons. Metal inclusions that included sulfur had also reacted with the wet Oregon woods to create an acid bath that ate away parts of the rock.

Its overall shape came, though, from its fall through the atmosphere. The conical form of the Willamette means it's what scientists call an "oriented" meteorite—a rarity. Its alignment when entering the atmosphere was stable; the meteorite stayed pointed in a single direction instead of spinning and tumbling. Thus, air and heat shaped the Willamette in a rather regularized fashion. It is the largest oriented meteorite in the world.

In space, the Willamette had been buffeted a bit more. At least twice it was heated and melted because of what meteoriticist Vagn F. Buchwald calls "violent cosmic shock." No other meteorite is known to have been melted twice. The Willamette may have wandered for a billion years before landing on the Earth.

All this makes the Willamette a rare specimen in what is a relatively rare category of meteorites to begin with—the irons, which make up about 7 percent of all meteorites. Of the several thousands of recovered meteorites, only a few hundred are irons. (The largest meteorite in the world is an iron, the 66-ton Hoba, which is still embedded in the ground in Namibia but partially visible in an outdoor display.) These space rocks are made of 85 to 95 percent iron, with the remainder in nickel. Dark, brooding, often showing regmaglypts—those pressure-and-heat formed depressions—iron meteorites look as though they *should* come from space. Far denser than terrestrial rocks, irons have a seemingly unnatural weight. In your palm, an iron meteorite suggests that gravity has taken on a new and unsettling authority.

When cut open and etched with acid, irons show a cross-hatching intergrowth of two nickel-iron alloys, high-nickel taenite and low-nickel kamacite. Meteorite expert O. Richard Norton calls this accidental design "the most remarkable and beautiful structure to be found in any class of meteorites." This lovely metallic weaving is called the Widmanstätten pattern, after one of its discoverers, Alois Widmanstätten. No earth rock shows anything like it.

The cooling of asteroid interiors was very slow; every million years an iron core might lose about 200 degrees Fahrenheit. Deep inside an asteroid, the cooling allowed kamacite to grow onto cubes of taenite, whose structure featured not corners but flat triangles. That's where the kamacite clung, and the growth pattern created octahedrons. Most iron meteorites formed this way and are named octahedrites. The pattern in octahedrites is subdivided from coarsest to finest. Technically, the Willamette is a medium octahedrite IIIAB. Its polished interior looks a bit like speckled linoleum or subflooring. The relative dominance of taenite or kamacite also leads to other classes of iron meteorites, and irons are further classified by the abundance of trace elements such as iridium and gallium, traces that allow researchers to link the meteorites to more than a dozen parent asteroids where the iron was either in the molten core or in a partially molten pocket elsewhere in the interior.

The appearance of these patterns depends on which plane is cut against. Each time what's revealed is a surprise.

In November 1903, as visitors stared at the heavenly monster sitting under Ellis Hughes's shed, perhaps only the scientists cared much about the complex interior of the rock. But even they would have known little of the Willamette's past, earthly and cosmic, and in any case, a different form of ancestry was being traced in the offices of Williams, Wood and Linthicum, attorneys at law.

ON NOVEMBER 27, 1903, Oregon Iron and Steel sued Ellis Hughes in Clackamas County Circuit Court, claiming he had stolen its property. Whereas

Oregon Iron and Steel now valued the meteorite at $10,000, Ellis's attorneys, the Latourette brothers, estimated its worth at a mere $100—the highest offer the company had made to their client. As the cases were being prepared, Henry Ward wrote that "sympathy lies mainly with Hughes." Of course it's impossible to know how the jurors felt when they convened on a 50-ish, drizzly April 28, 1904.

The transcripts reveal Ellis Hughes as evasive, unable or unwilling to answer many questions, in particular those having to do with cardinal directions. "I don't understand the compass," he testified.

When he first took the stand, Ellis claimed not to know who owned the land on which he found the iron. But he did admit that it wasn't his land—and that the meteorite was at his house. He also lied, saying that the meteorite was "set right on the top of the ground . . . it wasn't embedded in the earth at all." Finally, Ellis was forced to admit that the rock had been on land "said to be the Iron Company's."

After company counsel C. E. S. Wood asked Ellis Hughes about the road he had cut, he admitted he'd cleared the swath, saying he didn't want people to "know what I was doing," then adding that the area "was already cleared," then offering that "there is lots of roads in there," then saying that he wasn't trying to conceal his activity, then, in response to further questioning, explaining that he "didn't know which way to take it."

Wood continued to trip up Ellis Hughes as he asked detailed questions about land claims and property lines, with Hughes at one point confessing, "I don't understand this," to which Wood blurted, "You understand your own land?"

"Yes, sir," Ellis replied.

When Wood noted that the defendant had valued the meteorite at $100, the attorney offered Ellis the money then and there to settle the case. He didn't take it. Wood moved on to ask if Ellis had used wire or rope in moving the iron. Typically indirect, Ellis responded, "Yes, sir, sometimes I did."

After this bizarre testimony came someone whose testimony would carry weight not in this trial but in deliberations to come decades in the future—a Klickitat Indian named Susap, who, when asked his age, replied, "I am pretty old." Claiming to have known Wachino, chief of the local Clackamas, Susap said he had seen the rock many times before when "I was running around with this man that was hunting deer over there." As a boy Susap had listened to Wachino's story. It went like this: "That rock standing there is iron and there is a hole in there and when it is raining and the water drops in there and fills that rock and the Indians go in there and wash their faces there and put their bow and arrow in that they have got for war." The rock was called Tomanowos—various sources translate the name as Sky Person—and Wachino explained that the medicine men believed it had come from the Moon. Alas, Susap didn't help his credibility when he first denied and then admitted that he had been to Ellis Hughes's house to see the rock. He said that the miner had paid him to renew acquaintance with Wachino's moon creature.

After Susap came Sol Clark, a forty-seven-year-old Wasco, who said his mother talked of the meteorite in similar terms, saying it was magic.

The testimony over and closing statements complete, Judge Thomas McBride told jurors to ignore the defense's main arguments: that the meteorite had somehow been transported to the discovery site and that the object, having been venerated by Indians, was an abandoned relic and thus belonged to its discoverer. That it was, in short, a really big arrowhead. McBride's exasperation was evident in his jury instructions. "Did Hughes take this off of the Oregon Iron & Steel Company's land?" he asked. "If he did, then you ought to find . . . for the plaintiff." As to Ellis Hughes himself, McBride didn't quite call him a liar but drew attention to his motivations, his evasiveness, and the trial's mounds of "contradictory evidence."

That same day, the jury reached its verdict. Ellis Hughes was to return the meteorite—worth $150—to Oregon Iron and Steel.

In late November 1904, McBride denied Ellis's motion for a new trial and agreed to allow Ellis Hughes's attorneys, the Latourette brothers, to withdraw from the case; G. E. Hayes took their place. The Latourettes were soon working for men named Koerner and Meyers, who were claiming the meteorite belonged to *them*. They said the meteorite had been found on their land, then moved to Oregon Iron and Steel property, then moved to Ellis Hughes's. The meteorite had fallen on Koerner and Meyers's property—the men insisted on this point—and the jury in the January 1905 trial even went to view the purported crater. Had Ellis worked out a deal with Koerner and Meyers? Were the pair working their own scam? It was true that Ellis Hughes had crossed Koerner and Meyers's property with the meteorite.

A newspaper article datelined "Oregon City, Or., Jan. 19," and noted as a "Special," reported the second trial's outcome with the following stacked headlines: "FELL FROM SKY," "Meteorite's Possession Is Fought in Court," "FINDER LOSES HIS SUIT," "Infinite Toil of Ellis Hughes in Moving Mass," "LAY ON CLACKAMAS HILLSIDE," "Owners of Property Adjoining Oregon Iron & Steel Company Are Also Defeated in Attempt to Obtain Big Piece of Iron." One of the subheads was "The Rape of the Meteorite." Oregon Iron and Steel won again.

The story also reported an additional detail of the Koerner and Meyers ruse. The men had blasted or dug out a crater, then convinced nine people to perjure themselves in saying the crater had been there all along.

And the meteorite, this jury decided, was worth $10,000.

ALTHOUGH THE COMPANY HAD the meteorite moved from Ellis Hughes's land, the haul stopped just down the road at a place called "the Johnson farm." Ellis had rehired the Latourettes, who, with G. E. Hayes, appealed the case to the Oregon State Supreme Court. There would be a *third* trial. The appeal papers were filed with a clerk whose last name seems appropriate: Sleight.

Meanwhile, the rock remained under the care of farmer Johnson, to whom the guardianship proved an irritation. In 1938, Johnson's son recalled how his father jumped out of bed when he heard the distinctive bell-like ringing of the meteorite being hit by hammers and chisels; he'd run outside with his gun to chase off the hopeful thieves. But Johnson, like Ellis Hughes, probably allowed scientists and others to cut samples from the meteorite—for a price, of course. At least 100 pounds of fragments have been removed over the years.

The case appeared before the State Supreme Court on June 21, 1905. While conceding that Ellis Hughes had found the meteorite on property other than his own, his attorneys focused on the idea that the stone was "an Indian relic" and "abandoned property." Finders keepers, they argued. Ellis's attorneys repeated the assertion, an accurate one, that the meteorite had not fallen on the Oregon Iron and Steel Company's land; rather, nature had somehow transported it there. But with flourishes shading into falsehoods, the attorneys asserted—incorrectly—that the meteorite had not been buried when their client found it and that the meteorite was "fashioned, erected, [and] maintained" by the Clackamas. The tribe had, they said, "gouged out its interior into those fantastic pot-holes (*no other reasonable theory of their existence has been advanced*) . . . [and] *they erected it to a standing position upon a prominent knoll*" (emphasis added). This, they concluded, demonstrated that the rock was property and thus subject to salvage. The meteorite "stood there erect, like a sentinel, like a Tomanowos. (The court will take . . . notice of the meaning.)" Obtaining the rock, they said, had required no excavation.

It is hard to believe that Ellis Hughes's attorneys would not have been aware of Henry Ward's article about the meteorite, which had been published a year before in two journals, including the prestigious *Scientific American*. It was there Ward offered evidence that the cavities had been created naturally, not by human hands.

Oregon Iron and Steel's angry brief replied that the meteorite had been on its land and that case law in the United States showed that meteorites belonged to the owners of the property on which they were found, a principle still recognized today.

So on Monday, July 17, 1905—a hot, dry day in Salem—Ellis Hughes lost the meteorite again, this time for good, though if anyone thought the Willamette's legal tussles were over they would have been wrong: as the Oregon supreme court noted, "Nature does many fantastic things." For Ellis, fear of deportation was enough for him to let the matter drop.

Quickly, Oregon Iron and Steel hired men with horses to move the meteorite along the Willamette River at the junction with the Tualatin, where it was loaded on a barge. The iron would go on display at the Lewis and Clark Exposition in Portland as a major attraction. Did Ellis Hughes watch the meteorite disappear downstream? Did he even speak to stepson Edward? Edward, who was working for the crews moving the iron to the barge. Was Edward going to earn some extra money that might help his stepfather pay legal fees? Or did Edward's new job feel to Ellis like betrayal?

· *Chapter I.3* ·

Tomanowos

Unveiled from beneath an American flag at the Lewis and Clark Exposition, the Willamette meteorite seemed destined to stay in Portland, given the public's enthusiasm for the huge iron and its story. But among the usual dignitaries on that August day was a patron of the American Museum of Natural History, the eminent Mrs. William Dodge. She was the kind of woman who could contemplate buying a 15-ton meteorite because she was the kind of woman who got asked to buy a collection of table spoons for $16,000.

Henry Ward had already tried to secure the iron for the museum, when Hughes still had it, and he'd paid Ellis and Phebe Hughes $340 for samples of the meteorite. He exclaimed that the couple was therefore friendly to him, but he soon backed out of attempts to gain the entire specimen.

Through a Portland financial agent, the museum offered $20,500 for the iron, but when Oregon Iron and Steel approved the sale at its January 16, 1906, board meeting, it wasn't to the museum, it was to the estate of W. S. Ladd—which happened to control the company. A month later, however, the company reversed course, selling the Willamette for $20,600, cash provided by Mrs. Dodge. She gave the iron to the museum.

While Oregonians bemoaned the loss of the meteorite, museum officials were trying to snatch up all the fragments that had been chipped off in order to protect the value of exchange specimens—those traded with other institutions for different items—and to keep the meteorite from being further defaced, which was part of the condition of the sale. Associate Curator E. O. Hovey complained bitterly that Ward himself had taken more samples from the meteorite than he had documented and that he was selling them through Ward's Natural Science Establishment. The issue strained relations between the museum and the company for a long time, but eventually the company turned over its fragments to the museum.

The giant rock was put in a box and shipped by Union Pacific Railroad to Chicago, then by Erie Railway to New York. The meteorite got weighed too—31,107 pounds, minus the box. A receipt of the Oregon Railroad and Navigation Company for March 14, 1906, drily noted cargo received from Northwestern Transfer as "1 Car Meteorite (Boxed)."

In New York, spectators lined the streets to watch a horse team owned by Schillinger's Reliance Express and Storage Co. haul the Willamette to the museum, where, once it arrived, a nameless official recorded in longhand, "Willamette Meteorite arrived at Museum April 14, 1906." Director H. C. Bumpus told Mrs. Dodge that the floor would have to be strengthened before the rock could be displayed.

Later that year the museum made an effort to compensate Oregon for the loss of its fantastic specimen by offering a replica to Oregon Iron and Steel, which the firm accepted. This replica now stands on the University of Oregon campus in Eugene.

In West Linn, near where the meteorite was found, in front of the quaint, rambling Methodist Church on Willamette Falls Drive, one can today find another replica. Smaller than the real Willamette and lacking its impressive holes and caverns, the concrete replica at least recalls Ellis Hughes with a plaque. The replica looks worn, and despite the attractive landscaping—a trim hedge, sprays of purple cosmos—the fake Willamette attracts trash from passersby.

For years, the replica resided in front the fire station, just down the street. A photo from the 1962 dedication shows the two tight-lipped Hughes sons—Edward and Clark—standing by the monument. When the city remodeled the fire station, the replica seemed fated for the dump, until resident Ben Fritchie lobbied to have it saved and moved in 1985.

Fritchie, who once led a local parade under the theme of "It Came from Outer Space," has more than a passing interest in the story and the replica, for across the way is the hardware store he owned and at which shopped one of Ellis Hughes's sons—he can't recall which now. "A very quiet man," Fritchie says. "He would find his purchase and come to the checkout . . . with a sock full of money . . . then . . . say, 'Take what you need.'"

A FEW YEARS BEFORE the West Linn replica was dedicated, Oregon officials made an attempt to bring the real meteorite back for the state's centennial. They were quickly rebuffed. Decades later, though, two schoolgirls sought a permanent return—and made a lot of noise doing so.

In 1990 Annie Campbell and Stephanie Corey, nine-year-olds from Forest Hills Elementary School in Lake Oswego, gathered 40,000 signatures on petitions urging the meteorite's return. The third-graders gave a slide presentation for then-senator Bob Packwood, who promised to help HEWMAC—Help End Willamette Meteorite's Absence Committee. The students argued that because the American Museum of Natural History already had one of the world's largest meteorites—Tent, which explorer Robert Peary had taken from Greenland—the Willamette ought to come home. Government bodies passed resolutions of support. There were articles and TV news reports and parades. Engineers and truckers offered their services. There was the "Meteorite Rap," a video of which was sent to *The Tonight Show.* "It's the meteorite, we want it here to stay / It's not like the others that get all the fame! / It's in New York now so let's bring it back to town / Bring it back, bring it back / Bring it back now!" In early 1991, Annie and Stephanie got an appearance on television with Johnny Carson.

Meanwhile, an attorney drafted an analysis for his grade school clients, suggesting that Native American claims against the meteorite might complicate the museum's presumed ownership. One of the teachers who helped the students recalls that the museum "took our challenge very seriously and would have nothing to do with us."

In the end, there wasn't a chance that the American Museum of Natural History would give up the meteorite to some pesky pupils. No, nearly a century after the first arguments over the heavenly monster, the most serious claim yet would come from the ancestors of those who believed in Tomanowos. Susap—"I am pretty old"—and Sol Clark were long dead, but their testimony would be part of the reason why museum staff in 1999, as one person put it, "freaked."

In 1990, Congress passed the Native American Graves Protection and Repatriation Act—NAGPRA—which was intended to recover sacred remains and artifacts from museums. Nine years later representatives of the Confederated Tribes of Grand Ronde were visiting museums looking for objects that might qualify for such return. In September 1999, they saw Tomanowos in the middle of the construction area for the Rose Center, then being built around the iron. And they asked that Tomanowos be returned.

The group included several individuals, including Ryan Heavy Head, who worked as a NAGPRA consultant to the tribe. The contingent spent six days at the museum and took hundreds of photographs. Heavy Head told the Associated Press that museum officials were "civil but not necessarily cooperative. . . . Eventually they gave us hard hats and let us in, but when we started taking pictures, they freaked."

The tribe's bid to bring Tomanowos home was a bold step in a rather stunning revitalization of a people who had almost gone extinct. Just decades after Lewis and Clark's arrival, most of the Clackamas were gone, and in 1855, the sole eighty-eight remaining signed away their lands for cash they never received. They were sent to live on the Grand Ronde Reservation. In the 1950s, the Grand Ronde tribes—made up primarily of the Umpqua, Rogue River, Chasta, Molalla, and Kalapuya—lost their federal status. Tribal lands that had been thousands of acres in extent were reduced to a 5-acre cemetery. After years of effort, however, the Grand Ronde regained federal recognition as a tribe in 1983 and is now thriving. After such a recovery, regaining a meteorite that their ancestors had worshiped hardly seemed impossible.

According to a National Park Service official, the tribe had to demonstrate the sacred past and current ceremonial necessity of the rock. Use of the rock had diminished toward the end of the nineteenth century, but the meteorite's legends and its sacred songs had been passed down. An anthropologist noted in the press that meteorites "often have great spiritual or powerful magical significance for the tribes." Indeed, the Comanches, Apaches, and Kiowas worshiped the Wichita County meteorite, just as the Blackfoot and the Cree worshiped the Iron Creek octahedrite. The Blackfoot even considered puffballs—"dusty stars"—to be meteorites. Meteorites have been found in Hopewell mounds and at other Native American sites in Nebraska, Montana, Mississippi, and Arizona. For the Pawnee, meteorites were the children of the Creator. Sometimes, however, shooting stars could terrify, as when the Dakota were frightened by one especially intense meteor storm. Such storms suggested catastrophe. Native American beliefs are, I will learn in the months ahead, analogous to many in other cultures across the globe.

Despite the rich cultural associations between first peoples and shooting stars and despite the clear evidence of the prior worship of Tomanowos, the *New York Times* editorialized in favor of keeping the rock in New York, noting both limits to NAGPRA and the public interest in the meteorite. Neil deGrasse Tyson, the likable and well-known planetarium director, said that over the years something like 50 million people had viewed the meteorite. Then there was the not-so-small matter of the Rose Center having been built *around* the meteorite.

The tribe remained unmoved, stating that the only acceptable outcome was the meteorite's return. And that's when the museum sued. The lawyers asserted—echoing trials long past—that the Willamette was a "feature of the landscape, rather than a specific ceremonial object." The lawyers asked the court to declare the museum the rightful owner, once and for all.

Then, rather astonishingly, the tribe and the museum were thrust headlong into, well, mutual understanding. Tribal chairwoman Kathryn Harrison sat next to museum communications vice president Gary Zarr at a show in the new planetarium, the Hayden Sphere, which is plunked inside the Rose Center. They watched a 3-D rendering of the cosmos based on data from the Hubble Space Telescope. "You fly out into space to see the limits of the observable universe . . . our long address," Zarr tells me during my museum visit, when I ask him to recall how a compromise was reached. Zarr recalls how, after the show, Harrison told him, he says, "Our people call the Earth a blue pearl." Somehow the planetarium show softened attitudes. Somehow the Hubble helped set the stage for political compromise.

The tribe and museum agreed that the meteorite would remain in New York and that the Grand Ronde would have private access to it for ceremonial purposes. The museum agreed to tell the story of Tomanowos's spiritual importance and start internships for Native American youth. When asked if the meteorite still had powers, Grand Ronde Council Chair Kathryn Harrison said, "Look what it's done; it's brought people together."

There were still critics who wanted the meteorite returned, and even some of the interns expressed disappointment. "There were kids crawling over it," one intern was quoted as saying during her New York stay. "It was out of place. It should be here." Here meaning back in Oregon.

A FEW MONTHS AFTER touching Tomanowos in New York City and working at the American Museum of Natural History, I travel to Oregon to see what I can find. This trip I'm alone. It's the autumn of 2001. I left my wife in April.

The site where Ellis Hughes found the meteorite is in West Linn, not far from the intersection of Grapevine and Sweetbriar Roads. There, beside a driveway, is a meager swale where the Willamette once abode. A diminutive magnet can fetch up rusted bits of former meteorite out of the hole, a dirty, grassy scrape. Meteorite hunters still come by and ask—or not—to poke around.

Ellis Hughes. I've felt a distance between my interest in him and who he actually was, which is mostly lost after all, but in looking through documents kept by various owners of the discovery site, in the house of the current owner, I'm stunned to find that they include the Hughes divorce papers. In the court transcripts, I had learned that Hughes for a time was living with the elusive William Dale.

In a month, my wife will be granted her divorce from me.

On July 8, 1910, Phebe Hughes filed for divorce. She claimed her husband had "been guilty of cruel and inhuman treatment." His swearing and physical threats had begun less than a year after their marriage in 1893. She asked for property, custody, attorney's fees, and child support—$60 of her husband's monthly $90 income. He denied the charges, saying in his court papers that "if or whenever [he] has spoken otherwise than most kindly to the plaintiff, it has been because of the exasperating conduct of the plaintiff herself." The court dismissed Phebe's suit.

Eventually, Ellis himself filed for divorce, saying it was Phebe, not he, who was abusive. She had slandered him with allegations "of improper conduct with his own children." Despite this, he still protested her refusal to live with him or to let him see the children. Further, when Phebe had her estranged spouse arrested for failing to provide income for the children's necessities, he was "greatly humiliated." (The charges were dismissed.) Phebe, Ellis Hughes claimed, had "nagged him until . . . life [had] become burdensome."

There were issues with land and money, and their retorts appeared with the clockwork predictability of disputes between two people who had married themselves to other feelings.

He got his divorce on May 10, 1913. She got the kids and $30 a month from her ex.

Regardless of the truth of the allegations, the affidavits record little but heartbreak. One wishes, of course, that it had turned out differently. In the Hollywood remake of his story, Ellis Hughes would win the meteorite back and he'd be a good guy.

Three years after the divorce, on April 3, 1916, the sickly fifty-six-year-old Phebe Hughes died.

In a 1938 article about the still-bitter Ellis, then seventy-seven-years old, J. H. Pruett, of the American Meteor Society, wrote a condescending tribute. "If this sturdy, intelligent Welsh miner . . . will . . . foresee that distinguished scientists will be speaking of him long after his own neighborhood has forgotten him, he may realize that the honor thus acquired through the discovery of this country's finest meteorite overbalances . . . loss of its mere material possession."

At least Ellis might have taken pleasure in the demise of Oregon Iron and Steel, which collapsed in the 1920s.

Standing by a rough wood doorway in a black-and-white picture, Ellis Hughes looks past us. Born November 1, 1860, Ellis, at eighty-two, died at his home in Willamette on December 3, 1942. He left behind four children. One obituary noted his meteorite caper. If one looks carefully enough at a photo of Ellis taken while he was moving the meteorite and another taken decades later, he seems to be wearing the very same striped shirt, the same suspenders.

For a time, Clark Hughes, one of Ellis's sons, posted a homemade, carved wooden sign that, in 1978, read "WELCOME TO DRIVE IN RECORDED HISTORY. THE WILLAMETTE METEORITE FROM OUTER SPACE REPLICA and 500 FEET, HISTORICAL SPACE SATELLITE." At eighty-one, Clark with his dog Kirk lived on a remnant of his father's farm and one day took a reporter to a barn to show off his own replica of the meteorite. It could be viewed for a quarter—still the same price. The door opened to reveal a large replica "of concrete and plaster" twinkling on a turntable under the lights. Clark proclaimed, "For the good Lord, having no more use for the meteorite, caused it to plummet into our backyard," not quite getting the property lines correct. Just like his dad.

AFTER LEAVING THE SITE of the meteorite's discovery, still stunned by news of the Hughes divorce, I find Oregon City's Mountain View Cemetery but can't locate Ellis Hughes's headstone, which, for some reason, I absolutely must see. So I rush back to the cemetery office, past four high-schoolers cutting class and smoking, and the sexton comes along with me. Walking on the just-rained-on grass, she shows me, at my feet, the grave of Ellis Hughes. Twelve plots down from a big oak whose September leaves flutter in this breeze-after-storm. Two of those leaves rest on the flat pink granite marker that reads "HUGHES," and I gasp to see a kind of morbid reconciliation: Phoebe—spelled with an "o"—and Ellis are buried next to each other. Sunlight silvers different curves in the letters while a Steller's jay calls nearby. The sky clears to big cumuli. I bend over the grave, the sun warm on my back, and I see how rain has collected in the shallow cavities of their names.

I REMEMBER THOSE WET LETTERS the instant Brent Merrill tells me that they had poured water on Tomanowos. At the end of a long day in Oregon, I'm sitting at a conference table at the new tribal headquarters of the Grand Ronde.

Kathryn Harrison had shared with me in person what I had read of her feelings in the press. *Blood recollection*—that's what Kathryn Harrison felt. *You are my people*—that's what she heard the meteorite speak.

Brent, the reservation's congenial newspaper editor and public affairs officer, tells me that at the first ceremony the tribe conducted, just a few months ago, they had anointed Tomanowos with water brought from the lands of the Grand Ronde. He still has some of the water in a plastic container in his office. A Cheyenne River Sioux, Travis Benoist, led the ritual at dusk with the museum closed. Of what was uttered there, no one would say, but Brent shows me the picture he took of Travis, whose head is bent in prayer, whose left hand holds an eagle feather and whose right touches the meteorite. The ceremony was solemn, even after it had been interrupted by a ringing phone somewhere in the museum. Brent laughs about that.

"I'm not a religious person," he goes on. "But I do believe in a Creator. I felt complete reverence in its presence." Brent pauses. "It's *not* a feeling of being more special." He looks hard into my eyes. "I feel *lucky*. This reaffirmed my sense of family." I nod.

While Brent gives me copies of the tribal paper *Smoke Signals* for me to take home to my temporary apartment back in Kansas, I consider what Kathryn told me when I asked her what she thought of Ellis Hughes. "He was there to rescue it," she said, almost whispering.

In the midst of small talk foreshadowing the end of a visit, Brent retrieves the container of ceremonial water. He pries off the plastic lid, saying nothing. Then he takes my right hand, turns the palm up, dips his fingers in the water, and dabs my skin. I had told Brent of my findings, and of the divorce of Ellis and Phebe Hughes, and I had told him, briefly, of my own, which was impending. I look at my wet palm then I look up at him, quizzical.

"In case," he says quietly, "you need healing."

Book II

What Breaks Out Entire: The Eliza Kimberly Story

The fireball bore down fast toward the east, splitting the air. Shooting out of the southwest through a clear sky, it was as bright as the sun, if noisier. At five P.M. on Saturday, May 10, 1879, in Estherville, Iowa, people heard one, two, three, possibly four explosions and their echoes—the first like "the discharge of a cannon," the second like "a heavy blast," according to eyewitnesses. "When it burst there was a cloud at the head of the red streak which darted out of it," an account reported, "like smoke from a cannon's mouth and then expanded in every direction." The fireball had become a bolide: a meteor that explodes in midair.

Next came the growling, as though the sky were clearing its throat to speak. One man looked up. Sun-blinded, he couldn't spy a thing till he looked down and "saw dirt thrown high into the air at the edge of a ravine, one hundred rods from . . . where he was standing." A wandering chunk of the cosmos had just clobbered the ground, and a fraction's difference in orbit, in trajectory, or in the man's position would have been his death.

No one died in Estherville that day from getting coldcocked by a meteorite but large masses did fall, including one that punched through 6 feet of clay soil. It measured just 2 feet by 2 feet by 15 inches, yet the stone weighed *more than 400 pounds.*

Even the smaller pieces were a cause of consternation. Thousands of "pellets" blown off the main mass laid down a metallic rain. A boy saw a lake "peppered like in a hail storm," while milk cows panicked. High above, a cloud of pulverized space rock dissipated.

Estherville was a mesosiderite. One researcher called the interior of such specimens "a sort of fruitcake . . . consisting of a matrix of very brittle stone in which are imbedded metallic nuggets." On their exterior, the mesosiderites presented a "fearfully rough" surface with "ragged projections of metal," some Esthervillians recorded. "Rough and knotted" was how one scientist described the stones. It's easy to imagine more than one cut hand that Saturday night.

After the fall, a man named J. W. Lucas managed to cut off one of those "ragged projections of metal" and fashion it into a ring, rendering chance into ornament, into symbol. For weeks throughout 2001, from the spring to my autumn trip to Oregon, I think about the meteorite ring and where it might have gone, especially when I notice the absence of my wedding band—while drying dishes, say, or dusting. Did a descendant still keep the Lucas ring? Had it been slipped into a velvet pouch and kept in a drawer? Was it lost? Eventually, I write to the local paper asking for information, though later I give the ring little thought.

Estherville was one of a series of remarkable nineteenth-century meteorite falls in Iowa. In 1875, a meteorite had landed in Homestead. In 1890, Forest City witnessed a fall. Back in 1847, Marion had been hit. "It began to look as if the entire universe had its guns trained on the State of Iowa," wrote a meteorite researcher named Harvey Nininger in his 1933 classic *Our Stone-Pelted Planet.* Some Iowans fancied that a weird, subterranean pith in their crop-rich country attracted the attentions of outer space. Was there some corn-belt magnetism that lured meteorites to their demise? No, just chance, most unloved.

It would be understandable, however, if Eliza Jane Bennett believed in her later years that some design had been involved. A meteorite specimen in Iowa once attracted her attention, and this fact became, arguably, the central event of her life. She just didn't know it at the time. But it's unclear which meteorite she saw, where and when. Eliza may have seen a meteorite as a young girl. If it was, as some claim, an Estherville specimen, she would have been thirty years old. I know this is trivial, but the loss of detail regarding Eliza's first view of a meteorite strikes me as poignant. Time has erased that particular actuality.

Though I knew the bare outline of Eliza's story before my marriage ended, it is months after my separation and divorce that I begin to research Eliza's history. It will be years before I realize I began to keep a more assiduous journal when I had not only put my marriage in jeopardy but started to see how many moments from any life are always irrevocably gone. This is not earth-shattering news, but it pains me in ways I don't expect. I wonder, though, if I also savor such loss, just a little, because it means that the task of recording and re-creation is never complete, which further means that I can't—I shouldn't—imagine my own conclusion.

But whatever meteorite Eliza saw—and whenever she saw it—at least she never forgot. Or at just the right time, she remembered.

DARK, HEAVY STONES—lumpy, irregular in shape and size, like huge potatoes that have been punched, like hopeless loaves of bread. That's what Frank and Eliza Kimberly, née Bennett, kept finding on their farm in Kansas after they moved there from Iowa in 1885.

One sweltering day Mary tagged along behind her father—she was planting corn after his plow—when he "jerked the [horse] team to one side. He was awfully hot tempered and he cussed and said, 'We left Illinois and Iowa to get away from rocks and here I've knocked the edge off my damn plow with one of them.'"

Frank had to stuff the plowshare into a gunnysack to get the blade resharpened. The rocks wrenched arm sockets and neck muscles. A hard whack could make your teeth ache. Frank Kimberly hated the rocks. Eliza did not.

"Frank, do you see that rock? Do you know what it is?" Eliza asked, according to meteorite hunter Harvey Nininger, who met the couple in their old age. Imagine them under the sweep of Kansas sky, Frank in overalls, Eliza in a long, dirty dress. "Well, it isn't any ordinary rock," she said. "It's a meteorite." With seasons balanced on a blade—one sometimes broken by the rocks—Frank didn't give a spit.

In another version of this conversation, also written by Nininger, Eliza spoke in the plural: "Those are meteors."

"Meters!" Frank blurted. "What's that?"

Eliza explained that they were "shooting stars."

"Well, from the way they've been driving the plow handles into my ribs," Nininger has Frank declaiming, "I'd rather believe that they'd come from the other direction."

Like a novelist, Nininger entered the mind of the dubious farmer, who was said to have thought, "A rock's a rock. It's plenty of grief these have caused me already, without packin' 'em up to the shack to stumble over the rest of my life."

Which is what Eliza wanted Frank to do. She believed that if she could contact a scientist, he could confirm that these were "meters," which, she must have reasoned, could be sold to museums and collectors. A new crop for the farm!

The meteorite that Mary saw her father curse, she, with her mother, scraped out of the ground "with a hoe and a butcher knife," she said. "It was awfully heavy. We took a bucket along to gather up the little stones: There were several gallons of them all the way from the size of the end of my little finger and on up. Mother thought they were so queer and believed they were valuable. Daddy hitched a chain on the big one . . . and drug it to the house. He said, 'Liza, don't let that get away.' He was still mad about hitting it."

Liza probably knew she wasn't the only person to notice the rocks. The year the Kimberlys arrived, a wandering cowboy walked onto their land and stole three of them. He surely thought they might make him a profit, though it seems he didn't know what they were. His haul weighed more than 200 pounds, and his sweating pony couldn't make the eight miles to Cannonball Green's Stage Station. So the cowboy deposited the three stones in a gulch or badger's den, took ill, and on his deathbed fessed up about his discovery.

One of those rocks, which weighed about 100 pounds, sat outside an office in nearby Greensburg for three years. The owner, a Mr. Davis, called these stones meteorites. Naturalist Henry Ward wrote that various Kansas farmers knew the rocks were meteorites but didn't care. Only Eliza, it seems, laid careful plans to put them on sale.

None of this piqued Frank's interest or lessened his ire. He knew that not all years would be good ones—in fact, in 1887 and 1888 Kiowa County had meager growing seasons—so he had more important things to do than rassle with a bunch of damn "meters."

At least some could be put to use. They weighed down a cellar door, a haystack, a loose stable roof. Over on Jud Evans's land, one of them filled a gap in a pigpen fence. Sometimes the men used the heavy stones in sporting contests. Frank continued to move the rocks from the fields and out of the way of the plow. Frank and Jud even exhumed a 500-pounder, which they named "Moon Rock," a perhaps unintentional echo of the (incorrect) theory that meteorites originated in active lunar volcanoes.

A photo of Eliza Kimberly—she looks to be in her sixties or older—shows a large woman with a downturned mouth and clear eyes. She looks imposing. She must have been patient too, because Eliza spent five years putting up with jokes, laughs, insults, and curses. "Her pile of black stones," according to nature writer Edwin Way Teale, "became a standing joke throughout the region." A granddaughter recalled that those doing the ribbing—they called her "the rock woman"—"didn't do this too loudly as she had a mind of her own and did not hesitate to use it."

If Eliza was ever hurt by the doubts and gibes, there's no record of it now. She tended to the needs of the household—cooking, sewing, cleaning . . . rock-hauling—and she made time for one more chore: writing letters to professors, explaining that right outside her door in Kiowa County were visitors from space and would they care to see them?

ONE SUCH MISSIVE ARRIVED in the hands of a young professor of natural history at Topeka's Washburn College, the Harvard-educated Francis Whittemore Cragin. While no letter from Eliza Kimberly to Cragin seems now to exist, I learned from a variety of sources that he took a trip to Greensburg on Thursday, March 13, 1890, to personally inspect the alleged meteorites. He may have had alternative plans, for if the rocks were uninteresting specimens of this planet, as doubtless they were, then Cragin could do some shooting and collecting, maybe discover something new and write it up so he could make a name for himself.

The day Cragin traveled was "a spring-like day in south-central Kansas" and the geese were flying, according to a 1979 self-published book by farmer Ellis Peck. Peck had become the owner of the Kimberly land and, fascinated by its history, threaded fact through invented scenes in his charming homespun chronicle. But Peck's neglect of Mary Kimberly's words—which had been published in the local paper before he wrote his book—suggests that he disregarded sources when they didn't suit him.

So from Peck we imagine, correctly or not, the entire Kimberly clan—Frank and Eliza, Mary, now a young woman married to Jud Evans, and her brothers Myron and Ed—waiting in the house for the professor to arrive. Outside stood a pile of about twenty brown-and-black rocks. With inertia we might personify as patience, the rocks had waited beneath snowdrifts, thunderstorms, hot sun, and the prairie itself. A few more hours before Cragin arrived meant nothing.

Off the train in Greensburg, Cragin "hired a horse and rig," Peck wrote, got directions to the Kimberly place, then stopped to make a hotel reservation.

From atop Moon Rock, Myron proclaimed that the dust cloud on a distant road had to be a visitor driving a horse and wagon. Cragin had been prudent. If the rocks were indeed meteorites, the wagon could take back whatever he managed to get.

"The buggy stopped near the group in front of the home. The stranger who stepped out . . . was a muscular man of medium size, about thirty-two years of age. The visitor's friendly, confident appearance, his candid and direct manner evoked the . . . approval of the farm folk," wrote Peck. Cragin doffed his bowler hat upon greeting. A studio photo in the Washburn archives shows him in a group pose, holding just such a hat. He has large ears and receding black hair. Perhaps it's the tilt of his shoulders as he leans that gives an impression of both energy and awkwardness. He sports a mustache, and his eyes are deeply inset, though that may be a trick of shadow.

"How do. Mighty glad you come, Perfesser," Eliza said, according to Peck. "Yes, I'm the one that wrote you about these rocks because I'm purty sure they meterites and are worth some money. We was real anxious for you to look 'em over." Eliza violated frontier etiquette and showed Cragin the rocks before offering a drink, for which he had to ask. While he sipped, Myron tended Cragin's horse.

Soon enough, Cragin squatted beside the pile of dark stones. He hefted them, chipped them with a hammer, studied them with a lens. He asked Jud and Frank to help tilt the Moon Rock so he could see underneath. "Cragin slowly rose from where he had been kneeling near 'Moon Rock,' and laid the chipping hammer aside," reported Ellis Peck. "He removed his hat and took a handkerchief from his pocket and deliberately wiped the perspiration from his brow. The Kimberly and Evans families moved closer, tensely awaiting the verdict."

When F. W. Cragin announced that the rocks *were* meteorites, Peck said that Eliza blurted out, "Hah! I knew it!"

Cragin may have discerned that these were not only meteorites but rare ones. But to have said so would have weakened his position, for the "perfesser" now faced a negotiation more daunting than he might have guessed.

THE METEORITES OF THE so-called "Brenham fall"—named for the post office in Brenham, Kansas—constituted an astonishing find of pallasites, which, with mesosiderites like the one that fell on Estherville, make up the category of stony-iron meteorites. It is an odd poetry of coincidence that Eliza may have seen both kinds of stony-irons, the first, a mesosiderite, leading to the other, a pallasite.

The stony-irons—rarest of all, making up one percent of meteorites—are amalgams of different materials. In pallasites, a lacework of nickel-iron is dotted with olivine, so when cut open and polished, pallasites reveal the green-gold olivine—it has the translucence of prehistoric amber—set within a shiny metal web. The olivine, whose cousin on Earth is the gemstone peridot, seems to glow from within. Slices of pallasites look a bit like the coat of a metal leopard with green spots. Because sections of pallasites are so gorgeous, they are popular among those who can afford them. The well-known meteorite dealer Robert Haag once offered a hand-sized, polished slab of the Esquel pallasite for a cool $3,800.

Their appearance suggests that the opaque olivine, which looks so recently viscous, had oozed into the harder metal lattice, but just the opposite is true. Pallasites formed when the mantle of an asteroid began to cool. The mantle is that semiliquid zone between the crust of a planet and the first layers of the deeper core. Rich in olivine, the cooling mantle began to crack, and into those crevices crept up underlying liquid metal that eventually hardened. Impacts might also have pushed the metal up into the olivine zone. The metal webbing occasionally shows the Widmanstätten pattern, so familiar as the distinguishing feature of the interiors of irons. Pallasites get their name from Peter Simon Pallas, a mid-eighteenth-century naturalist who either saw in the field or had sent to him a strange mass—found in Siberia—though at the time no one was sure what it was or where it came from. It would eventually be known as the first sample of a pallasite meteorite.

The other type of stony-iron, the mesosiderites, have interiors with none of the easy glamour of pallasites. Complex tidbits, chunks, and strands of various materials indicate a messy past. Composed of metal, glass, and chips of igneous rock—and thus sampling core, mantle, *and* crust—mesosiderites were "subjected to transient and variable high temperatures and fragmentation, followed by slow cooling," write Robert Hutchison and Andrew Graham. But some mesosiderites—the ones that were very hot—cooled off very quickly, and strangely, all mesosiderites have significantly fewer samples of the mantle—of olivine—than they have of the core and crust materials. Within a single mesosiderite one can find blebs of nickel-iron with different Widmanstätten patterns, which indicates separate histories for these blebs, followed by impact shattering and later conglomeration.

One of the things that so excites scientists and collectors about meteorites is linking material that falls on the Earth with specific bodies in the solar system. Having matched the spectra of some asteroids with the spectra of meteorites found on Earth, researchers think that the "mesos" come from Vesta, that large minor planet with a diameter of some 300 miles. Harry McSween argues, however, that "the very slow cooling rates of mesosiderites argue for deep burial, so Vesta would have to have been destroyed to liberate the mesosiderites." A more likely candidate, he suggests, would be the impact of "an iron meteoroid . . . with a Vesta-like surface," a surface that was once volcanic. A 2001 research team wrote that mesos originated when an asteroid large enough to have a molten core was hit by an impactor of between about 30 and 90 miles' diameter. As for the pallasites, researchers think that they sample "A-type asteroids," whose composition appears to contain a goodly portion of olivine. Other research supports the Vesta origin for mesos but not for pallasites; the pallasites appear to have come from the same asteroid as some iron meteorites.

It would be decades after Eliza's finds before science could conjecture much about the Brenham meteorites themselves. By determining the age of the soil, scientists now know the fall took place 10,000 years ago, which means that paleo-Indians may have seen the blazing crash. It's thought that Brenham may be the largest pallasite fall yet known, and that the mass totaled about 6 tons. The meteor having exploded in the air, the different fragments thundered in supersonic flight but still weren't going fast enough at impact to form large craters; they just punched into the earth and, over time, were covered by dirt.

The insides of Brenhams: the hatchings and lines of metal—the Widmanstätten pattern, evidence of exquisite growth—but shot through with olivine, that green-gold, yellow-brown translucence. Brenham olivine is autumnal and ethereal, like an October forest and sky in a luminist painting. The curves of metal look like sinuous paths connecting lakes seen from on high. A slice of Brenham? It's a silver sponge that soaks up honey light.

I don't know if Eliza Kimberly ever saw the interiors of her "meters." If she did, she must have gasped. Who'd have guessed these ugly rocks held such material grace? It's nearly clichéd: how homely are the Brenhams and how lovely. It's reassuring, though, to be reminded that sometimes the beautiful does lurk.

ELLIS PECK FOUND NO RECORDS of that front-yard rock sale, no ledgers of cash received from a young professor anxious to gain every specimen he could. There can be little doubt he wanted them all. Cragin could make a profit—and a reputation—off the Brenham pallasites.

Eliza told Cragin to select any five of the meteorites, and he chose well, picking a 345-pounder, a 75-pounder that had weighed down the lid to a rain barrel, a 36-pound specimen from Jud and Mary Evans's land, another that weighed 125 pounds and may have adorned the roof of the Kimberly's barn, and, not surprisingly, the huge "Moon Rock."

Cragin and the men loaded the three smallest meteorites, then shoveled out a trench, according to Peck, so the wagon's end was even with the ground. Then the two biggest meteorites were loaded. Cragin apparently left the wagon with the Kimberlys, to be delivered the next day to a Topeka-bound train.

That night, Peck imagined, the Kimberlys celebrated. But just how the party unfolded, whether with Praise-to-the-Lords or uplifted bottles of whiskey (or both), we can only speculate. At the festivities, did someone coin a new name for the Kimberly land, a name that locals and meteorite hunters alike would use in the future? To heck with wheat. The Kimberlys were owners of a unique operation: the Kansas Meteorite Farm.

DESPITE HOW CHARMED I am with Ellis Peck's narrative, I find myself vexed by a chemist's claim in an October 1890 issue of *Science* that he had identified "a small fragment . . . as being of meteoric origin, and steps were taken . . . to obtain some of these masses." Had Eliza sent off samples *before* Cragin's visit? Had various scientists been in touch by mail? Did Cragin know ahead of time what he'd find? If so, had Peck made up the tense scene of Cragin's visit, despite Peck's assertion that he wanted his book "to be true to historical and scientific facts"?

It's after I find and reread the 1961 *Kiowa County Signal* reprints of Mary's recollections that I learn just how much liberty Peck had taken. According to Mary, her mother "took a match box and cut it in two. . . . She told me not to tell about sending the rocks away, but she wrapped . . . two pieces in some new patches left from a dress, and put one into each half box." One was bound for Cragin at Washburn, the other for naturalist F. H. Snow at the University of Kansas. Mary took a horse to Brenham and mailed the boxes for 25 cents a piece. She later mailed a sample to another scientist for the same price. "That was an awful lot of money, when we were so short we didn't always have enough for bread," Mary said.

She remembered how a rider sped on horseback toward the Kimberly house a few days later. While she hoed outside, her mother was in the house with son Ed. The rider approached so swiftly that "mother was scared and thought it was a death message, and I was sitting kid fashion waiting to see what it was about."

"Hold everything," said the telegram from the University of Kansas, as Mary told the story. "We will be there immediately."

Soon another rider appeared with a telegram that said "sell nothing. It's a meteorite." Mary recalled that it came from Tiffany expert George Kunz, but it might have been Cragin. In either case, the Kimberly women were beside themselves.

Eliza said to her daughter, "How are we going to tell Dad?"

"I'll go tell him," Mary replied. "He's out to the barn. He can't do anything more than lick me."

Mary told him the news.

"Now she's played hell, ain't she?" said a surprised Frank Kimberly.

Soon enough more than one professor visited.

"That time the professors came to look at the meteorites they had no way to weigh it, so one said, 'We'll test its specific gravity by dropping it in this barrel of water.' Before we could stop them they socked it into the barrel. It was all the water we had," Mary remembered. "They were awfully sorry—and went with daddy to get another load of water."

In a typescript manuscript, Harvey Nininger also wrote of more than one scientist arriving at the same time. Without citing sources, he told a humorous story of Cragin and Snow eyeing each other on a southbound train before they each admitted they were after the same goal: Eliza's meteorites. At that point, they joined forces. In this version, it's a "doctor named Blank" who alerted scientists to Eliza's meteorites and who even met the scientists and showed them various specimens.

Long after I think that these were the only competing narratives of the Brenham discoveries, Jessy Randall, a librarian at Colorado College, shows me the story from Cragin's own hand. Cragin, who went on to teach at Colorado College in 1891, had written to his mother the spring of his visit with the Kimberlys. It's the only clear, direct account from Cragin that I know of. Cragin says it was Eliza's father who "thought [the stones were] meteorites and on various visits took some away with him (one 211 lb. one) to Iowa." The Kimberlys sent samples to both Kansas geologist Robert Hay and to Cragin, though Cragin made it there first. He'd been authorized by Washburn College to "buy all at 25 dollars"—which he could not do—so instead he borrowed $400 back in Topeka, returned $25 to the college president, and bought the five meteorites. Later, the president tried to claim the entire purchase for Washburn, but the school's board and some local citizens rose to Cragin's defense.

So, of course, it wasn't a matter of awestruck discovery and mutual fulfillment. The costs of wonder in this case were the literal bill of sale written out by Cragin to Frank Kimberly (not Eliza) on March 13, 1890—for the five pallasites weighing 1,051 pounds and costing $400. The costs of wonder were also, in time, Cragin's job at Washburn.

He paid Kimberly $25 with a promise to return with the remaining $375 two days later. He did. Cragin also told his mother he sold "926 lbs. for $1,583" to George Kunz at Tiffany, thus making a very handsome profit. At the time of the letter, Harvard University was offering $525 for Cragin's 125-pound specimen, the latter offer being more than enough to make up his $400 credit. Cragin planned on keeping a half-pound "original sample" as "a memento."

Eventually, I ask archivists at Colorado College if there's anything else in their holdings concerning Cragin, and they unearth an unpublished study that quoted a former student of his. The former student remembered the plot of a story published in the magazine *Youth's Companion.* But I'll never locate its original publication. According to this version, a professor named "Cragstone" had learned of the meteorites in a newspaper article, for it seems the meteorite had just fallen—a fabrication. After his train ride to Kiowa County, Cragstone walked to the farm, where the owner had one meteorite he would sell to clear his $400 mortgage. Cragstone couldn't get the money from his college, so he went back home, borrowed the funds, and on his second train trip saw a rival geologist. After they both arrived near the Kimberly residence, the apparently quite athletic Cragstone *ran* through streams and fields and got to the farm first. In this version, he donated half his find to the college where he worked. And a further retelling of this version concludes with the meteorite sale allowing Mr. and Mrs. Cragstone to afford a long-delayed honeymoon or "honey-star," as the writer said Mrs. Cragstone called it.

Ellis Peck would have loved all this invented drama.

Reading these myriad stories reminds me of the theories of some cosmologists. They say that in any given moment, an infinity of outcomes takes place, a splintering of realities into parallel but distinct universes, where all things are possible and all things happen. In one of those universes, Cragin—er, Cragstone—runs through muck and gets to the farm too late. In another, he tries to cross a stream but drowns. In another, Eliza grabs the $25 Cragin has and sells at that price. In yet another, Cragin loses his prize stones to another scientist but it doesn't matter because he's fallen in love with Mary Kimberly and steals her from Jud Evans. In yet another universe, Eliza keeps a long journal detailing her finds and her feelings about Frank's insults over the stones; after the sale, she divorces him and moves to Tulsa or Los Angeles. In one such universe, I find her journal and I'm brought closer to the person I expected to know better. In this universe, the one I'm in, it seems I can pretend any of those possibilities are true and call them fact. But I won't. I'm left with splintering narratives, uncertain contingencies. Suppositions. Fancy.

Two—no, three, four, five—narratives of scientific confirmation. They all speak to something wider, to the emotional truths of the story. The noticing, the obsessing, the collecting, the desiring, the dreaming. The desire for possession and profit, for possession and legacy. Variations on a Brenham myth, an Eliza archetype.

And I suppose there's a small-minded meanness in me that wants to correct Mr. Peck, even as I rely on him and enjoy the tale he spins. I love how he paces Cragin's revelation that the rocks are meteorites. It's a perfect little scene, accurate or not.

Then too I contemplate the delicate conspiracy of mother and daughter, those images of Eliza and Mary covertly cutting up a matchbox in order to post little nuggets of space. And in this version, Frank seems a darker, harder figure, spiteful even, drunk maybe, who knows?

Then there's Harvey Nininger—a man who will loom in importance for me and who misremembers Eliza's name as her daughter's, Mary. He said Eliza was young when she saw the Estherville specimen, when she would have been a woman; he misspelled Cragin's name and in one account got the date of Cragin's visit off by five years. More than once I'll come to think of these oversights when I read of his own life with rocks from space.

And there's Cragin's letter to his mother, which tells a bit of the story from Cragin's point of view but reveals nothing of Eliza or Cragin's observations or feelings. The letter ends with his decision to stay in science and teaching rather than become a Christian missionary.

The "Cragstone" narrative shows just what suckers we still are for a bit of excitement—a professor outrunning a carriage? It sounds like a cartoon.

There's more to all of this. Ellis Peck wanted to tell a story based on his chance association, because he owned that old Kimberly land. And this fact triggers recognitions. For I love how stories emerge from the accidental associations of a person with a place. It's one of the ways in which I bear myself in the world, it helps me feel connected. The competing accounts also force me to acknowledge again the subjectivity of memory and the effacing power of time. Does part of me want to walk into the gaps between these memories and accounts? Not to invent details but to experience emptiness?

As my marriage collapsed—thirteen years dissipating in the course of a month—my wife remembered what I had forgotten, and I remembered what she had forgotten, and where we remembered together it was often with distinctly different images and story lines. Memories have gaps between them the way planets do.

History is its own space, and I never found a letter of Eliza's. Her distance from me seemed somehow intimate. It was familiar.

THE MARCH 19, 1890, *Topeka Daily Capital* included what seems to be the first public notice of the remarkable discovery in Kiowa County. A small item reported that Cragin had made a presentation on the meteorites to the Topeka Society of Natural History. He shared the stage with a man showing pictures of his trip to Arizona and California and another who displayed some white maple blossoms. Cragin wrote a separate article himself but didn't reveal how he acquired the meteorites. He described their pittings as "shallow cupules, resembling the marks of finger-tips on a body of putty-like consistency." The 125-pound specimen could be viewed at the Rowley Brothers pharmacy, he concluded.

That summer Cragin wrote to the trustees of Washburn College, hoping they'd buy one of the Brenhams. He offered a 30-pound Brenham "at cost" for $110. "Pallasites are so rare that only one other fall than the Kiowa has occurred in America," Cragin said, adding—and this seems like conscious hyperbole—that "this may be the last that will be derived from that field." He also mentioned that "some little doubt" had come up regarding "the title of the Kiowa County meteorites," but downplayed the risk. A walk among display cases in today's science building at Washburn University reveals no pallasite, whole or cut.

After his notice in the *Topeka Daily Capital,* Cragin more or less dropped from the scene. Other scientists began to publish material about the meteorites in scientific journals.

The geologist Robert Hay called on the Kimberlys the day after Cragin. It seems that he bought no specimens but he did sketch a simple, evocative drawing of the homestead. The piece is strongly horizontal, with the prairie broken by a few rough bushes, sandhill plum perhaps. There's a windmill, a house, two low outbuildings, a white sky. Hay also mapped where specimens had been found and shared the data with Tiffany's George Kunz, whose June 1890 article in *Science* was the first major research article on the discovery. Hay eventually obtained some of the meteorites and reported the discovery of clusters of small meteorites that ranged across portions of the fall site.

The third man on the scene, F. H. Snow of the University of Kansas, visited the Kimberlys on March 17 then on March 22 and again on March 29. Snow obtained five specimens, including those that the sickly cowboy had nabbed. He sold all to Kunz except for an "irregular plum-shaped mass," which he kept. He estimated that the total weight of specimens thus far was more than 2,000 pounds, all having dropped in a zone about a mile long. He also reported that the meteorite sales had "cleared the [Kimberly] farm from a heavy mortgage, and placed the family in comfortable circumstances." Indeed, Eliza Kimberly became, wrote Edwin Teale decades later, "the richest woman in Kiowa County."

Though Snow merely mentions how one specimen he obtained was found "only after a long and anxious search," other writers have managed—somehow—to elaborate on various aspects of Snow's activities. According to Harvey Nininger, Frank paid $50 for the meteorite that a local attorney had owned, then resold it for more than $200, presumably to Snow. According to Teale, Eliza—who seemed now to be treating the meteorites as a business enterprise—bought this meteorite for one dollar and sold it for $500. A *Kansas City Star* reporter wrote that Frank bought the stone from the lawyer for $3 and sold it to Snow for $150. The paper also claimed that "Kimberly refused to consider Dr. Snow's solicitations, but finally . . . led [him] around the end of an old hay stack and . . . revealed a small iron meteorite which was immediately secured by Prof. Snow."

This collecting frenzy kept Frank looking for meteorites on the Kimberly land and in town, where, like Eliza, he bartered and bought rocks from locals. He was suckered at least once when a colorful rock he bought turned out to be burnt coal cinder. As the money poured in, Frank regretted how he had dumped one of the "meters" in a field years before instead of bringing it to the house. "I had a fortune in my hands and threw it away," he said.

The frenzy pitted rival parties against one another, as the *Kansas City Star* made clear. After paying Jud Evans $500 for a 218-pounder, Snow and Evans found that Tiffany had telegraphed an offer of more than twice the price. "The agent threatened litigation," the article said, "but Dr. Snow finally convinced him that the $500 sale was closed." When a museum in Austria offered Snow $1,500 for the specimen, he wouldn't sell.

The one original field document I uncovered in my Kimberly research—Snow's notes about his visits to Kiowa County—gave me cursive glances into the collecting. Snow noted weights, ownership, appearances, uses, and which ones Cragin had managed to acquire. For example, the pallasite kept atop a stable roof was nicknamed "The Scraggly Meteorite." Snow briefly told the story of the now-dead cowboy and his stones, wrote of a Mr. Harmon's meteorite (this was the one that had held down the haystack), and a Mr. Tritafore's meteorite (which had weighted down the pickle jar).

I don't know why I had expected Snow's notes to be anything other than what they were: documentary, with nothing recorded of negotiations, conversations, gestures, the highlights of the day. No portrait of Eliza. I guess I'd thought that a naturalist like Snow might have mentioned singing birds, flowers in bloom, the clouds in the sky. But no. When I held his pages, so small within their leather binding—and written of the places and people that so interested me—my disappointment only widened the gap between me and Eliza Kimberly.

So far as I could tell, Eliza's obsession wasn't that different from Ellis Hughes's. The rocks mattered for the profit they might bring. This made me sad. Then again author Edwin Teale wrote that Liza "was long interested in the study of the stars and the constellations. Her eyes, I was told, were unusually far-sighted." So maybe she had a touch of wonder about the sky after all.

One day, glancing in desultory fashion at George Kunz's article on the Kiowa meteorites, rereading the facts I'd already absorbed, and still feeling stung by all I'd never know, I stop. Kunz noted how "the olivine crystals are very brilliant and break out entire." Lacking the contact with Eliza Kimberly I had so hoped for, I thought instead of the meteorites themselves and understood, suddenly, that the mineralogical was also metaphorical. Here was another form of wreckage, the sliced-open pallasites that sometimes dropped their olivine crystals, leaving a gap where the sheen of gem had been. Yes, the beautiful breaks out entire.

I REACH KIOWA COUNTY with Edwin Teale in mind, for his story of the Kimberlys was one of the first things that excited my interest in meteorites. When Teale visited Kiowa County, it was part of an adventure undertaken with his wife as he traced the passages of the four seasons across the continent. By the time I drive to Kiowa County, I'm not only divorced, I'm with someone new. I'm touched again, as I hunt the stories of the Brenham pallasites and the personalities associated with these remarkable meteorites, by the necessity and occasional frailty of what binds us to the world and to those we trust with our lives and feelings. Kathe sits in the seat beside me as we drive south out of the flat-topped Flint Hills and toward the flatter prairies of south-central Kansas. It's spring 2002.

Kathe's been back with me for a few months. Not long after we fell in love—I was still married, she was ending her own marriage—she had left to teach in China, giving each of us distance in which to think about our future. She was there when I had visited Oregon, in my search for the history of Ellis Hughes. After a late summer and autumn overseas, Kathe had returned. Her father had fallen ill and soon passed away. The past few months have been ragged and difficult, but the answer to Kathe's question was yes. The question was, Did we belong together? The day we arrive in Kiowa County, a sun-warm, breeze-cold March day, about a year has passed since I left my wife and house. I did so with neither grace nor honesty.

This day of paintbrush clouds is one of the last I'll spend in southern Kansas, where for years I've come during migrations, to see thousands of pelicans at Cheyenne Bottoms or rafts of sandhill cranes above the marshes of Quivira. Later this year Kathe and I will move to Utah, where I'll start a new job.

I keep thinking of that summer decades ago when Teale and his wife watched the Perseid meteor shower from near the very ground that yielded meteorites to Eliza. But it's not his descriptions of the Perseids or the Kimberlys that have haunted me. It's how, at the end of his story, Teale watched a hungry stray cat as it ate graham crackers he'd set out for it. Then how, under the moon, a jackrabbit appeared and how for Teale it was a moment from "an age that antedated fear . . . [a] moment of silence and moonlight, of strange and lonely companionship . . . a moment in another world." Then the "very small kitten filled with courage, ambition and graham crackers" began to stalk the diffident rabbit. It was only after falling in love with Kathe that I understood this cat was emblematic of all hungers, of how they might, in fact, possess dignity.

This late afternoon we stand near, not on, the old Kimberly place. The prairie is bathed in honey light, and wind shakes cedar shelterbelts and gives the hawks easy soaring. It's 112 years to the day that F. H. Snow last visited Frank and Liza, but I can't cross the property lines he did. The current landowner has been coping with ill parents and couldn't tell me when he'd be there. I'd explained that I wasn't hunting meteorites—though if I'd found one, would I have kept it?—but just wanted to see the land and the old farmstead. He wouldn't allow me to walk the place alone.

Squaring a map from an 1890 issue of *Science* with a 1968 topographic map, I've marked an X, and we've driven to an edge of the land lined with cottonwoods. Red-winged blackbirds and meadowlarks call and sing. Beside a sandy road, I find no space rocks but bend to take a locust pod that I'll keep near at hand for years.

In the drive leading up to what was the Kansas Meteorite Farm, robins chirp and flutter. I see a Quonset hut, a house, a trailer, a large antenna, slumping piles of hay, and rows of trees. Light keeps slipping toward its amber end. So where Teale had watched Perseids, we watch a landscape at sunset—though I also think back, chagrined, to last November's Leonid meteor shower, in 2001. Kathe had seen it while in China. Shouting students brought her outside to witness a flurry of shooting stars. Alone in Kansas and some three years since seeing a few Leonid fireballs with my ex-wife in our front yard, I dragged my sleeping bag to the apartment-complex lawn to find cloudy skies. I went back inside and slept. The next day a friend told me that the clouds had briefly cleared and through that gap he'd seen a spectacular outburst of meteors. Great, I thought, *great,* here I am writing a book about the damn things and I had slept through the sight of a lifetime. I'll never stop berating myself for missing them.

We drive along the edge of the old Kimberly property when I'm stopped by the sight of a stunning gray, black, and white bird, a loggerhead shrike that swoops from a tree, catches something in its beak, and spears it on a tree thorn. Shrikes earned the nickname "butcherbirds" for this behavior, something I've read about but never seen. After the bird flies off, Kathe spots a green caterpillar oozing on the thorn. Shrikes will spear their prey thusly, producing caches that the birds sometimes forget they have, as if the collecting itself were sustenance.

Somewhere back behind the shrike's flight, Cragin is stooping over rocks, Hay is looking up from his sketch, and Snow is shaking Frank's hand, the fence between past and present tense is broken, Eliza counts bills, and I scoot my boot beneath the lowest wire strand to cross the line with just my toes. The parallel universes waver in and out in the wind.

Eliza and Frank retired in 1920, but their youngest son—Edward, who had been sickly as a child—died just a year later, so they returned to run the farm. The couple passed away in the early 1930s.

Others walked the Kimberly land long after the visits of Cragin, Hay, and Snow. Harvey Nininger excavated the shallow Brenham crater in the 1920s and found more meteorites. In the 1940s, an electrician named H. O. Stockwell rigged up a metal detector on a wheelbarrow and found a 1,000-pounder. A few years after my visit, local landowners, meteorite hunters, and scientists will discover more Brenhams, including Steve Arnold's spectacular find of what will be claimed as the largest oriented pallasite in the world, a 1,400-pound mass that will draw a $200,000 bid in 2007, three to four times less than expected and not enough for an immediate sale. The finds will spur plans for new meteorite attractions in the area and T-shirts that read on the front, "World's Meteorite Farm Kiowa County, Kansas" and on the back, "466 LBS—Found 1886, 740 LBS—Found 1947, 1000 LBS—Found 1949, 1400 LBS—Found 2005, Next?"

In Greensburg, Kathe and I see Stockwell's 1,000-pounder on display. In a concrete building that serves as a tiny museum/gift shop for both The World's Largest Hand-Dug Well and this "Space Wanderer," the pallasite sits in a back alcove past knickknacks and postcards. "The Space Wanderer" is much smaller than a bale of hay and hulks in a glass-and-wood display case. The black–rusty-red–dingy-gray rock has one large grapefruit-sized pit in its top and many other dimples and in a few places some bits of old opaque goo—olivine extrusions? On the end, a section's been cut away, etched, and polished to show the metal matrix with its few dark lines. The rock is rough and dusty and, in sum, a bit scrofulous.

When I ask the woman staffing the museum if most visitors come to see the well, she says yes, but adds that "old-timers come in who saw the meteor shower this one's from," nodding toward the display case. I smile, happy to have seen the ugly ol' Space Wanderer and unwilling to explain that such visitors would be old-timers indeed.

Years from now, the meteorite will be buried again, under debris, after a tornado destroys most of Greensburg. In the recovery of the town, the meteorite will be found once more.

MONTHS AFTER VISITING KIOWA COUNTY, I hear from a descendant of J. W. Lucas: the Lucas ring. It *had* been handed down over the years, and the current owners send me a photo of the simple gray band that had been cut from a fragment of the Estherville mesosiderite. The woman who writes tells me the ring is rarely worn, since it nettles the skin. And then I remember a different ring, one never made. Harvey Nininger had told a jeweler to use some Brenham olivine for a ring he'd give to his wife, but the crystal cracked beneath the jeweler's tools. I think of these two rings a good deal now, because I'm keeping my old wedding band in a drawer, unsure what to do with it.

And though I've failed to come any closer to Eliza Kimberly, I've come to know F. W. Cragin better than I could have expected, and it troubles me.

Most of what I've learned comes from historian Dorothy Price Shaw, who had the bewildering job of organizing the vast accretion of Cragin's papers, books, notes, and maps. She explained in an article that at forty-five—"well on the way to becoming a renowned geologist, married, with three children"—he had quit his job at Colorado College. Cragin then pursued an independent and grandiose plan to publish massive volumes of Western history. He traveled incessantly, read voraciously, interviewed pioneers and their descendants, and scribbled cramped notes on the backs of long-kept student tests. He sought patrons too, though not very successfully.

He never published his books of history, though he's known today to Western historians for his huge corpus of notes and manuscripts at the Pioneer Museum in Colorado Springs, the town where Cragin made his final home. He had made his mark as a kind of collector, not as a scientist, not as a historian.

F. W. Cragin also slowly lost his mind.

Not long before he died "one of his acquaintances . . . declar[ed] him 'balmy,'" Shaw said. Part of the difficulty was that, in Shaw's words, "his brilliant mind . . . burst like a skyrocket in a hundred aimless paths." Another author said Cragin wasted "incredible amounts of time investigating trivial historical details." Ill and nearing eighty, he lived his last months in a poorhouse, having lost his dreams, his two marriages, his money. He had loitered among friends' houses in order to be asked for dinner. He had sold off parts of his collection to pay off debts. Cragin died in 1937, not long after the Kimberlys.

I've learned more about the Brenham transactions than I had expected, and the multiplicity of narratives underscore so much—the randomness of actions, the difficulty of memory. But Eliza remains pretty much a cipher, and Cragin remains a warning, about the dead ends of obsession, whether a scientist's or a meteorite hunter's or a writer's.

It turns out that before donating the papers that Shaw had to arrange, one of Cragin's children, a daughter, faced what she called with italicized emphasis "a *hopeless* task"—sorting out the "great disorder" of her father's "thousands of papers." She "worked many days trying to save what he would want saved, but . . . finally we had two *truck loads* . . . hauled to the city dump."

Perhaps more histories of the Kiowa County meteorites have rotted there ever since, discarded facts, discarded stories, the words of lives left to waste.

Book III

Higher Latitudes: In Search of Peary's Meteorites

· Chapter III.1 ·

Please Bring Your Wu Wei to the Upright and Locked Position

Cramped in my airplane seat, I read this passage from Zen writer and translator Alan Watts: "Yet as you get older and wiser it is not just flagging energy but wisdom that teaches you to look at mountains from below, or perhaps just climb them a little way. For at the top you can no longer see the mountain." I sigh then read on: "And beyond, on the other side, there is, perhaps, just another valley like this . . . and you must not mistake this for a kind of blasé boredom, or a tiring of adventure. It is instead the startling recognition that in the place where we are now, we have already arrived."

Where I am now is in an airplane stuck on a runway in Dallas, because we aren't being allowed to pull into the gate, what with all the lightning. Where I have arrived is not-at-my-gate, which may mean I'll miss my flight from Dallas to Baltimore. Which will mean I'll miss my flight from Baltimore to Greenland. Which, for the second year in a row, will mean I'll have failed to arrive at a remote U.S. Air Force base called Thule, which perches on the rugged coast of northwest Greenland high above the Arctic Circle.

It's been about four months since Kathe and I moved from Kansas to northern Utah's Cache Valley, and with a couple of weeks to spare before I begin teaching at Utah State University, I'm hoping to make all my flights so I can meet up with biologists and volunteers with the Peregrine Fund. The P-Fund works summers in Greenland, motoring a boat alongside wave-battered sea cliffs to survey for falcons. If all goes according to plan, I'll be aboard, and while they look for birds, I'll be looking for meteorite history.

From Thule, where the P-Fund has a research station, we're to travel some 150 miles on Arctic waters to Cape York and camp along the shore, close to where ambition pushed a troubled man to cross what he called "splendid distances," even as such passages placed him in mortal danger. By crossing those distances—then stealing meteorites out of rock and ice—polar explorer Robert Peary saved his career from certain disaster. If I'm lucky, I'll be one of the few non-Inuit to see where he found the saviksue, or iron mountains, which, before his discovery, had been the stuff of legend. These were the ghost rocks of Greenland, venerated by the Inuit and used by them as a source for iron points and blades. These three "star stones," as Peary called them, these Cape York irons, are meteorites so large and so unwieldy that he almost failed to nab them all. The tale of the saviksue is a mystery, an adventure, and a story of crime all at once, and in my pursuit of its details I'll find myself thinking not only of Peary, but of Ellis Hughes and Eliza Kimberly and F. W. Cragin, and just where my own actions fit on a scale between theft and discovery. I'll wonder what distinctions can be made among varieties of profit and legacy.

If the P-Fund's boat—nicknamed the *Barb,* for "Big-Assed Red Boat"—can thread her way through iceberg-clotted Melville Bay, if we can match up old photographs with the present landscape, if we can find hammerstones the Inuit discarded while chipping off bits of astral iron, if we can find the natural pier where Peary's ship, the *Hope,* sidled up to take the biggest star stone ever moved on this planet, if—

Well. The closest I may get to Cape York is sitting in this muggy metal tube reading a Zen honcho's insights into the nature of journeys while lightning spangles the evening sky. Quite frankly, my paperback Thoreau isn't helping either. "Be rather the Mungo Park, the Lewis and Clark and Forbisher, of your own streams and oceans; explore your own higher latitudes," he writes. "Be a Columbus to whole new continents and worlds within you. . . . Every man is the lord of a realm beside which the earthly empire of the Czar is but a petty state, a hummock left by the ice."

Fair 'nuff. Provocative thoughts all around. All good. My own higher latitudes would like to stretch their legs. I twitch, and the plane does not move.

LAST SUMMER, WHILE STILL LIVING IN KANSAS, I tried to get to Greenland after a terrible loss. A few months into the separation from my wife, a few months into my new relationship with Kathe, she and I had driven to the Nebraska Sand Hills for a "star party." This is where nerds such as myself deploy their telescopes and stay up all night looking through them. After our first night, when I went to check my voice mail at a pay phone, I crumpled, holding the receiver to my ear as I stared at a sunny general-store parking lot. I was listening to message after message from my sister imploring me to answer. My mother had died.

Kathe and I drove back to our separate apartments in Kansas. From Kansas, I flew to Indiana, where I helped scatter my mother's ashes while reading aloud lines from the *Tao Te Ching*. Some of the ashes fluttered in the air, some descended quickly.

Then, stunned and raw, I returned to Kansas in time to begin packing for a drive back to my sister's house in Indianapolis. From there I'd drive the next day to Baltimore, where I'd get on a flight bound for Thule—my first try at Greenland. After, presumably, finding the sites of the saviksue, I'd return to Kansas, then say goodbye to Kathe, who was about to leave for China to teach English at a university.

So when I found myself again in Indiana in July 2002, it was with two losses framing my drive to Baltimore and my polar plans. Behind me was my mother's death. Before me was Kathe's departure. That morning in my sister's driveway, I stood by my car, uncertain in the heat. My sister had gone to work, and I could leave anytime. Instead, I went for a walk.

I often take my time before leaving, a long-held habit, this reluctance to embark. I am wedded to final views of rooms. Is this just a hesitation to journey? Sometimes, yes. But do I linger also because I love the inherent poignancy of departure? I mean, I sometimes walk about a room before going to the grocery store, as if seeing the place for the first and last time. Maybe it's just a way to avoid running an errand. But slow departure suggests regrets and maybe lengthened potential. Departures—on one's own terms—can last and last. For those being left, of course, departures can wrench, they can devastate. And arrivals? Arrivals can feel like epiphanies. They too are full of potential. Having gotten to where you are means a new thing is under way, though an arrival ends a journey. Arrival can be demise.

After my walk in a park, shirt clinging to my skin, I wanted nothing more than to sit on my sister's porch step for perhaps the rest of my life. I missed my mother. I was going to miss Kathe.

Not long before her death, I had told my mother over the phone that I had been unhappy for years and that I had found someone new, someone I loved and with whom I felt completely myself. My mother, who had been depressed for decades, was still suffering from aphasia caused by a stroke. But in her last months she'd been on an antidepressant, and she was like, I imagine, the young woman in photos from her days in nursing school, smiling and bright. She told jokes, she laughed, she stopped smoking. After I had told her my news, my mother stuttered into her phone, "Happiness is . . . happiness is . . . " She sighed with frustration, and I told her to take her time. "Happiness is—is *okay*." Nothing else she'd ever said to me meant more than that. I wept.

In my sister's driveway, my car was stuffed with drybags full of cold-weather gear, wet-weather gear, boxes of Clif bars provided by their manufacturer, a camera, notebooks, and an expedition-class tent that I had practiced setting up in the Kansas heat. The tent stakes had burned like narrow cautions in my hand. I'd take the tent down, bring it inside, and leave it on the floor in my apartment, where, on the kitchen counter, I'd unfolded the best maps of Greenland in the entire Midwest. The maps were dotted with place names—Inuit and Danish—as clotted with consonants as the waters are with icebergs. Akuliaruseq. Niaqornarssuaq. Arfeqarfik. I would trace my fingers along the Greenland coast, and as I imagined sailing past the island of Conical Rock, on beyond the Crimson Cliffs, where algae stain the snow like blood, past white interiors marked "unexplored," I knew that I was trying in my own bumbling way to reconcile splendid distances with higher latitudes. What had Robert Peary—the long-dead explorer I would follow to Greenland—what had he found, really? What had he hidden? In gaining what he did, what had he left behind? In other words, what do we hold dear?

. . .

A YEAR AFTER THOSE PORCH-STEP meditations, I'm sitting in the stalled airplane in Dallas and still thinking about those questions. Are we little more than fancier versions of that hungry cat Edwin Teale fed as he paused on his journey to the Kansas Meteorite Farm? Is hunger all it's about? Does hunger have the grace I think it can have?

By the time the plane reaches the gate, other, more pressing questions are on my mind, such as when do I leave for Baltimore and will they hold the flight from Baltimore to Greenland? Eventually the answers are "in a while" and "yes."

Finally boarded on the night flight to Baltimore, I begin to relax. Thule, Greenland! Hundreds of miles north of the Arctic Circle! Chill, I tell myself, keep your hopes down. Even if I get to the Air Force base, the weather will be in charge and the *Barb* many hours away from Cape York. I may get to Thule and end up meditating on Alan Watts and Henry David Thoreau for a week in a Quonset hut being pelted by wind and snow.

After all, last summer when I left my sister's and reached a motel in Baltimore, I learned that storms at Thule had gone from bad to hopeless. I had a choice. I could wait in Maryland and take a later flight, though there was no guarantee I'd get to Cape York anyway and a good chance I'd miss being back in time to see Kathe off to China. Or I could bail.

I bailed. And I decided that if I couldn't get to Greenland that summer, I'd go to the next best place: Cresson, Pennsylvania.

SO IT WAS THE PREVIOUS AUGUST that I found myself at the cash register of a convenience mart in Cresson, where, without meaning to, I was irritating a teenaged clerk, a girl who, all acne and hurry, told me I was *already* on Admiral Peary Road or Highway or *Whatever*. No, she didn't know where the park was. I had crossed out "arrive Thule" in my planner and was instead looking for a statue. Under warm, cloudy skies, I drove past discount stores and smoke shops.

Looking for the Arctic, I got a city park. Looking for adventure, I got irony and sculpture: a birdshit-festooned Robert Peary and his loyal dog on whose head a prankster had placed a straw hat, one with, just above the brim, a ribbon printed in colorful seashells. Metal-frozen, Robert Peary and his dog gazed outward from the town of his birth. Across the way were the mobile homes of the Laurel Woods Community, and though they looked spiffier than the one I had lived in as a kid, the trailer park seemed an appropriate neighbor to the Admiral Peary monument, because the explorer had grown up in humble circumstances.

The rippled bronze was green with age, and the fur-clad admiral less than life-size. The tongue of the bronze dog beside him drooped. This meant that the statue didn't show Peary at the North Pole—getting there had been his life's dream—because when, he claimed, he had reached 90 North, on April 6, 1909, he was all alone. No dogs, no human companions, no one but himself. Now many believe he fell short of the Pole. Still, the statue-Peary's thick mustache connoted unfathomable time in the wild. The explorer's eyes looked toward the horizon, beyond the evergreens hiding the trailers, and I saw that a spider had attached ends of its web to the dog's hat and to Peary himself, as if commemoration must include a trap.

I read the statue's plaque—a list of achievements, including his traverse on the Greenland ice cap in 1892 and his later, presumed conquest of the Pole. *Fame sure keeps engravers busy,* I thought. And then I decided to join him, hoisting myself up beside the husky. From my perch, we three saw that a road sign for the park kept shaking in the wake of speeding trucks.

OF COURSE MY AMBITIONS are far more modest than Peary's. Right now, a year after I hung out with Peary's statue in Cresson, the only goal that matters is to run down the concourse of Baltimore-Washington International Airport. Just in from Dallas, I and other Greenland-bound passengers hustle toward the Air Mobility Command counter so we can check in and board the flight to Thule Air Base. Panting, we haul up with papers and luggage, then I take a moment to introduce myself to Bill Burnham, the president of the P-Fund. I apologize for the delays, but from under his wide-brimmed felt hat, Bill Burnham says nothing to me, not a thing. His silent gaze is my first hint of the ice to come. We don't speak for the entire flight.

Aboard, alone in a row, I'm wired, I'm irritated at Bill's cowboy quiet, I'm disbelieving.

Greenland? Greenland? Ultimate north. Land of polar bears, land of visiting snowboarders bagging unnamed mountains, land of traditional seal hunts and of adventuring CEOs looking for wrecked B-17s. Place of storm, glacier, aurora, mosquito, icemelt. An island more than 1,000 miles long, five times larger than California, with an ice cap that hoards 10 percent of all the fresh water on Earth.

Living on the fringes beside the ice cap and moving across the top of it, the Inuit are Greenland's first people, having migrated from North America 5,000 years ago to the area that now holds the Thule base—"Pituffik," as it's otherwise known. There were perhaps 500 of them, and they endured the harshest living conditions on the planet. Their ancestors almost succumbed later to far greater forces: traders, missionaries, whalers, and the Danish government.

Somehow, though, the Inuit have persevered, cleaving to their subsistence traditions and blending aspects of their animistic religion with Christianity. Home rule came in 1979, just as environmentalists were about to wreak havoc on Greenlandic culture by trying to ban all hunting of marine mammals. Today pollution and overfishing are threats to the well-being of Greenland's environment and its 56,000 inhabitants, who also must cope with dizzying changes brought on by, among other things, increased access to electronic media and alcohol.

Greenland is a nation without highways where fishing and hunting remain crucial to survival. It is a country in two worlds, the ancient, still-abiding world of the hunter-gatherer and the modern, metastasizing world of the cybershopper. It's the weary fringe of Danish social policy and the high, quickened heart of global warming. Greenland's a frosty temporal nexus of dogsleds and cell phones. And it was, for Peary—well, let's call it his stage.

On May 6, 1856, in Cresson, Pennsylvannia, Robert Peary was born into a family of uneducated business people. According to biographer John Edward Weems, after the boy's father died—Robert was not yet three—his mother moved them to Gallitzin, where her son would watch trains vanish into hillside tunnels, "waiting anxiously for the plumes of smoke to reappear on the other side." They next moved to Maine, where Peary's mother tried to cure her son of boyish wildness. She did this by dressing the profane, sometimes cruel young Peary in a sunbonnet and by making him sew. Still, he retained a sense of adventure in nature, and when he was six, read a newspaper article about the American polar explorer Elisha Kent Kane. It made a mark.

In college, Peary continued to live with the three crucial facts of his life: His depressions. His mother, with whom he boarded. And his dreams of polar glory. After graduation, Peary became a taxidermist, then a draftsman, then joined the Navy. He also thought very deliberately about becoming the first person to reach the North Pole. In his private writings, Peary compared Arctic expeditions to the body itself—the leader as head, the men as arms—and with bodies he was literally concerned, for Peary believed that men must have access to women in order to remain "contented."

In 1886, Peary reached the Arctic, traversing farther and higher on the Greenland ice cap than anyone had before. He tested equipment. He wondered about the constrictions of marriage. Twice he almost died, once by falling in a crevasse, once by falling into rapids. He returned, lectured, and was inducted into the American Association for the Advancement of Science. He was thirty years old. Peary thought, "I am writing my name before the world."

Into this grand questing came Josephine Diebitsch, whom Peary married in 1888. Peary's mother accompanied them on the honeymoon.

On his next Greenland expedition, Peary broke his leg, but he still retained a tough sense of command and barked out orders while his underlings built Red Cliff House, a headquarters for him and his wife, who, to the amazement of the public, was with him in the wilds. At a time of deep-rooted prejudice, Peary also had brought to Greenland his right-hand man Matthew Henson, an African-American. While his men cached food north for a possible run at the Pole, Peary observed the Inuit and professed to like them. Certainly he did not mind how the women sometimes went topless and how the husbands swapped wives. Blue-eyed, tall, volatile, smiling, and muscular, Peary impressed the natives, but Inuit writer Kenn Harper says that his "eyes were devoid of emotion." In the spring of 1892 Peary saw the ice of the Arctic Ocean for the first time. Having proven himself a capable navigator and a man who could endure storms,

Peary returned home, gave countless more lectures, and raised more money. Success seemed assured.

But in 1894 Peary's attempt to cross the ice cap was turned back by storms and temperatures that plunged to 60 below. This Second Greenland Expedition was, in Weems's words, "doomed to a wretched existence on the ice, and eventually to utter failure." There were blizzards, insane dogs, lice, frostbite, and snow blindness. Back in camp with his wife, in the summer of 1894, Peary saw no one in his rooms but Jo and the doctor. The couple ate well while the men subsisted on "seal and walrus." So when Peary's relief ship arrived that summer, his men were thrilled. Peary, says Robert M. Bryce, was embarrassed by "having accomplished so little" and so elected to stay with two of his most trusted men, Henson and Hugh Lee. And Peary sent a letter back for publication critical of those returning home while he stayed on. Meanwhile, Jo was terrified of the dangers her husband had faced. She wanted him to quit Greenland for good.

She would not have her wish.

IN THE PLANE, I wake to brown land below, land the color of mud, with ribbons of valleys, inferences of rivers, glaciers as dirty as city snowbanks. It looks like a premonition of Mars. Wrinkled, terraced, ridged. Gullies running straight down steep slopes. My first glimpse of Greenland makes me appreciate even more the lie of Eric the Red's name for this country. I look out the window, glad to have a civilized breakfast of French toast, sausage, and cantaloupe instead of a Second Greenland Expedition repast of rotted walrus in a blizzard reduction sauce.

Last summer's Cresson sojourn seems mighty far away now.

Ice islands and icebergs speckle the water. Around some icebergs swirl bergy bits and growlers, looking like spume or a trail of white crumbs. Icebergs and their just-under-the-surface flanks glow a preternatural blue-green, just a shade greener than the color of a ship's log I'd seen from one of Peary's expeditions.

Beyond the water and the rock is the great heap of the Greenland ice cap.

What a curious admixture: my stupor and the epic open weirdness of this place, the oddity of musky cantaloupe in my mouth while ice and water and rock pass below. Hundreds of icebergs, like a sky with clouds that have sharp edges. The freakishness of a single one below—it looks like an Arctic stingray with three spikes rising from it.

Suddenly I sit up. There's Bushnan Island amid a flurry of icebergs. I grab my map. *Yes,* Bushnan Island and behind it is—*Meteorite Island,* sunlit, and icebergs aplenty, an aorta clotted with them on the side where the biggest of the Cape York irons was found. And there, the isthmus off the mainland close to where the other two meteorites rested for centuries before Peary "retrieved" them. At least I've seen the places. Breakfast settles warm in my belly.

The sky darkens. Through breaks in the cloud deck, a strip of sunlight renders the sky into a Rothko hanging above barren brown and scoured land. In the distance a black peninsula patched with snow brims with metal buildings. Thule Air Base. The water wrinkles as we descend, the plane is buffeted. Above and beneath green-blue water are the plunging red cliffs of the shoreline. We're flying through snow now, but we land without incident 700 miles north of the Arctic Circle.

Almost there, I think. Almost. Now it's time for close quarters with frosty Bill Burnham and my real challenge: traveling by open boat beside that staggering coast.

· *Chapter III.2* ·

Thule and the *Barb*

Thule Air Base festers like an industrial blister on brown rock. It's a sculptured gravel pile sprouting metal buildings that sit on pilings because of the hard permafrost. There are power lines, streetlights, stacks of drums, dirty buses, construction equipment, blue hangars with orange doors, the runway, and a snakes' convention of aboveground steam pipes. Everywhere there is construction, an atmosphere of diligent impermanence. Dirt roads scrape past sheds, jet-fuel pipelines, and storage tanks, where great flakes of metal sheeting lie here and there like rusted mica. Built after the Second World War, Thule served first as a refueling station for U.S. bombers and later as an outpost of the Ballistic Missile Early Warning System. Today the 600-square-mile complex is home to the 12th Space Warning Squadron.

The base evokes in me a kind of spooky nostalgia. Though I came of age at the end of the Cold War in the 1970s and 1980s, the black-and-white movies of the 1950s and 1960s—from science films in grade school to the sci-fi flicks on late-night TV—inculcated my affinity for fighters, rockets, and radar domes. The part of me that fondly recalls my cadet days in the Civil Air Patrol, when I wanted to be a fighter pilot or a missile launch officer, finds Thule captivating.

Beyond the main portion of the base, cliffs rise, and in North Star Bay icebergs float. A butte, Dundas Mountain, looms nearby at the end of a peninsula. Across the bay there's a scattering of wooden buildings painted red and green with windows boarded white—a trading post founded by Knud Rasmussen in 1910 and the village that grew up around it.

After I retrieve my luggage and meet Kurt Burnham, Bill's son and the manager of the Peregrine Fund's summer work here, I head to lunch at the dining hall with several soon-to-depart guests of the P-Fund, most of whom know one another well. They are back from the field, where they've been locating the nests of peregrine falcons and gyrfalcons, temporarily capturing birds in order to take measurements, drawing blood for later analysis, and attaching transmitters to track their movements after the birds are released unharmed.

For the next two days, before I join Bill, Kurt, and Regan Haswell on the *Barb* to head south along the coast, I have some time to myself. I decline to embarrass myself at basketball and take walks instead. On my first, I saunter past battered warehouse-like buildings and beneath the steam-heat pipes that arch above the roads. I see snow buntings, arctic hares like tiny hunchback albino kangaroos, and gray arctic foxes that cavort amid rubble and stacks of pipes. And I worry about the Military Police. I have my papers with me, just in case. "You're on base now," Kurt had said, gap-toothed and friendly. "You can walk anywhere."

I duck under a huge, rusted pipeline and find willowy cotton grass and willow brush at my feet. Here is a black lichen shaped like a fan. Glaucous gulls and black-legged kittiwakes rest on a spit and in brown shallows, keeping out of the brisk winds. There are rocks everywhere—black, brown, pink, red, speckled. I see a few yellow-green flowers in blossom, the arctic poppy, which is abundant, though most are now withering to dark fists. My walk ends at a chipped gravestone dotted with bird dung. The stone reads, "SACRED to the MEMORY of Wm. Sharp OF H.M.S. North Star who departed his life NOV. 1st 1849 AGED, 26 YEARS." The headstone rises from a pile of orange-lichened stones. In cold August wind, I sit beside the rocks, listening to trucks, watching a cargo plane land, watching icebergs. No one arrests me from my reverie.

Back at the P-Fund building, the High Arctic Institute, or HAI, Kurt quizzes his friends from *The Pointing Dog Journal,* and I tell lame jokes during a lame James Bond flick (Roger Moore–era). After pizza, I do the dishes and go to bed, peeking at the midnight light through the blinds, unable to believe I'm here.

The HAI, host to civilian volunteers and scientists working on bird conservation and other research projects, is just one of the many narrow metal buildings on base that look like souped-up mobile homes. The HAI includes workshops, sleeping quarters, a brown-carpeted living room, and a heavy, latching door of the type used for walk-in freezers. The shotgun hallway reminds me of a submarine; you put your back against the pale wall to let another pass. Nature photographs provide relief from ugly-comfy furniture that must have been airlifted out of a 1973 Laramie bachelor pad. The photos are the work of base meteorologist and HAI caretaker Jack Stephens, who lives here year-round. The books are his too. Greek classics, biographies, Greenland reference works, and Stephen Mitchell's translation of the *Tao Te Ching*.

The next day—a sunny, 50-ish Friday—Jack takes me for a long drive in an HAI pickup truck and says, "An asteroid whack would be good for the human race." With his jeans and flannel shirt, white hair and beard, aquiline nose and tinted squarish glasses, Jack seems both down-home and professorial. For about three decades, the Georgia native has lived at Thule, and his job—along with the truck—give him a mobility few residents enjoy.

We jostle on the rocky roads. He elaborates: not a planet-killer sort of asteroid but one big enough to wake up the human race. Jack doesn't think we're taking the threat of globally catastrophic impacts seriously, and he returns to a favorite topic, the possibility of moving the industrial economy and its deleterious environmental effects off-planet. I equivocate as we drive by fell fields and tundra, fast-running rivers and calm lakes, secret military installations and smatterings of buntings, past yellow oil drums with sawed-off poles stuck in them and more cables, then past a blue sign for Kap Atholl (where musk oxen roam, though I do not see them) and a dashing ringed plover. Jack takes me to the tops of mountains. On one, above Wolstenholme Fjord, four glaciers empty their water and ice: Salisbury, Chamberlin, Rasmussen, Moltke. I gape.

"It *is* like living on another planet," Jack says.

Then we're off to P Mountain, with its ruins of metal posts, broken concrete, and a few rusted screws as thick as my arm. There, beside a traffic sign of peeling yellow paint—"Slow" in black letters—I stare at the ice cap mounding the horizon like white iron and feel a slow-burn scare.

On the way down, we stalk an arctic hare on the side of the dirt road. I get within 18 inches—we have our cameras out—and I can hear its muffled snorts. It flinches but stays. The eyelids flutter. "You're a really good sport," Jack says to the hare, and I think that of Jack too.

TONIGHT, after stowing my gear in vinyl drybags, after catching a bit of the entertainment—*Tango & Cash*—I talk with Regan. Tall, bearded, with intense eyes, a tight haircut, and a Durham, England, accent, Regan, like all of the P-Funders, is wound up and easygoing all at once. We share names of favorite bands. We talk of the trip to come. I even make Bill laugh when he asks me if I have much experience at sea. Bill is a lean man, with gray and black hair and a face creased from sun, wind, cold, and salt spray. His reserve toward me has begun to lighten. Experience at sea?

"No," I reply, "there's not much ocean in Kansas," later thinking that odd, as I had already moved to Utah.

Hours later, unable to turn in because I'm feeling so fidgety, I walk the base again and stare at sunlight streaming over long, flat-topped Dundas Mountain. It's 11:30 P.M. The light feels like late October, soft and beneficent, and I see how it makes two silver slashes in the earth—water-filled tire tracks reflecting light. They look like quotation marks in an open-ended sentence. Steam galumphs in the pipes, sounding like a bittern caught in a commode, while down the road, rock 'n' roll plays. The wind blows around the pipes to helix with the rock songs and trapped steam, to make a sound that reminds me of coyotes calling, a sound from my new home in Utah.

BAD, BAD, BAD: The Saturday we're to leave I miss my alarm.

Regan wakes me—I can't remember the last time I slept through an alarm—and I dress faster than I have in years, then head to the galley for breakfast, where Danish contractors, Inuit employees, Canadian pilots, and American personnel gorge on pastries, eggs, sausages, fruit, raw fish, and coffee. I treat my crewmates to the meal as partial amends for my faux pas.

Afterward, I get my gear together right away, making a point of being seen waiting outside for the others and feeling absurdly overdressed in countless strata of long johns and sweaters and rain shells. I wobble in my waders and play with my little wristbands designed to prevent seasickness.

"Do those work?" Kurt asks, throwing gear into the pickup that we'll drive to the pier.

"I don't know. We'll see," I answer. "If not, I've also got two kinds of drugs."

"Boy, you've got all the stuff," Kurt replies, smiling, a bit of sun glinting off his glasses.

I wander back inside, where my scholar of the ice, genial Jack Stephens, is dressed in his robe. He isn't coming along. I had noticed a translation of *The Epic of Gilgamesh.* We both admire the story, and often when I thought of Peary I was reminded of Uruk's resident bad boy. Both wished for fame. Both were full of lust that defied conventions. Both resented mortality. Both were drawn to meteorites. *Gilgamesh* isn't a classic because it's about gods and heros, it's a classic because it's about humans and failures and the nearness of solace.

Soon it's time to go, and Bill surprises me at the dock by announcing this as my maiden voyage and taking a snapshot of me next to the *Barb.*

My maiden voyage is cut short, however, by the boat's mushy steering. We turn around after a bit and head back to Thule. I watch aft of the steerage, where Kurt and Bill stand behind a large windshield. Despite the steering problem, the boat is stout, built to withstand the bumps and scrapes of rock and ice, though at 25 feet it's no longer than many leisure craft one might see on a local reservoir. The boat has a supply hold, small seats facing forward, and some facing to the rear. Its most distinctive feature is the red foam "collar" that runs along the sides and wraps around the blunt front above the metal hull. Good for bumper-car with icebergs.

The next day finds the weather turning stormy, and I begin to worry if we'll be socked in for the rest of the week. To kill time, Kurt says—and this is not a shock—that we rent a lot of movies. I wait a beat and suggest that we could listen to the audio version of the *Iliad.*

"No," Kurt blurts out. "No, no way!"

"But it's action-packed," I respond.

Jack opines softly, "It is."

That afternoon when I try to walk all the way to the old trading post, I'm stopped by rising winds, slippery rocks, and dark skies, so for a few minutes I sit on a boulder black as my mood. Have I come this far only to come up short?

IN THE SUMMER OF 1818 the British mariner John Ross arrived along the coast of Greenland in search of the Northwest Passage. The 200 Inuit Ross found were, to say the least, surprised. They had believed that they were the only human beings in the world. Ross registered surprise too, especially at the discovery that these isolated people had tipped spears and had edged into bone handles a few rough blades of iron. How could a culture with no smelting technology fashion iron tools? The Inuit told Ross of "saviksue," an iron mountain. The tools Ross brought back showed that the iron contained a high amount of nickel, a sure sign of meteoritic origin. Finding this iron mountain became, if marginally, one of the goals of many polar expeditions. None succeeded, including the *North Star* expedition, on which served William Sharp, whose grave I'd visited.

In the spring of 1894—after the failed ice-cap crossing but before the departure of most of his men—Peary decided to find the iron mountain. He needed a spectacular feat to save this expedition from failure. Without some success, his polar career could be over. He was desperate.

On May 16, 1894—Peary calls it "a glittering wintry day, with fresh south wind, the temperature 25 F., and abundant cumuli casting cloud shadows on the white expanse of the bay and distant ice-caps"—he and comrade Hugh Lee left by a sledge hauled by ten dogs. Having promised a significant prize—a gun—to the Inuit who would show him the iron mountain, Peary stopped at Castle Cliffs, where two young men were camped in their sealskin tents, or tupiks. One of the men, Panikpah, agreed to guide Peary and Lee, but, for some reason, perhaps common sense, he abandoned the attempt, and a man named Tallakoteah replaced him, claiming that there were not one but three irons.

They set out again, with two sledges, heading south under clouds but on good ice and making rapid progress. The enticing conditions would not last. In three hours, Peary, Lee, and Tallakoteah found themselves in a blinding blizzard, listening to narwhals breathing amid the lapping waves beyond the pack ice. In camp, their snow shelter collapsed when a dog walked on top of it, so they got moving again, often traversing on melting sea ice. Conditions were so nerve-racking that, according to Peary biographer John Edward Weems, Tallakoteah "began singing an impromptu song describing the deep snow, telling of numerous cracks in the ice southward toward Cape York, warning of the ice breaking up behind them . . . detailing the pain in his legs, and predicting a deep covering of snow that would hide the iron stones." At least once Tallakoteah tried to pass off snowdrifts as the meteorites. He left the party, but returned to lead Peary again.

Eventually, the weather improved. This was Peary's chance. The party soon came upon another Inuit, Kessuh, of whom Peary was fond. He joined the trek, which now made good speed toward a small inlet that Peary would later call Saviksoah Bay and beyond which was a then unnamed island—it would be called Meteorite Island—and the smaller Bushnan Island. It was well past midnight, and the sky showed cold stars. Stopping for a meal—Kessuh had killed a seal—the party arrived at the small bay after four A.M.

"After passing some five hundred yards up a narrow valley, Tallakoteah began looking about until a bit of blue trap-rock, projecting above the snow, caught his eye," Peary wrote in his book *Northward Over the "Great Ice."* "Kicking aside the snow, he exposed more pieces, saying this was a pile of the stones used in pounding fragments from the 'iron mountain.' He then indicated a spot four or five feet distant as the location of the long-sought object." Tallakoteah cut away at the snow, revealing a massive meteorite the size of a couple of barrels, which was far too large to move with sledges.

"At 5:30 Sunday morning, May 27, 1894, the brown mass, rudely awakened from its winter's sleep, found for the first time in its cycles of existence the eyes of a white man gazing upon it," Peary wrote. Tallakoteah, Lee, and Kessuh got to work removing more snow and ice, and once that job was done, the explorer sketched, took pictures, and recorded measurements.

The brown mass was more than iron, more than a prize. It was called Woman, and she had been flung from the sky by the god Tornarsuk. Why Tornarsuk exiled Woman is not clear, but at least Woman arrived with her iron Dog and her iron Tent. Dog came to rest close to Woman, though the Tent was a few miles distant on the unnamed island. While the tale may have been made up for Peary's benefit, shooting stars were a familiar part of Inuit lore. John MacDonald, in his book *The Arctic Sky: Inuit Astronomy, Star Lore, and Legend*, notes that the Inuit have considered meteors to be the feces of stars. One Inuit tale involves a wry and naughty Moon-Man who spat on the earth through a hole in his floor, causing those below to exclaim that "the stars are shitting." Knud Rasmussen was told of a woman who was "physically entered by a meteorite," transforming her into a shaman.

After watching his guide demonstrate knapping iron off Woman, Peary then made a mark of his own on the meteorite, "a rough 'P' on the surface" (which has never been found) as "indisputable proof" of Peary's discovery. And he built a cairn to hold a note confirming his find.

Peary paused, looked back again at the brown rock ("the celestial straggler"), and walked back down to his sledge, where he rested before the clouds and wind picked up. After rest and a hasty meal, the party set out to find the largest of the saviksue, but the snow proved too formidable. Tallakoteah pointed to where Tent was, but the team moved on and had a excruciating return journey. The ice broke up, they ate rotted walrus, and they rocketed on sledges down a glacier at 60 miles an hour before reaching their home camp.

That summer was the season of discontent among Peary's men, eventually leading the officer to stay in Greenland with Matt Henson and Hugh Lee. Peary had found the iron mountains but didn't have them in hand. Their size and the weather meant moving them was impossible. He'd have to wait for the next summer to take them. Robert Peary still had little to show for his work in Greenland, and the next year's assault on the ice cap would be nearly fatal. His grasp on legacy was beginning to loosen, and he knew it.

KURT IS BLASTING EIGHTIES rock 'n' roll from the *Barb*'s speakers, "Everybody's Working for the Weekend." I tap my fingers to it, and we go farther than our first attempt, passing icebergs and smaller chunks of ice—bergy bits, growlers, and brash bobbing in the wake of our boat. At Parker Snow Bay, thick-billed murres are thick in the air beside a cliff, and Yes sings "Owner of a Lonely

Heart." Kurt stops the boat 100 feet off the rocks to scan for falcons and points at the cliff on the north side of the bay.

"This is the eyrie where Minik got his gyrfalcon," he tells me. Minik was an Inuit boy, and the falcon was for Peary, who was once here, right here among columned, sun-touched, cloud-wrapped cliffs all cracked and shit-streaked. All along the coast are such high cliffs, with their interruptions of glaciers, their parentheticals of grassy slopes and hummocks, their spills of babeled talus. Up and down the coast, grass seems to cascade like waterfalls on rock faces.

As we speed on, the boat sometimes slamming down on the water, lurching, the sun emerges, and I praise this development. Bill gives a thumbs-up. When I say I still can't believe I'm here, he jokes, "Jump in the water, then you'll believe." The crew of the *Barb* is, today, a happy one.

When we reach a small nameless bay fed by a small nameless glacier, Kurt eases the boat to rocks, where Regan and Bill jump ashore in order to find falcons. In a flash, a wave sends the boat toward Bill, who clambers up a rock to get out of the way. We almost smash into him. Moments later, Regan radios us, since we can't talk over the sound of the surf and the motor. Bill has gashed open his hand.

"He's bladin' on me," Regan says. "May need a few stitches in his hand."

"For Bill to say he needs eight to ten stitches . . . that's bad," Kurt says, staring at the shore. My heart is pounding now. "I was worried about crushing his leg," Kurt adds.

"It's not so much that it's long but it's deep. You can see the ligaments," Regan says.

Still, Regan and Bill continue to try to tempt in a distant falcon so they can capture it, do their work as fast as possible, then set it free. Kurt directs me to a white speck on the cliff, my first gyrfalcon. But queasy with rollers, I can't keep the binocs up. To distract us from our stomachs and Bill's injury, I ask Kurt how last night's movie ended.

"Oh," he says, "Hannibal escapes."

Before fetching Bill and Regan, Kurt has trouble with one of the two Yamaha 130 motors, so I lean over the big white coolers set in the back between the aft and the seat, stretching out over a platform just above the water to choke the engine. This will become my regular task. Bill and Regan soon jump on board with packs and poles and nets, and Bill's bloody wrapping is at once brighter and darker than the pastel pinks of the snow patches of the Crimson Cliffs.

When we reach the harbor, Bill is rushed to the hospital, I help Kurt refuel the boat and tend to other chores. Despite today's accident, I feel a sense of ease, finally, a sense of belonging. Kurt and I chat about nature writing, about football, and then, when Kurt asks me to tie off the boat, I think, *Dear God, knots.* I manage, though once Kurt has the motors cut and things stowed, he steps onto the pier and reconfigures my work without comment. This is when I tell him. I dropped out of Webelos. This is when he confesses. Kurt Burnham—falconer, captain of the *Barb,* Oxford University doctoral student in ornithology, Westerner by birth, savvy hunter, classic rock 'n' roller—carries a book of illustrated knots.

"It drives Bill crazy." He laughs.

THE YEAR AFTER HIS DISCOVERY of Woman and Dog, Peary had his ship the *Kite* in Melville Bay. It was the summer of 1895. After some two years in Greenland with nothing to show for it, Peary was depressed and anxious. The ship made its way to the isthmus of Saviksoah (or Saviksue) Bay, where, on a slope among a dizzying array of rocks and boulders, the sky-exiles Woman and Dog had lain for centuries. Peary wanted to get both these meteorites on board as booty.

"Standing here," Peary wrote, "the eye roams southward, over the broken ice-masses of Glacier Bay, the favourite haunt of the polar bear; eastward, across the glacier itself, to the ebon faces of the Black Twins, two beetling ice-capped cliffs, which frown down upon the glacier; northward, to the boulder-strewn slopes of a gneissose mountain; and westward, over the placid surface of Saviksoah Bay, which presents a striking contrast to the berg chaos on the opposite side of the isthmus."

Peary directed the crew of the *Kite* and his little fiefdom of Inuit to wrest Dog and Woman from the ground. Dog weighed 900 pounds and looked like a wad of gum as large as a beach ball. Woman weighed a serious 3 tons.

Peary surveyed the situation, happy that the isthmus's little valley retained a good hard crust of snow and ice, useful for sledging, though he noted that between the meteorites and the snow there occurred nothing but rocks and boulders.

"The next day, [Emil] Diebitsch [Peary's brother-in-law] began work with the ship's crew and the Eskimos," Peary wrote. "The 'woman' was lifted out of her bed with jacks, and a rough sledge of spruce poles made for the 'dog.' On the second day, the 'woman' was blocked up ready for transportation, and the 'dog' rolled upon its sledge and dragged by the combined force of the ship's crew and my native allies over the boulders and down the snow-drifts to the shore."

The men rolled the Dog onto an ice ferry—"a cake of ice"—and floated it on open water until it was man-hauled again on hard ice beside the *Kite,* "where it was hoisted on board and deposited in the hold."

It did not go so well with Woman.

Diebitsch commanded the Inuit to make a "rude road-bed" that the natives "roughly graded" from the local rocks, according to Peary. Meanwhile, Peary wrote, they built a "heavy timber drag" so that the Woman was "slowly transported upon iron rollers over a plank tramway" set upon the Eskimo road. So far, so good.

The Inuit talked among themselves of the last time someone tried to take Woman or at least a part of her. Wanting their iron source closer by, a band of Inuit had come from the north to take the head of Woman, which had separated from the rest of her body after years of hammering. But the pack ice, unable to endure the weight of the iron head, broke open, and water engulfed the sledge with its heavy head. The party didn't drown, but the survivors often told this cautionary tale. Peary's Inuit were dubious about this new attempt.

Despite the warning, the Americans and Inuit were able to haul, sledge, and float all of Woman, all three tons, on an ice cake that they positioned next to the ship anchored a mile from shore. The men worked with ropes and chains to bind Woman to the *Kite* and lift her in place for final loading.

That's when the ice broke, Woman began to sink, and the ship listed hard.

Men shouted, ice ripped, water rushed. The *Kite* groaned in protest.

But some of Woman still showed above water, and there was just enough support on the remaining ice to allow the sailors to affix more line. The winch chains tugged and tugged till "the meteorite slowly warped up to the rail and swung inboard," Peary wrote.

At last. He had two of the three irons in his grasp.

"Everyone breathed a sigh of relief," he said, "when the sulky giant was safely deposited in the hold."

The *Kite* would steam home. But Robert Peary was far from done.

· *Chapter III.3* ·

The Isthmus and Meteorite Island

Today Kurt has the *Barb* barreling beyond the Crimson Cliffs—it has become a spectacular day—and I recall how Bill turned over his shoulder to yell at me, pointing out the Peary Monument on the snowy-clouded cliff of Cape York. The dark obelisk sits on Cape York like a needle stitching land and sky. It also pokes up like Peary giving the finger at his critics.

Now past Cape York and into Melville Bay, I suddenly think, *Peary, you magnificent bastard, I read your book!*

The goofy outburst comes unbidden but I know its source. Earlier in the week, I was sitting in the High Arctic Institute living room with Kurt as we watched the movie *Patton,* a childhood favorite. My words echo the general's just before he battles his nemesis Rommel in North Africa.

My exclamation, like an iceberg, has much hidden beneath the surface. I puzzle over it. In part, I blurted this declaration to myself from the sheer exhilaration of the day, of going farther than before. It's anticipation. The boat is running smoothly, no one's been hurt, and we are closing in on the sites of the Cape York irons. Standing behind the steerage, resting a knee on the padded back seat, I hold on to the rail and happily let the wind blast my face and hair. I know that we can make it today and I believe we will.

Of course it's also become a contest, though not with Peary—my God, how could I hold up against the travails he so often overcame? No, the contest is within, with the part of me that still considers myself a quitter. The one who walked off the junior high football practice field. Who stopped short a journalism career, who dropped out of graduate school, who brought his marriage to a crashing halt.

And there's another reason for my outburst. When I first read shorter accounts of Peary's retrieval of the star stones, I was full of admiration for his daring. The more I learned of him, however, the more dubious I became. For example, on one of his journeys back from Greenland, he brought not only meteorites but Inuit friends who would go on display in New York City and who would die while Peary turned his back on them.

I shake my head at all of this, at history, and at this place.

On my right is the horizon line of ocean, punctured by icebergs of myriad shapes and sizes, crossed by flying murres and dovekies. On my left, shore cliffs rise craggy and steep, brown and gray. Here and there, grass grows, and dirty, crevassed glaciers reach toward the water. Snow patches in among the rocks. Broken sheens of orange jewel lichen vibrate with light. The sky lightens or darkens as clouds and sun jostle. The *Barb* zips along.

My mood has changed from the morning, when I refrained from taking notes because I was tired, because I was seeing again the sights of the past two trips, because I expected failure. Clouds glowered. Regan had surprised me at the dock by declaring that he admired the birch paintings of Gustav Klimt, then saying, out of nowhere, "I never saw water like that yesterday. It was like mercury." This was when we brought Bill in for his injury. Regan's comment put me on edge.

There had been one good omen: a rainbow arching over Wolstenholme Island, not far from Thule. Despite that, and another rainbow along the Crimson Cliffs—the green in the rainbow like the green of the grasses—the ride had been cold, rough, and windy.

A little while ago, we dropped off Regan and Bill at the bay where Bill had gashed his hand, and the procedure was by now familiar. Regan and Bill would troop off with nets while Kurt and I tried to steady our binocs against the rollers as we looked at specks (birds) or swarms of specks (flocks). I mixed apple cider in a mug, the hot water steaming up while I munched a trusty Clif bar, my mind wandering.

After Bill and Regan returned and Kurt had us under way again, I was startled to hear choking. Regan stood beside me in the aft and his eyes widened—I froze—while his Adam's apple moved up and down, up and down, till he waved his hand to indicate he was okay.

"Death by Pringle," he said, coughing. "What a way to go. I'd prefer something more dramatic, like death by narwhal."

I exhaled.

Now, at the northern end of Melville Bay, I see more icebergs than ever before. The views are, as Josephine Peary had once said, "surpassingly fine." At one point, the bergs are so many and so close that one entire horizon is nothing but a vast riprap confusion. A few appear to be the size of five-story buildings, some the size of football fields, spaced in varied intervals, 20, 30, 40 yards apart, Pluto brought down to Earth.

Regan tells me to cover my ears when he fires off a harmless round to scare up falcons that might be nesting on the cliffs.

The sun hides again as we cross the bay, the static flames of icebergs flaring against a dark sky, and they are white, they are blue, the blue of airless ice, they are blue and white and smooth, they are banded gray with trapped dirt and pebbles, they are pock-marked and rise at every degree of slope, grooved and claw-marked and faceted by a crazy jeweler and they look like pearl melted, then flash-frozen into chipped-tooth hoodoos and eyelash arches. They are bowls and buttes and boulders. They are odalisques. Sometimes we see holes in them. Here is a perfect black circle tucked in one intrados. We glide by their flanks, which, when sunlit as some are now and seen up close, slip bluey-green beneath blue water as if buoyed only by color.

We are so close. Just miles away. I contain my frustration that now, at four in the afternoon, we continue to go, stop, shoot, look, dropping Bill and Regan off then retrieving them, and, as if sensing my mood, Kurt tells me, "It's not just data to us. We really care about these birds." They've never lost or injured a falcon, he says, and as a birder, I understand. Though my crankiness subsides, my feet are freezing and I just hunker down against the wall that divides the steerage from the seat.

"You don't need to sleep do you?" Kurt says with a grin.

"I didn't come here to sleep," I say, smiling back.

So Bill and Regan get on board once more, and Kurt motors the four of us in the late-day light of the 24-hour sun, the wake stretching out behind, rippling hand-sized clumps of ice.

And as if it's the most natural thing in the world, we are pulling into this small inlet, across from the isthmus of which Peary wrote and near which he found Woman and Dog. The peninsula extends from the Ironstone Fjeld. Here, Kurt eases the boat up to the rocky edge, and this time it's me and Bill stepping off. We're just yards away. If I can find the divots where those meteorites once where, amid all this smooth round rock and boulder . . .

Before us the land slopes up in a wide, rocky draw, on either side of which is a small hill or ridge. As we top the middle of one rise, I see a tremendous confusion of icebergs on the other side of the isthmus, the most we've seen so far. The icebergs are packed in, mites in a bird's nest. I see no holes in the ground, no piles of hammerstones.

Kurt and Regan motor across the bay to set up camp. I hold up one of the laminated copies of Peary's pictures that I have with me as guides. This is the right place—but shouldn't we be able to see the depressions where Woman and Dog once sat? I bite my lip, anxious not to fail or appear inept. Bill, with a shotgun slung over his shoulder, and I, sweaty in my parka, spread out in the cobbled gloom and boulder-hop.

As Bill moves up toward the northerly side of the divide, I head farther down the sloping draw and see—cairns built by Robert Peary. Unmistakable! They still stand on the north ridge there, those small towers of stone. Then, at quarter till seven, Bill Burnham looks down at a mossy circle and sees—*the site of Woman.* He calls me to it, and I run up as fast as I can and I breathe hard, astonished breaths. Showing Bill, I match my copies of Peary's pictures to the view before us. We had at first gone up the wrong side of the valley. But, dammit, here we are. I'm stunned. We've made it.

Here is the utter actuality of this remote mossy circle. I do a quick measurement—about 11 feet wide. I stand before it, cold-cheeked, heart pounding, full of thanks. It's surrounded by a speckled confusion of rocks punctuated here and there by grasses. Then a bit higher and half as wide, another mossy-orange rough divot in the ground, the site of Dog, I conjecture. (Later I'll remember that Dog's site was supposed to be farther away, though the hole in the ground seemed the right size.)

Descending from the circles like a cape in the wind are the hammerstones, the blue traprocks fanning out and into a huge, partially conical pile. The round, gray basalt stones show breaks where the iron of the meteorite had defeated their pounding. It's said that there are about 10,000 hammerstones, all carried in by Inuit to knap iron from Woman and Dog.

Of these "trap-cobbles," Peary wrote, "The circumference of this pile of stones at the base is some 60 yards, and its height from the toe of the down-hill slope to the top is 18 or 20 feet. The contrast between the smooth rounded greenish trap-cobbles and the rough, angular lichen-covered grey gneissose rocks of the vicinity is very striking." *It is indeed,* I think.

So too is the history of the meteorite itself.

Researchers Vagn Buchwald and Gert Mosdal write that the Cape York fall took place in about the time of Christ. Thus the Inuit may have witnessed the fall, though label text at the American Museum of Natural History says the Cape York shower took place 10,000 years ago. The fall of 60 tons took place inland, and the movement of the ice sheet deposited the irons here on the coast. The Cape York fall—with about 275 meteorites recovered—is second to the 1947 Sikhote-Alin fall, the largest iron meteorite shower known. Sikhote-Alin dropped 100 tons of irons.

After forming more than 4 billion years ago, the asteroid that would become the Cape York meteorites suffered a collision; this was about 650 million years ago. Because there are so many samples of Cape York—including the "largest meteorite slice in the world," at about 6 by 4 feet, according to Danish researcher Henning Haack—scientists are able to determine when different materials crystallized; this means they can tell what was at the center of the meteorite and how the meteorite varies chemically from side to side, things difficult to determine with other irons, given how small they are.

What they've found is worthy of a good science-fiction novel. The meteorite "crystallized as a single crystal," Haack says, "cooling . . . a few hundred degrees per million years." The slow cooling allowed "the crystals time to grow to kilometer size." That is, Cape York's parent body contained supersized crystals more than a half-mile long. It's by studying the decay of various elements in meteorites (we know the ancestor of lead is uranium, for example, and how long it takes for the change to occur) that scientists can determine the "formation age" of a space rock—when it was born or, at least, when the rock registered a temperature of 1,300 degrees Fahrenheit. Dating can reveal the "cosmic-ray exposure age," or how long a body has been grazing through space, as well as the "terrestrial exposure age," which is how long a meteorite's been on Earth.

Exposure to conditions in space also weathers asteroids and meteoroids, as micrometeorites and the solar wind alter surface particles into a veneer of iron; recognizing this allowed scientists to match all those common ordinary chondrites with so-called S-type asteroids. Until space weathering was discovered, researchers thought those asteroids too rich in metal to be the sources of ordinary chondrites. As to terrestrial exposure ages, some meteorites have been sitting for a half-million years, though most, like Cape York, have been on the planet on the order of thousands of years. It's all very Jules Vernes–ish.

"There's some coal over here," Bill hollers, a few yards downslope. Almost tiptoeing around the edge of the excavation hole, I go to look, and Bill shows me a small flake of burnt wood—wood that must have been brought in, for there are no trees here.

"I wonder if they were cooking or heating or doing something else," he says, placing the flake in my hand, and what I feel so delicate on my skin isn't a piece of charcoal but a portal. Years later I'll take that small bit of burnt wood and smudge my face in a ritual of thanks.

I step back up to the mossy circle and stare at a single quartz rock in among greeny-orange moss and white-and-black lichen, like a real star made small and inset into the earth. It's beautiful, and the silence is overwhelming. I'm surrounded by boulders and basalt trapstones, perched beside the former home of cosmic iron, Peary's cairns just up yonder, icebergs cram-packed beyond, and a single gull flying over. It's almost an out-of-body experience.

My telescoping magnet pulls up a jot of rusted Cape York half the size of my pinkie fingernail. It's tiny, brown, unobtrusive, but I put it in a sample bag. Then I sit back on my haunches. Should I keep it? Technically, it's not a meteorite, having been rusted and changed by terrestrial weather. Still, is it, like that tiny charcoal flake, another piece of my puzzle about obsession? That to possess an object makes one's dreams, one's achievements more tangible? Of course—we show, then we tell. It's a way of believing, to have these things we find along the way. Something found is fixed in time, then in a story, if we can take it with us. In this way, can hunger bring us home?

Bill comes closer, and I offer him the magnet, and he finds that some of the hammerstones are magnetic too, perhaps from the iron dirt on them. Or could the hammerstones be Disko basalt, with Earth-iron trapped in them? Before wandering off, Bill takes a couple of pictures of me, and I see that in leaving me here beside the former site of a saviksue, he's granting me privacy. I had misjudged Bill at the airport and I'm glad for his help and companionship.

So on the edge of this mossy circle, I sit and draw in my breath, deliberate and calm, satisfied and bewildered. A light rain sputters. Kurt and Regan, having set up camp, are coming up the rocky valley.

Now I discover flakes of weathered wood that must be from Peary's expedition. I take one long sliver that is gray with age and two small flakes that are a light sand color. I add these to the charcoal Bill found. Then, well away from the site of Woman, I take examples of the local rock, a piece of quartz and a small piece of white granite studded with a tiny bit of lichen—rock tripe, I'll later determine, a lichen that on other, less fortunate expeditions has kept men from starving to death.

The P-Funders soon find rock walls built up waist-high around narrow chambers, which the Inuit had used as quarters when they were whacking iron from the meteorites.

I look out at the hulking mountains—sudden thunder twice, no, not thunder, the sound of icebergs calving—I look across the bay and see our yellow tents, I look out at the shards of icebergs, a glacier in the distance, the unreal blue hues, boulders everywhere, and I look down at my feet, standing where Tallakoteah stood scraping snow off a meteorite. Every detail is crisp, clear, and present, rendered by the Arctic air into such definition it's as though the place were too real.

We see how a crude "road" comes up to the side of the site, and Bill suggests that the body-high cairns on the ridge are channel markers. A good supposition, I think. With Kurt, I unspool a measuring tape and find that the pile of hammerstones is 27 feet high, different from Peary's figures. Still. "The word 'bunch' is too small for this," I mutter. Years later, I'll wonder if I had in fact come to the site of Dog and had missed the site of Woman in my excitement.

But in this moment I'm climbing to the cairns, and just as I reach one, I see no one from my party; Kurt, Bill, and Regan are heading back to shore where the boat is tied up. I'm alone on Cairn Ridge, my name for it now, above the bay with calving iceberg and wave crash going on and on, and though I don't really think heavenly creatures exist to hear me, I nonetheless thank Sila—a kind of Inuit version of the Tao—then thank my crewmates and even Peary himself.

I wheel around—the cairns! Could Peary's note of discovery still be there? He had put one in a cairn. But no, there are just flat stones shimmed with more flat stones, one atop the other, and of course, the note would have been retrieved or else it disintegrated over time.

Reluctantly I walk away, leaving the cairns behind, past the former site of a saviksue, past the pile of 10,000 hammerstones, down that slope of them to find my backpack, which I had left behind for the ridge climb. I heft it on, dallying a bit, and toting my tiny bag of lichens, local rock, also a seal bone, three flakes of wood, one flake of charcoal, a tiny fleck of rusted former meteorite, and two large hammerstones. I hop and peer and trundle across the fell field to the water's edge. Kurt sees me and turns the boat around, and in a few minutes, I step aboard for the short ride to camp.

Tonight Kurt makes dinner, and I head off over brownish hummocks of tundra and rocks to fill bags with glacial melt from a stream, which I lug back to camp. There, drinking wine, eating beans, corn, and sausage, we talk about the day. I offer a toast to my good companions. Bill berates himself for not inviting an Inuit to our dinner; he'd motored up to camp a while ago, a large seal he'd shot in the forehead draped across the prow. Regan says, "Think of how few people have been here." We eat standing up as the sun sends gold-pink light to the cliffs on Meteorite Island a few miles distant. Again, a crash. I see waves from the calving, ripples reaching the shore, splashing up.

"You'll hear this," Kurt says, "all night long."

And much of the night I do. And much of the night I check Regan's watch, for I have misplaced mine, left it behind along with the two hammerstones I meant to take with me. Before I had gotten on board to head from the isthmus to camp, I was discombobulated by all my stuff and set the watch and hammerstones aside. I forgot them. My watch will sound an alarm at 7:20 A.M. for a very long time. Will foxes turn it into a salt lick or toy? I hate that I left it there.

My chest thumps and thumps and I try to relax by snuggling into my sleeping bag, but I'm both exhausted and unable to stop thinking, riding a flow of images like ice bobbing on ripple upon ripple upon ripple upon ripple.

Tomorrow comes a new adventure. To find the site of the biggest Cape York iron of all, Tent, over there on Meteorite Island.

AFTER RETRIEVING WOMAN AND Dog in 1895, Robert Peary found that same year that he could not budge Tent. His men had dug three feet down in "scant turf and moss on the crest of a terrace on the eastern side of Meteorite Island, eighty feet above, and some three hundred yards distant from high-water mark." Most of Tent lay under the surface, and despite the excavation, Peary's party still had not found the bottom of the meteorite. The top of the iron jutted up out of the ground like "a dorsal fin," he wrote. Not even "two ten-ton screw-jacks" could move the mammoth star stone. Peary estimated its weight at 100 tons—in reality it was 34 tons and the size of a small carriage—and this last meteorite would have to wait. After four days of struggle with Tent, he gave up, and the *Kite* left Melville Bay as pack ice began to freeze. According to Kenn Harper, Peary wrote in his diary in 1895, "I have failed." He couldn't remove the meteorite. And he hadn't come close to the North Pole.

In the archives of the American Museum of Natural History in New York, I found the ship log kept in 1896 when Peary went after Tent again in a new ship. Dirty, grayed, creased, with foxing and torn edges, the oversized, iceberg-blue log of the SS *Hope*, commanded by John Bartlett, was kept by mate William Smith. Its understated entries record difficult summer days on Meteorite Island in 1896.

After leaving St. John's, Newfoundland, the evening of Friday, July 10, 1896, the *Hope* crossed northern waters and arrived at the Greenland coast at midnight on August 1. In thin cursive, William Smith noted weather, winds, direction, landmarks, miles traveled, ship's speed, the boarding and unboarding of passengers, crew's duties (which involved much shifting of coal and washing of decks), and, of course, the presence of icebergs.

On Saturday, August 22, 1896, Smith recorded, "Begin with a light breeze & clear. 4 p.m. full spd bound for Cape York. Midnight strong breez & clear. 4 a.m. Stopped at Cape York & sent boat on shore for Esquamauz [Inuit]. 5 a.m. left for Kite Harbour to ship the meteorite fresh breez & snow showerz. Noon arrived and moored alongside of clift." The pack ice was still thick, but Captain Bartlett found a lead to steer into.

The next day Smith ended his sea log, and began the harbor log, while "Esquamaux on shore clear[ed] rubbish from Meteor. Being Sunday crew unemployed." According to Peary, the natives used picks and shovels, and by day's end "the brown monster stood out in all its immensity as to length and breadth." On Monday, the "Crew employed landing timber & various apparatus for lifting Meteorite," Smith wrote. During the day, in clear air and in rain and fog, the crew of the *Hope* and the Inuit worked at digging out Tent with jacks. At night, Peary and his own expedition party took up the task.

During the next days, the men dug out the huge meteorite, jacked it up, and rolled it downhill on a crude roadbed.

"The huge brown mass would slowly and stubbornly rise on its side, and be forced to a position of unstable equilibrium," Peary reported, "then everyone, except the men at the chain blocks down at the foot of the hill, would stand aside. A few more pulls on these, then cable and the chain straps would slacken, the top of the meteorite move almost imperceptibly forward, the stones under the edge of revolution would begin to splinter and crumble, then, amidst the shouts of the natives and our own suppressed breathing, the 'Iron Mountain' would roll over."

The meteorite crushed wood and stone alike, punching holes in the earth as it lurched toward the natural pier the ship was docked beside. Once it was at the bottom of the slope, the jacks—which had a hard time finding purchase on the irregular surface of Tent—lifted the meteorite up so that beams and rails could be set beneath it. The meteorite forced the steel railings to pinch into the wood, and, more dangerously, if the meteorite shifted under the pressure of the jacks, "the head of the jack, like a melon-seed pressed between thumb and finger, flew out with serious risk to adjacent legs and arms," Peary wrote. Three of the four jacks gave out—though no one was injured—leaving a single 100-ton jack that couldn't do the job.

The meteorite, which was shaped like an overgrown, legless pig, deeply vexed Peary.

"Had the matter been a subject of study for weeks by the celestial forge-master, I doubt if any shape could have been devised that would have been any more completely ill suited for handling in any way," he wrote. To compensate, the men had placed wood blocks as shims beneath thick chains wrapped around the iron. At least the meteorite was now beside the ship.

According to William Smith's log, the crew gathered ballast and began to prepare the ship for taking on Tent and leaving the snowy bay. On September 1, Smith noted, "Tested timber & found it of sufficient strength to carry Meteorite. Lieu. Peary & party working to night."

Of that night Peary nearly sang, virtually praising the difficulty that he and his men faced. It was, he wrote, "a night of such savage wildness as is possible only in the Arctic regions. . . . The wild gale was howling out of the depth of Melville Bay through the *Hope's* rigging, and the snow was driving in horizontal lines. The white slopes of the hill down which the meteorite had been brought, showed a ghastly grey through the darkness; the fire, round which the fur-clad forms of the Eskimos were grouped, spread its bright red glare for a short distance; a little to one side was a faint glow of light through the skin wall of a solitary tupik." Peary could see his men beside the "black and uncompromising" bulk of the meteorite itself.

"While everything else was buried in the snow, the 'Saviksoah' was unaffected. The great flakes vanished as they touched it . . . as if the giant were saying, 'I am apart from all this, I am heaven-born, and still carry in my heart some of the warmth of those long-gone days before I was hurled upon this frozen desert.'" Peary saw that "if a sledge, ill aimed in the darkness at wedge or block, chanced to strike it, a spouting jet of scintillating sparks lit the gloom, and a deep note, sonorous as a bell . . . the half-pained, half-enraged bellow of a lost soul, answered the blow."

It was for naught. On a windy and rainy Wednesday, September 2, Robert Peary decided they could not get the meteorite aboard. He'd been foiled again, and all prepared to leave. For days, the men had been singing a ditty they'd composed: "On my Johnny Voker/We will turn this heavy Joker/We will roll and rock it over/We will turn this heavy Joker/Oh my Johnny Voker, Haul!" The heavy joker hulked on the edge of land, with Signal Mountain in the distance showing between snowy billows.

THIS MORNING—SUNNY, STILL, AND cool enough that the bay is skimmed with patches of fresh ice—we're heading to Meteorite Island. To get gas. On one side of Meteorite Island is the site of the Tent meteorite's discovery, on the other is a tiny Inuit settlement called Savissivik. I continue to obsess over my lost watch and hammerstones. We've come 140 miles.

As we approach Savissivik—a town of about forty buildings, of which about half are tiny homes—Inuit men in jeans and jackets appear by the dock, smiling, curious about the boat but speaking no English and we no Greenlandic. They help haul empty fuel bins up from the boat to fill them, then hand them back down while I run a portable pump to empty the gas from the bins into the *Barb*'s tank.

"You're becoming a good hand," Bill says. We'll have to hire you on."

"Let me finish my meteorite book and I'll send in my résumé," I say. Bill nods, laughing.

After showing the villagers pictures of falcons and maps of the coast—the Inuit point to places worth surveying—we all walk into town on broken pallets and shingles that are thrown down on the mud as makeshift sidewalks. One of the most remote towns in Greenland, Savissivik is haggard, though the bright colors of the buildings—yellow, red, green—try hard for cheer. I stay clear of the sledge dogs, some running free, some chained, and I wish I could interview the villagers about the meteorites.

But I remember what the Danish researcher Holger Pedersen had written to me. "People in the Thule district, and in particular in Savissivik, have no recollection of the meteorites' presence," he said. "This is easily understood from the size of the population and its migration. Just a few . . . may have an awakening feeling that the meteorites really belong to Greenland."

We buy a few things at the local store—Kurt warns me that this is part of their supply of food and not to buy too much—then we head back to the boat and motor out for another round of falcon searching. At about noon, we're on the move to the backside of Meteorite Island, where I match my copies of Peary photographs to the saddle shape of Signal Mountain, our best landmark.

Kurt has us in a ragged, jagged jubilee of ice, a blocked horizon of bergs, melt runnels everywhere. At times we have just a couple of yards of clearance between icebergs. I could stop and watch their undulant soffits for hours.

"I'm not sure I've been in big bergs like this, not this dense," Kurt says, without taking his eyes off the narrow lead. Shoosh and thud, the hollow thud of brash hitting the metal hull. Bill and Regan have the photo now, the one that shows the approach to the natural rock pier. We're all looking for the cleft in the cliff that is the "pier" itself. Kite Harbour.

Regan stands on the forward deck, his mouth open. "Don't you feel like explorers?" he exclaims.

"It's like we're looking for treasure," Kurt says.

In the early afternoon on a sunny-cool Wednesday in August, the crew of the *Barb*—three falconers and one space-rock eccentric—pull into the small inlet where Peary's ships had landed more than a century ago. As we do, Regan is asking questions astronomical, as in "How old are meteorites?" When I tell him, he repeats the numbers.

"Four and a half billion years . . . Jes-us Christ."

Bill is out first, over a short, vertical rise of rock, then he ties the boat off and beckons me ashore. "Just go hand over hand and haul up . . . that's right," he says, and here I am where Tent had rested for months awaiting its final journey.

THE STEAMER *HOPE* CAME BACK on August 12, 1897, to find Tent still on its pilings—what could have moved it, after all?—and snow everywhere. The ice had been thick, again, and the ship survived a minor grounding on the rocks to dock alongside the shoreline cliff. The situation was dicey. Next to the squat cliff, the ship—whose railings were just lower than the flat edge of the land—was exposed to winds that could push pack ice toward the vessel. The "open" water was narrow and icing up. Snow lay thick on the freezing water and could harden fast. Peary had just hours to lay a span of a few feet between ship and shore to secure what was then the largest known meteorite in the world.

"I proposed to construct a very strong bridge," Peary wrote, "reaching from the shore across the ship; lay the heaviest steel rails upon this, and then, after depositing the meteorite upon a massive timber car resting upon these rails, slide the huge mass across the bridge until it rested directly over the main hatch; remove the bridge; then lower the meteorite with my hydraulic jacks through the hatchway to the ship's hold."

And that's exactly what his crew did. They built the bridge—which featured massive timbers and huge iron bolts—and erected a lattice of beams to support the iron and keep it from moving in heavy seas should they get the thing on board. As if to underscore this precaution, an iceberg shattered during the construction, sending waves that shook the ship and unnerving men who were convinced the meteorite was cursed.

When the jacks lifted Tent off its pilings, the cart was slid beneath, the meteorite was lowered again, and, as with Woman, the weight of the iron forced wood, rail, and bolts to pinch together. The rock was chained, the ship moved, and boulders set upon the bridge as a counterweight. Draped in an American flag, the meteorite was ready to be loaded, but first Peary's child Marie, who was with him and Jo on this trip, christened Tent with a nonsense name, "Ahnighito." Then they got the little girl out of the way.

The winch pulled the cart with its meteorite till it was over the ship's hatch, and Robert Peary waved his hand to stop the operation. Just then, he claimed, clouds parted and sunlight broke from beyond Signal Mountain and fell upon Tent, "changing it into molten bronze" and lighting mountains and icebergs "in unspeakable tints of rose and yellow." It was, Peary thought, "as if the demon of the 'Saviksoah' had fought a losing fight . . . and yielded gracefully."

They loaded the iron in the hold, and hours later the ship made way for home.

WHILE KURT, REGAN, and I walk up the hill that rises above the natural pier—the hill down which Tent was moved to its pilings—Bill's on a satellite phone calling P-Funders back in Boise, taking care of business. On this side of Meteorite Island, the water is open for a few yards, but then come the whetted edges of bergs, the sharp white clutter and the reflections of sharp white clutter in water nearly the color of cobalt and maybe just as tricky. The sky is clear, the sun warm on my skin.

There are no piles of hammerstones, but the hillside depression where Tent had been partly buried for centuries is as obvious as the rough road that runs from the excavation site to the shore cliff. The road resembles a tear-shaped gouge, with softened divots where Tent had punched into ground on its slow roll down, and the ground itself is lichen-speckled white gneiss, clumps of brown grass and bumps of moss.

Before I know it, Regan, Kurt, and Bill are readying to leave. They'll spend the afternoon puttering along cliffs and landing at beaches on their falcon survey. Me, I'll be all by my lonesome.

"Y'all come back now, you hear," I yell.

"See you next year," Kurt yells back.

My spine does in fact tingle, as the boat backs away from the cliff. I look at the gear they've left for me. A pencil flare gun. A radio. A shotgun for polar bears.

"God," Kurt had said, "don't waste shots in the air. If that thing's close enough to see"—Kurt and Regan indicated a ledge a few yards distant—"shoot it. Aim for the center of the mass. No eye or ear shots. You don't want to piss it off." No, I think, I don't want to piss it off.

Frightful, beautiful, this crenellation of ice. It's so quiet that I can hear icebergs melting many yards away in the water, a trillion, trillion rivulets. To be here?

Soon I discover a long rusted metal strap. Good Lord, it must be from Peary, I think, perhaps used to bind timbers together. Now, here, a section of rusted rail from Peary's bridge. And more recent detritus, cigarette butts and a green bottle from the islanders. I sit on a mossy rock considering all these things, objects on display in the private museum of the world.

Gunshots and calving icebergs punctuate the quiet and then I hear a sigh, *a sigh,* a very loud and very long animal sigh. I stand and look to the crest of the ridge and retrieve the shotgun in one motion. Nothing. I look around. Nothing. Must have been an animal in water. Yes, an animal in water, nothing more.

Still nervous, I decide it's time for my planned small ritual. Though I'm not a believer in gods or design, I had decided to give a small piece of the Willamette, of Tomanowos, to the waters of this place. It was one of the flecks I'd found by the driveway in Oregon. I had begun to feel that Peary had committed a great wrong here—the meteorites had not been his to take—though his operation was, to say the least, impressive. Before my trip, I even drafted a letter to send to a Greenland newspaper explaining why I thought Peary was wrong in taking the star stones, why I thought the meteorites—or at least one of them—ought to be returned. The story of Tomanowos seemed instructive. My words were too stark, too preachy, disallowing nuance in history the way I had disallowed nuance in my own life. I never sent the letter. Perhaps I just got lazy.

At the cliff I toss a piece of Tomanowos into icy water, wishing for the well-being of the people to whom the saviksue mattered so much and hoping that one of the star stones might come back home. I give thanks. Maybe this private ceremony is, in part, a way of acknowledging that these rocks from space should be more than prizes, more than a way to make money, more than a way to clear a mortgage, more than a way to make a career. I don't worship meteorites as gods, but they're more than lucre. I know, though, that their stories have become a kind of prize to me.

Also, I know this: The tossing of the fragment of Tomanowos echoes what I did a few months ago, prior to moving to Utah, when I stopped by a prairie lake on a cold spring day with my old wedding ring. After I learned of the Lucas ring—the one made from the Estherville mesosiderite that Eliza Kimberly may have seen—I began thinking about how to part company with my simple gold band. I didn't need it as a reminder of my failures; of those, I had plenty. Though all things on this earth are gifted from the cinders of former stars, I couldn't launch the gold back into space. I could, however, give it back to the world. In the arc it made over the lake, it described the elements of all journeys—beginning, middle, end. As that little piece of Tomanowos just did.

I see murres and maybe loons. I tape the sounds of icebergs calving. I take photographs. Along the old scraped road, I find a rusted piece of wire. Without consideration, I put it in my backpack, along with bits of plants, granites, gneisses, and years-silvered wood, next a fleck of rusted former meteorite my magnet pulls up from where Tent had been. Evidence and adventure, but are my hands alive simply to grasp? I find more wood. I take it. I measure where Tent had been in the ground—30 feet by 16 feet. Fixity, fixity, I could nearly chant, fixity, fixity, afraid of time that goes too fast.

Everywhere grow clumps of stunted grass, along with mossy tussocks, orange-green to gray-green, and a lime-green-and-white lichen that looks like a magnified view of veins. *Alectoria ochroleuca,* I write, though it might be the widespread reindeer lichen. And I identify brilliant red moss too. The grasses—northern woodrush or the same I saw at Thule, which was arctic-*something* grass. I'm angry with myself for having come so far to be defeated by the names of grasses.

Two ravens croak over the ridge. Ice cracks. Shadows chill me.

I can't recall when I first found Eleanor Lerman's poem "The Mystery of Meteors," but I have it with me here in Greenland. In the poem, the narrator walks a dog before dawn and keeps looking for November's Leonid meteor shower.

In the darkness, the dog stops and sniffs the air
She has been alone, she has known danger,
and so now she watches for it always
and I agree, with the conviction of my mistakes.
But in the second part of my life, slowly, slowly,

I begin to counsel bravery. Slowly, slowly
I begin to feel the planets turning, and I am turning
toward the crackling shower of their sparks

These are the mysteries I could not approach when I was younger:
the boulevards, the meteors, the deep desires that split the sky . . .

How are we privy to our hearts? I wonder at this. And is exterior doing enough to construct one's interior being? I didn't think so when I began to learn of Peary's life and I don't think so now. Yet Peary's drive to surpass the mundane and to leave his name behind is something I do understand, and certain of his private agonies resonate. A strong-willed and tormented mother, a fear of death, fidelities to passions beyond conventional marriage. Jo came to accept Peary's adulteries. Passion is its own fidelity.

The psychiatrist and anthropologist Melvin Konner writes of the "apparent inevitability of human dissatisfaction." To be alive is to be unfulfilled. For Konner, this sense of free-floating desire is a matter for everyday reflection and action. "I sometimes think that the more reckless among us may have something to teach the careful about the sort of immortality that comes from living fully every day." That's passion. That's a kind of obsession, obsession as cultivated recklessness.

The root of the word "obsession" entails a sense of a person being overcome by an evil spirit, but I can't accept that obsession is always a base or abject thing. An obsession that seems laudatory we call "ambition," and, if success comes, we name it "achievement." Failed obsession? Few things seem as pathetic. Think of F. W. Cragin's unfinished histories.

Standing here on the hillside where Tent had rested for thousands of years, standing here alone on Meteorite Island, I hadn't expected this welter of meditation, and I'm still troubled by my little bag of goodies—rocks, a bit of ex-asteroid, wood, metal strap. Do I disbelieve my own achievements enough that I have to carry something of them in the palm of my hand? As shadows lengthen over the road of Peary's conquest, a road I have in a sense claimed as my own, I walk down the hill and wait with my booty for the *Barb.*

There I find a few white sprouts of pixie cup lichen. Tiny lichens, such persistent lives, so small, beneath the weight of passing zeal.

THE *HOPE*'S HOMEWARD JOURNEY was rough. After Peary off-loaded his Inuit (with supplies as payment for their work), the weather turned awful at Wolstenholme Island. The wind blew water "as a graver's tool cuts metal and drove the liquid shavings in sagittate lines," Peary wrote. Waves tossed the ship. The compass pointing at the iron in its hold, the *Hope* made it through storms, fog, and berg-clotted waters, to arrive unharmed at the Dock Street Excursion Wharf in New York City on September 30, 1897.

According to records at the American Museum of Natural History, on October 2, 1897, the ship was towed to the Brooklyn Navy Yard, where the meteorite was unloaded by a 100-ton floating crane. Peary had telegraphed the museum from Nova Scotia to tell officials of the meteorite and his additional cargo, that handful of living Eskimos. Also in the cargo bay of the *Hope* were bones of Eskimos that Peary knew by name, whose graves he had plundered in the name of science.

For years Tent stayed at the Brooklyn Yard, until it was hauled by "wrecking barge, with a huge derrick," then loaded onto a wagon pulled by twenty-eight horses, according to *Scientific American*. Throngs watched as the team—long as a city block and passing the David S. Brown Soap Works—pulled Tent up Broadway and then on 77th Street, where, at four P.M. on October 1, 1904, it was unloaded at a foyer of the museum.

But it still didn't belong to the museum. It belonged to the Pearys.

In a deceptive and tactless letter to museum officials, Mrs. Peary emphasized that the Cape York irons were her property and that proceeds of any sale would pay for her children's education. She claimed she had no other way to fund their schooling and said a sale would not go toward her husband's future expeditions. The Pearys were pushing for the sale because their longtime ally, museum president Morris K. Jessup, had just died, and he had promised to buy the stones. Meanwhile, Peary had written his wife that the sale *would* fund a future effort for the Pole. Eventually, Mrs. Peary sold the irons for $40,000 to Mrs.

Jessup, who gave them to the museum. The sale price matched the cost of the Cape York expeditions, and the museum purchased the meteorite in 1906, the same year that the museum acquired the Willamette. All involved in the Tent dealings cultivated an image of serene gift-giving. But Kenn Harper says that Peary earned about $50,000 for the meteorites and that at the time his wife complained of being in poor financial shape she was buying $10,000 in bonds. He calls these various transactions a "long-time fraud."

As to the feat of retrieving the meteorites themselves, historian Richard Vaughan writes that earlier expeditions hadn't worked very hard to discover or take the meteorites and concludes that "Peary was assuredly not the genius who succeeded where others before him had failed, though the recovery of the meteorites was a product of . . . single-mindedness and resolve."

The meteorites remain on display in the impressive Arthur Ross Hall of Meteorites at the American Museum of Natural History. For a time, there was a mirror on the ceiling above Tent, and one could look up and see good-luck coins tossed upon its top. The mirror is gone now, but when the guards aren't looking people still tap the giant boar of an iron with their rings to try to scuff its surface. One can see where Peary's men had drilled holes in the meteorite. Tent looks like a boulder from a Magritte painting, big enough to fill a kitchen. Nearby are the smaller Woman and Dog, each worn smooth from centuries of hammerstones.

Other meteorites from the Cape York fall were found in later years; some are displayed in Denmark, and one, which was found on Meteorite Island, stayed in Greenland on display in the capital of Nuuk.

Peary earned medals, gave speeches, wrote books, and earned both money and fame. Many now believe his legacy prior to his purported 1909 conquest of the North Pole was the retrieval of the Cape York irons. Otherwise, Peary's work—especially from 1894 to 1895—was, in biographer Wally Herbert's word, "wasted." Further, Peary faced criticism that he'd taken away the sources of iron for the subsistence-hunting Inuit of the Thule region. Peary's defense was that the irons were no longer used, though Kenn Harper argues that "perhaps there would come a time when they would need again the Woman, her dog, and her tent." As to supplies that Peary had provided the Inuit, they stopped when the explorer stopped going to Greenland, after he claimed the Pole. But the most intense criticism Peary faced regarded his human cargo brought to New York, where most of the Inuit got sick and died. When Peary finished his book on Greenland, he wrote of the dead Inuit as though they were still alive in their tupiks.

From time to time, when I think of Peary's life, I recall a passage from a Greenlandic novel by Hans Lynge, *The Will of the Invisible;* in it, one of the protagonists "could not bear to be just half a person: Either there should be happiness alone or an unhappiness just as complete."

Peary died on February 20, 1920. He was a public hero. He was sixty-four.

"I DO NOT WANT TO GO." That is the last thing I write on Meteorite Island. The P-Funders pull up and announce a change of plans. They've broken camp, and we're heading back to Thule now, because there's no luck with falcons around here. I try to hide my disappointment. Bill lifts my spirits when he tells me he brought a hammerstone for me to replace the ones I'd left behind, and I'm so grateful for that, for all of this, that I could cry. I'm sitting on a boulder and put on my wind pants, parka, and shell, readying for the cold ride. I take one more picture and climb down the cliff.

We travel for hours through water becoming rougher and rougher, and for the first time on this trip, I feel a sustained nervousness. We drive into a fog bank, unable to see the bergs. The sun descends into this bank, though the clouds stop just above the horizon, so between water and the cloud deck glows a weak gold-like sunset. I'm wide-eyed till we get to the pier, where we unload coolers, drybags, and boxes. My 500 miles of travel by open boat in the Arctic is over.

ON MY DESK IN UTAH is all I brought back home: one hammerstone, cleaved in half with three main facets; an old bone; three pieces of wood, one gray, two tan; two rocks, one quartz, one granite, both covered in lichen; one piece of burnt wood or charcoal; one tiny chip of rusted former meteorite, all from the valley of the isthmus by Saviksue Bay where Dog and Woman once rested. From Meteorite Island I have a long piece of gray wood, a rusted coil of wire, six local rocks, another flake of rusted former meteorite.

These are possessions, but I want to believe they are also connections to story.

But had I taken them in the spirit of discovery or greed? Both? Had I been like Ellis Hughes and Robert Peary? Had I stolen? Or had I found, discovered, and kept, the way Eliza Kimberly had? Certainly they'd been thrills and still are. I won't sell these little tokens, but by being part of my story—one I'm being paid to tell—are they bound up not just in legacy (my hope) but also in profit? These questions, I think, are harder to weigh than a meteorite, and over the years I'll even wonder if the wood and the wire might be stuff dropped not by Peary's men but by the few others who have visited those places.

Not long after returning, I made a list. At the top was "Reasons to Send Back: 1) they should stay on the land; 2) I feel guilty for taking them; 3) I'll have story, more impt. than artifacts; 4) helps my cause that stones should be returned; 5) maybe they were given to me, or I took them, *in order* to give them back. *To let go.*" The second half reads "Reasons to keep: 1) inspiration to write; 2) very few people will get there; 3) I found them; 4) they are beautiful and historic."

A friend tells me that there's a distinction I'm failing to make. He says there's a difference between spiritual artifacts like medicine bundles and utilitarian ones like hammerstones.

"Were you raiding a historical site," one colleague at Utah State University asks me, "or were you just cleaning up some of Peary's trash? How long does the garbage of expeditions have to remain in wilderness before it acquires a special significance and attachment to the place and we start to feel it should be left there for historical reasons?"

"Perhaps Peary's abduction of three meteors licenses your abduction of Peary's wood scraps?" another friend suggests.

I had made small forays into the splendid distances of the Arctic and the past. But Thoreau had been right. The bigger journey was the exploration of my own latitudes, where I had arrived at these questions as well as some answers, even if they were tentative. The answers were yes—to the question of whether

I could find Peary's meteorite locales and to the question of whether I too believed in the importance of passion and legacy. But these artifacts, these prizes—over the years, I'll not know what to do with them. Trash or treasure? Mine or not? Inertia will set in, and rather than mailing them to someone in Greenland by way of easing my conscience, I will let the bits of rock and wood and wire remain on a shelf. I'll look at them from time to time, remembering.

Book IV

The Weather of Belief

Beneath a sky that once rained bacon, at a pull-out next to Road 201 outside the town of Ensisheim, France, I watch in broad daylight as a man in dress slacks, shirt, and tie hunches in a December wind and relieves himself. Traffic—whizzes by. Bruno, a young man with peach fuzz who works at Ensisheim's Regency Palace Museum, has driven me and Kathe to the edge of this field where now the man zips up, gets back in his car, and leaves. I want to laugh, but Bruno's English is not so good and I can tell already he thinks mine is a strange journey, though it took just three minutes to drive from town.

I get out of Bruno's Peugeot to stand beneath a sky that centuries ago was believed to pour forth not only bacon but blood, wool, trash, bad cheese, coins, and skin. Such was the weather of belief. The air and its fire-breathing dragons—what we'd call meteors—proffered strange gifts, and through eras named and unnamed, all over the planet, plebeian and pedagogue alike took this sky to be a dripping storehouse of portent.

And I look across a field, trying to imagine a boy who saw—astonishing—not a wedge of Roquefort but a *rock* land right over there.

At Ensisheim, in Alsace, the oldest known witnessed meteorite fall in Western history took place on November 7, 1492. In at least one chronicle, the stone that landed here is given equal prominence with Columbus's discovery of the

New World. No one could explain the sudden appearance of a rock falling from the sky above Ensisheim without resorting to the word "miracle."

A little more than 300 years after Ensisheim and about 300 miles away, not far from Paris, the sky would precipitate not one rock but thousands, and the distance between those two events is the story of how it became possible to believe in rocks falling from the sky not as weird tales of God-fearing and gullible peasants but as facts of nature.

It's the winter after my pilgrimage to Greenland, and I've come to France to trace this change in outlooks, a kind of flashback to times long before the exploits of the star hunters I first studied, Ellis Hughes and Eliza Kimberly. How was it that they and others understood their booty to be meteorites? I'm hoping my time here will help me answer that question—indeed, how all of us have come to accept stones dropping out of the air, and further, as I stand in the cold, I'm startled, even troubled, by the contrast between legacy and anonymity.

So of course we know the name of a king whose fate was claimed to be tied to the sky stone, but not the name of the boy. The king is history. The boy wavers, refused by time and station any singularity, but I try to imagine him anyway—in tawny britches and tunic, leaning on a walking stick.

Kathe and Bruno huddle nearby while I look at a slanting sunbeam and irrigation equipment in the south-southeast, the direction of the meteorite's arrival. At my feet, there's a sign noting the fall, and I slap a pedestal-mounted faux space stone rendered in grayish blues. I look at the plowed fields and fences, ditches and trees. This is a place so flat it seems that France has become the Indiana of my childhood—and God knows where the rock landed, exactly. Later I'll learn that the monument was placed here simply because there was room for it and because it might entice tourists to town. The fall site itself remains as elusive as the boy.

Bruno can't say what rock has been used to mimic the real one or what trees have been planted next to the pavement. Poplar, alder, and birch, I'll come to find out. The faux rock? It doesn't really matter. With the temperature dampening our enthusiasm, such as it is, and with Bruno needing to return to the museum, I feel a rushed grumpiness. I've experienced no cosmic connection to this place. What was I expecting? It's a roadside rest. I'm looking at a wet spot on the ground, trying to imagine a 500-year-old fire in the sky. I motion us back to the car, and we drive away, leaving the ghost of the boy behind.

He was, it seems, alone in a wheat field the day of the fall. Or perhaps he'd not yet arrived at the field. Perhaps he walked with a woven basket on his back and carried a pitchfork. A rope around his waist would have kept his tunic tight against the autumn chill. He was not, as legend has it, tending sheep.

Sometime between eleven A.M. and noon, the clear air burst with sound. He must have flinched. He must have seen a fiery cloud in the sky. Did he cower? Did he run toward town?

The aerial explosion was impressive enough that another possible eyewitness, the artist Albrecht Dürer, painted the event. His rendered sky is mostly black, and near the middle of the composition glows a pale-yellow object, more rectangular than circular, radiating with red spikes. Gray and black clouds churn, fire burbles.

Along the flight path the thunder diminished, the loudest noises having occurred earliest in the descent. At Ensisheim, there was a rumble. Scientists have concluded that the meteor's southeast-to-northwest approach took it over the Rhine until it burst at a point above and between the towns of Villingen and Luzern, where terrified residents prayed for safety.

Slowed by air, the meteor lost its original "cosmic velocity"—the speed at which it entered the atmosphere, typically at about 40,000 miles an hour—and at that "retardation point," a sonic boom sounded. Then the rock fell under the influence of gravity until, affected by aerodynamics, it was traveling at least 200 miles an hour.

After the stone hit—imagine the *whomp,* the plashing dirt—our nameless boy may have stood over a hole a few feet deep, the black rock at its bottom. Was he curious enough to crawl in? Was he afraid he'd be swallowed up? Did he pray? How soon did others come?

A 1513 illustration shows a fire-orange cloud surrounded by yellow, the meteor falling with red rays before and behind. In a field, a man in a blue tunic and white britches (there are no illustrations of the boy) is about to whip his horse, which is dragging a small harrow. The dark meteorite looks like a heavenly currant. It drops unseen behind the rider, who has no idea what's happening. Nor does the other man, who sows seeds from a bag slung across his neck and chest. Like the plowmen in Auden's poem, they go about their work. The village in the distance, above which the fulvous Vosges Mountains rise, is in for no ruin. The towers still point to paradise.

Another drawing, however, shows dead birds falling from the sky. Cerise streaks surround the stone as it drops. Creatures skulk or cower. A man points up, and from a perch, an owl—harbinger of doom—calmly takes in the prospect.

There had to be a reason for the fall, but just what the miracle foretold no one knew. One original account, included in a booklet available today to tourists, says that the fall of the rock was so obvious it was "as if God had wanted people to find it." That they did. Despite edges sharp enough to cut and a weight of about 300 pounds, the stone was hauled out, whereupon villagers began hacking at it till a magistrate—possibly one Guillaume de Rappolstein—told them to stop. They then took the rock to a church, whereupon it was closer to God.

When the German professor and poet Sebastian Brant of nearby Basel heard news of the fall (and he apparently heard the explosion but didn't see the fall itself), this man of letters didn't have God on his mind. He was thinking about politics. Brant dashed off a poem—he was a versifier with all the subtlety of a dredge—and his stanzas soon reached thousands of people. Brant wrote, "May the one who marvels at listening to weird stories/Remember . . . this report." He recited the usual litany of a sky jammed with "shields of blood and fire. . . . Tiles, flesh, wool, of the celestial wrath." He also claimed the Ensisheim stone had a crucifix and symbols written on its surface. Well, it didn't, not really, but that's beside the point. The tricornered stone meant the Trinity, which meant—voilà—"special calamities" for the French. Convenient, since Brant wanted Emperor Maximilian to attack them. A cardinal, meanwhile, believed the meteorite presaged a different attack, that of venereal disease.

Three weeks after the fall, Maximilian trotted into town with his troops, viewed the stone, considered it divine, and took two fragments, after which he ordered no further molestation of the rock. He proceeded to whip the French in battle. No word on whether he met the boy.

This rock was something special. According to scientist and historian Ursula Marvin, "Even as memories dimmed, the spectacular circumstances of the fall and the special aura of this legendary ruler must have clung to the stone for centuries helping to protect it from destruction."

The meteorite stayed in the air—hung aloft in church—until the French Revolution, when the putatively rational reactionaries could not abide stories about falling sky stones. Falling stones! *Merde!* So residents decided to move the rock to another town, to keep it safe from the hotheads. Returned in 1804, the meteorite was eventually displayed in the Regency, which would become the museum that now houses the largest remaining fragment. Its tenure since then has been pretty uneventful, though after the Second World War an American woman claiming to be in the military tried to secure the meteorite for the "Museum of Massachusetts" but was turned away. Locals say she offered so much money that town leaders realized again how important the stone was.

Kathe and I visited the Regency museum last night after we arrived from an all-day train trip. At the train station in nearby Mulhouse, we met a professor named Zelimir Gabelica, a chemist passionate about collecting meteorites and sharing their wonders with others. Zelimir graciously delivered us to a hotel, where we met our two kind guides, Jean-Pierre Bruyère, a businessman who volunteers with the museum, and Jean-Marie Blosser, a cherubic pharmacist who leads Ensisheim's Guardians of the Meteorite, a group that from time to time dons red robes and regalia to honor the stone and those with connections to it and to meteorites in general. Tilting his head and smiling, Blosser explained that Maximilian wanted the meteorite to be watched over by a secret guardian, though the group dates only to 1984. Over drinks, Blosser handed me gifts, including a booklet about the meteorite and a ribbon-tied reprint of Brant's curious poem. Then we strolled to the museum and saw the meteorite itself. But last night's glimpse came at the end of a long day.

Today, returned from our roadside adventure, Kathe and I now stand before the rock and really look. What's left of Ensisheim here—after various whacks over the centuries—is a mass about 10 by13 inches in dimension and weighing about 120 pounds. It sits in its case like a siren song made solid. Bruno even opens the vitrine so I can reach in to touch this historic meteorite, and I graze my hand across the slightly stippled surface. It feels like rough, dry soap. It feels like hard dough worked over with hobnails. I'm transfixed. The oldest known fall in Western history, a stone that landed the year Columbus landed in North America! It's crazy—I want to bear-hug the rock, I want to squeal. Kathe, who speaks French and is translating for me, asks how many people are allowed to touch it. Bruno says very few, so I give a look of grateful, melodramatic surprise. We all laugh.

Flat-bottomed and with a flattish back side, the worn and gouged meteorite is dull gray and, in places, the color of old pennies. Here and there is ancient black fusion crust. All in all, the meteorite might pass for the remains of a dinosaur's number two. Despite my metaphor, I keep touching it, but soon Bruno has to close the case and resume typing at his desk nearby.

I close my eyes. I see myself standing by a boy in a wheat field. I watch the meteorite land and land and land. Round, heavy, smooth, brown, like high-speed chocolate, and it whacks into Earth, the boy standing there, coughing from the cloud of dirt and dust. What can I say to him? That we'll forget you?

I open my eyes to see the red cloth on a platform beneath the meteorite, red as the fire of the sky that day. And I wonder about that tiny fleck of meteorite flake by the edge of a shadow, smaller than a broken pencil tip. I yearn to have it the way I yearned in Greenland—automatically. Is it just that ancient hominid grasping? Or is my desire to have these things a mistrust of the words I'll write?

Thanking Bruno for his help, Kathe and I head out in the dark where Christmas lights are strung above the village streets—two large white snowflakes with blue lines shaped liked stylized holiday meteors.

THE POWER ATTRIBUTED to Ensisheim—that way-back-when magic—is in keeping with how shooting stars and meteorites have been viewed by many people in many places: as messages from the divine, as signals of fate.

In ancient European and Near Eastern cultures, meteorites were often revered. Various Greek and Roman temples may have held them, and meteorites' origins were partly understood. Diogenes of Apollonia said that meteorites were "invisible stars that fell to Earth and died out." A rare form of achondrite—diogenites—is named for him. Egyptian hieroglyphs note irons from heaven, and some were even buried in pyramids. In Mecca, the Kaaba stone is enshrined in a mosque as a meteorite, though most scholars now think it's terrestrial. The Hittites believed that iron meteorites came from beyond the Earth.

In Asia, the Chinese long understood that rocks fell from the sky. During the thirteenth and fourteenth centuries, China's Ma Tuan-lin wrote a chronicle that recorded some 2,000 years of meteorite falls. The Japanese goddess of household skills, Shokujo, was thought to utilize meteorites "to steady her loom," according to Philip Bagnall. The world's oldest, well-dated witnessed fall took place at Nogata, Japan, on May 19, A.D. 861. The stone remains at a temple.

Different cultures have different answers to the question of what one does—and expects—when a meteor is seen. One widespread reaction, of course, is to "wish upon a falling star." In the Middle Ages, some believed that the number of seconds counted before a shooting star vanishes equaled the number of years in which you'll have good eyes. A Swiss tale suggests that meteors can stave off illness.

But not all welcome a falling star. The Baronga of Africa would exhort such light to "Go away, go away, all by yourself." In Siberia, meteors are bloodsucking fire worms. The Dobu of the Western Pacific believed meteors were "volcanic crystals from vaginas of flying witches." The !Kung use a term for meteors that means "evil thunder wizard named Mucus who sneezes blood." And meteor storms have long been seen as presaging the end of the world.

Meteorites themselves have been taboo in some places. Nigerian natives attributed a smallpox plague to the theft of a meteorite by British troops. Various tribal cultures have had strictures against touching or using meteorites or even speaking of them.

Still, from Greenland to Mexico cosmic iron has been worked into implements. Artisans and smiths have used iron meteorites to make everything from beads to blades.

I think the richest folklore of meteors and meteorites is that which connects them to the body. And not just human bodies; meteors have been seen as other creatures as well, such as dragons or bats. Just as the Inuit thought of excretory functions when they saw meteors or discovered meteorites, so too have other societies. In one Chinese dialect, the word for meteors means "stool of the thunder gods." In Greek, *meteorizesthai* means to have gas. A Baja California tribe saw meteors as "fiery urine," according to researcher Carlos Trenary. Trenary says that in some cultures the "idea of excrement" means more than what we would think of today. Among the Maya, it "includes . . . blood, urine, sweat, mucus, vomit, afterbirths, exhalations and semen."

The perceived sexual nature of meteors leads to some interesting metaphors. For some cultures, "meteors are actually produced at the moment of orgasm when the stars copulate," Trenary writes. The Tucano tongue in Colombia has a term for meteors that means "penis of the sun" or "semen of the sun." Seeing fluids in the sky, seeing the work of bodies up there, reinforces a sense of physicality down here. Historian John Burke notes a fetching Chinese story of a meteor called the Jewel of Ch'en-pao that sometimes flew above a town as a male pheasant, prompting the females to answer his passionate cries. In one version of *The Epic of Gilgamesh,* the king of Uruk doesn't just dream of a meteorite, he "makes love to it," writes a scholar.

We can view the world scientifically while keeping in mind or heart these enduring metaphors of a sensual sky. Metaphors are like the movies. We go to them suspending our disbelief so we can be swept up in a fiction that enlightens. We've made the sky a kind of body to remind ourselves of function, of necessity, of ardor. Perhaps in doing so we give the sky itself a kind of immortality that we ourselves lack; we invest the sky with timelessness as recompense for our own passages.

If the sky becomes a body that goes on and on, we with actual bodies have not escaped the wounds and deaths we despise. So we find almost everywhere the archetypal belief that shooting stars are dying souls or are somehow related to death. Fate breaks off a thread for each person, Lithuanians say, at death—a falling thread, a meteor. The Hindus believed that "meteors were the offspring of Rahu's severed body," according to a folklore dictionary I keep near at hand. Among some Mayans, a meteorite leads to an alligator-infested lake, which seems too bad, really. The Irish believed the Perseids were the tears of martyred Saint Lawrence.

Then again sometimes it's just a matter of circumstance. Ancient Middle Eastern texts emphasize that the outcomes of witnessing a meteor depended on the person's state of being, where the meteor appeared, and in what direction relative to the observer. All these factors could mean the difference between life and death.

For all the wonderful variety of beliefs associated with meteorites and meteors, the view of just one man came to dominate Western understanding of these phenomena: Aristotle, who taught that meteors were streaks of light created by earthly vapors, water, air, and fire. This belief and its derivations persisted for many centuries.

William Fulke in his *A Goodly Gallerye* presented a sixteenth-century natural history derived from such classical views. He wrote about everything from dragons to a lake whose water causes men to "abhorre wyne" (in Italy, of all places). Of shooting stars, Fulke claimed they were exhalations slipping about the Earth's atmosphere that presented the illusion of falling. For, after all, if stars could fall, would not the earth be glutted with them? Would we not walk upon stars instead of meadows?

It took the Enlightenment to prompt harder questions, questions not only logical but daring, questions that challenged the sway of rigid classical and Christian views of the world, questions that arose, slowly but forcefully, not from theological dictum but from careful observation.

ERNST CHLADNI ARRIVED TO demonstrate his new invention, a euphonium; he left with an idea—about rocks in the sky—that would shatter any last echoes of the so-called music of the spheres, that long-presumed harmony of the solar system.

If, by 1793—about the time of itinerant genius Ernst Florens Friedrich Chladni's visit with physics professor Georg Lichtenberg—the notion of a geocentric solar system had given way to the correct heliocentric model, and if no one seriously believed the Pythagorean notion that the stars and planets were attached to nesting crystal spheres whose geometric and mathematical relations could produce, literally, "a music of the spheres," then what remained of that wished-for symphony was an implicit desire for order, for a solar system and a universe characterized by divine regularity and accord.

Chladni's euphonium worked like this: one rubbed a wet finger across pipes of glass and metal, producing sound. The heavens worked like this: moons about planets, planets about suns, and from time to time, a comet. Disruption was kept to a minimum. There could be no random, zinging rocks. If we could not press our ears to the sky to hear God's song, at least He kept space as tidy as rows of pews.

And the once-prevalent belief that falling rocks formed high in the air was giving way to other explanations. It seemed possible that lightning struck matter on Earth, altering rocks on the ground and thus giving the illusion of falling stones. But the discovery of fossils and cultural material from earlier peoples suggested that some of the odd rocks or "thunderstones" purported to have formed from lightning were, in fact, animal remains or crafted artifacts. Some philosophers thought that volcanic ejecta might be thrown over very great distances.

Ultimately, we know, some of these ideas were disproven, but they did reflect advances in such fields as chemistry and archaeology. Numerous natural historians in the eighteenth century emphasized evidence, and the need for evidence meant a need for skepticism. Reports of falling rocks, by and large, were not to be believed.

So it was one thing to have banished the crystal spheres that a couple of centuries earlier Johannes Kepler had so enthusiastically endorsed (he even proposed a beverage dispenser based on his musical planets). It was, however, quite another thing to suggest that the solar system was a shooting gallery of rocks that bulleted toward Earth and even landed here.

Inspired by his own sighting of an impressive fireball, Lichtenberg told Chladni that shooting stars in the sky might come from beyond the atmosphere. The suggestion was as revolutionary then as the theory today that the speed of light can vary.

Chladni went to a library and emerged three weeks later with an impressive list of reports concerning lights in the sky and stones on the ground. His legal background came in handy, for the reports felt accurate to his lawyer's mind.

A handful of men had suggested that meteors and fireballs might come from outer space, even as they typically, though not always, discounted a connection between those phenomena and rocks falling from the sky. Part of the difficulty was that fireballs appeared to be monumentally big—as wide as villages—so if they were solid, wouldn't they drop more than just a mere sprinkling of stones? Further, a discovery by Peter Simon Pallas of strange iron in Siberia—which would turn out to be the first pallasite ever discovered—baffled natural historians. Things just did not add up.

Two years after Chladni's meeting with Lichtenberg, he published a short book with a long title: *Concerning the Origin of the Mass of Iron Discovered by Pallas and Others similar to it, and Concerning a few Natural Phenomena Connected therewith.* What was for Lichtenberg a conversational hypothesis was now for Chladni empirical fact. Fireballs resulted from melting and explosion of the solid forms, some of which survived to land as meteorites. This material originated not from atmospheric coagulation, not from volcanoes on Earth, and not from comets but from the detritus of unformed—or exploded—planets. (About this origin he'd change his mind more than once.) Such a cosmic birthplace could account for the almost unfathomable velocities associated with fireballs. Furthermore, he noted striking similarities in the reports of these phenomena: The sky was often clear but for one cloud, out of which fire was seen and an explosion heard. The stones that fell were often encrusted and black—evidence of heat.

Chladni's speculations predate Piazzi's discovery of Ceres—the first known asteroid—by just seven years. Chladni knew that his were shocking ideas. But he appealed to observation. Not authority, but evidence. After all, the complexity of the solar system had been underscored by William Herschel's March 13, 1781, discovery of Uranus, the first world found since ancient times. Maybe there was more up there than the old sages had thought.

Emboldened, others took up their pens to describe further accounts of meteors and falling rocks, but many critics attempted to refute Chladni and his allies, after they'd stopped laughing. (Chladni didn't get everything right; for example, he was partly wrong in suggesting that shooting stars—quickly lit meteors like the Perseids—were rocks from space that hit the atmosphere, lit up, then glanced off the air back into space, like stones skipping on water. Larger meteoroids will sometimes do this, however.) German reviews were disapproving. Even Lichtenberg reportedly regretted the book's publication. Critics resorted to the usual ridicule of peasants seeing funny things and to the powers of lightning.

But the critics couldn't quite square their views with a series of serendipitous witnessed falls. Siena, Italy, 1794. Wold Cottage, England, 1795. Krakhut (Benares), India, 1798.

Chemists began to study specimens from these falls and found some surprising results. William Thomson, an English scientist living in Naples, found materials in Siena that seemed not of the Earth. In England, two developments ensued. First, Edward King wrote a tract about meteorites that garnered wide attention, and second, Sir Joseph Banks, the president of the Royal Society who had collected natural history material on Captain Cook's voyages, stopped by a coffeehouse to have a look at the Wold Cottage stone on display there. Banks realized that Wold looked like the Siena stones and he put a chemist named Edward Howard to work examining them.

Howard, with the French scientist Jacques-Louis Bournon, examined several meteorites, unknowingly helped by the fact that all the stonies were ordinary chondrites. They saw black fusion crusts indicative of surface melting, though these rocks were not volcanic. They saw for the first time chondrules—those mysterious globes of glass, which don't occur in terrestrial rocks. In a pallasite and some irons, they found curious mixtures of nickel and iron. It's difficult to overstate the importance of their findings.

One commentator weighed in on the debates sparked by Chladni, Howard, and the others, noting that "what is striking . . . is that all of these stones contain nickel, a substance that is rare on the earth, and iron in a metallic state, which is never a volcanic product." That commentator was Jean-Baptiste Biot, and his words were published in August 1802. Just months later he would have the pleasure of securing one of the most spectacular confirmations of a theory in the history of science.

AND SO ON OUR FRENCH SOJOURN WE GO, to the Château du Fontenil, Kathe and I, to see a place where Jean-Baptiste Biot stopped, a country house by the village of L'Aigle some 90 miles from Paris. Now home to a few thousand residents, L'Aigle was, on April 26, 1803, something of a backwater town. But it was here that the last, most stunning fall of the late eighteenth and early nineteenth centuries took place, a fall that dropped thousands of meteorites, a fall that demonstrated without a doubt that even peasants could be believed sometimes—and that rocks do indeed drop from the sky.

On the train from Paris, Kathe translates a report Biot made in French while I read a summary in English. She tells me this is what Biot wrote about the château we're about to visit:

> The owner was absent, so I spoke with the concierge of the château, who seemed to me a man of sense and self-possession. He heard, like everyone, several violent noises, like cannon shots, followed by noises like that of a fire in the chimney. All of a sudden a very large noise was heard in the courtyard of the château, as if a great tree had fallen. The woodsmen ran at this sound, and the animals [were] disturbed. . . . A young man . . . said to have seen a rock fall, went up and found a rock . . . had made a hole in the earth of 18 thumbs deep. . . . I saw the young man who saw the rock fall, I saw the rock itself, I saw the hole made by the rock, and I took away a small piece that they permitted me to separate from the larger whole. . . .
>
> I also have a piece of similar rock that fell in a field near Fontenil: It passed just a hair's breadth over the head of a burgher, causing him much distress, and fell just 20 feet from him. The sheep were also very disturbed. . . . These details were given to me at Fontenil.

On the day of the fall, at one P.M., "a globe of fire" traversed a clear sky from south to north, exploded, then dropped its stunning load of meteorites. Those who saw the event included priests, former soldiers, and workmen who, Biot emphasized, were "possessed of strong natural sense and reason." There were so many witnesses that the report could not have been fiction. Indeed, the falling of some 3,000 rocks was "never mentioned without terror," for they had fallen "like hail," had bounced off the ground, had smelled of sulfur, had snapped off tree branches. The fireball had streaked along exploding for *five minutes.* The explosions sounded like cannons, then muskets, then drums. Smoke streaked outward from a cloud north-northeast of L'Aigle, a cloud stationed in the air like an eldritch balloon. Though Biot need hardly mention this, the rocks were different from the local geology. And there are no volcanoes near Paris. The largest specimen weighed 17 pounds. Biot found that the area where the stones had dropped was about 2.5 miles wide and about 6 miles long. This ellipse was the strewnfield, the term now used to designate the fall zone of a meteorite shower.

Biot traveled to L'Aigle in June 1803, stopping at houses, interviewing villagers, and taking notes. One of his stops was at the Château du Fontenil, where Kathe and I are about to arrive, having been driven from the train station past dairy farms and fields of rapeseed. Barbara Whiteman, an Australian living in France and a friend of the château's owner, pulls into the lane leading to the estate.

"It is not a practical thing for someone working for the Australian government in Sydney to buy a house in France," she had just told us. Barbara lives in a nearby village so small it has neither a doctor nor a pharmacist, though there is, she notes, "a painter with the worst taste in France. He has his shutters painted a violent purple." Dressed in a proper skirt and burgundy sweater, Barbara is polite and she holds herself erect, befitting someone with professional diplomatic experience.

We are met by the château's owner, a retired geneticist named Philippe Lherminier. Kathe and Barbara both translate for me as we make our introductions under cloudy skies. Philippe had joked on the telephone with Kathe that he looks like Sean Connery but not James Bond. Thick-chested and bearded, Philippe might be called "burly" if he were an American, but his tasteful plaid jacket, red turtleneck, charcoal slacks, and mud-flecked dress shoes give him the dignified appearance of a country gentleman. One who cuts his own wood, that is. Philippe is a L'Aigle native who knew, when he bought the château, that he was also buying a chapter in the history of meteorites and science. Near an ash pile, Philippe has posted a sign about the fall.

I gaze at the château with its black bricks crisscrossing in a diamond pattern through the predominantly redbrick walls. The château looms over us with a steeply slanted roof, a four-story tower adorned by a dragon wind vane and high chimneys. I shake my head. It's my first real introduction to a place that once housed such wealth. I've never even visited a castle. My mother, who hated that she had to raise my sister and me for a time in a trailer, would have loved this place, with the grounds of prodigious trees and hedges, the cedar of

Lebanon, the thuya, a fading hydrangea, and blooming roses. Though the house dates from 1544, the site itself has been occupied by a château since the eleventh century.

Inside, past the heavy dark door with its carved story of the Annunciation, we walk on a sixteenth-century stone floor and oak parquet. Biot himself must have stood in this foyer, with this staircase before him. When Philippe had agreed to meet us, he also had generously offered to let us stay here. *To spend the night at a château whose roof and grounds were once sprinkled with meteorites that changed our view of the sky? How cool is that?* I thought.

Well, cool indeed. I rub my hands in the chill of several unheated rooms. I reach out and touch the stone walls and feel their cold seep into my skin. There are fifteen bedrooms in the mansion—so many rooms, in fact, that in the summer Philippe hosts symposia of visiting musicians and artists. From the fireplace comes the comforting smell of wood smoke.

Fontenil is all lovely stone and dark wood and tall spaces and gray light and a kind of thickness in the air that is the season's dampness but also the accumulation of time, though I admit, only to myself, that the rococo parlor where we first sit feels less like 1803 and more like houses in any number of Stanley Kubrick movies. Maybe the stereo, with an amp worthy of a good bar band, sparks that connection. Under the yellow walls sits an orange vinyl chair, a tiny rug made from (I'm guessing) a fox, and neoclassical-style furniture.

Philippe and Barbara go out of their way to make Kathe and me feel comfy, welcome, even pampered. Philippe retrieves a yellow folder of copies and papers and shows me a log from an accession register for the natural-history museum in Paris noting newly arrived samples of the L'Aigle fall 200 years ago. Feeling foolish but wanting to be thorough, I ask if there are holes still visible, holes in the ground, holes in the roof . . . I anticipate the response. He laughs. Kathe translates, "I could make a facile one," meaning he could fake it, then he makes picture-taking gestures. At Fontenil, no preserved holes. And no meteorites. Philippe speculates, "When the grandmother died . . . 'Here's this old rock, just throw it out.'"

We have a rich lunch of duck breasts Philippe roasts in the fireplace. The meal includes honored guests, L'Aigle's mayor and his wife. The mayor and Philippe discuss plans for the bicentennial of the fall, then afterward Kathe and

I tour more of the château and grounds. I stifle a desire to sit alone in the attic with its original joists that would have reverberated with the explosions on April 26, 1803. My stomach warm with duck, I could also use a nap.

But there's place and history to see. After a stroll outside, wet oak leaves pressed to my shoes, the lunch guests departed, I wonder about the man who opened that thick front door for Jean-Baptiste Biot. The concierge would have had a cottage of his own, Philippe says. When I ask what his duties were, my host jokes, "Watch TV." We laugh, and I say that Fontenil was quite advanced for its day. Nothing's known of him. The concierge of Fontenil will remain like the boy of Ensisheim, a faceless ghost.

In the evening, Philippe begins to speak at length. Darkness appears over the high windows. He needs no prompting, and I relax in my chair and just listen to Barbara and Kathe. We have all eased into one another's company, and our hosts have made us feel like old friends. Biot should have been so lucky. Philippe says that it's not really the "L'Aigle" fall, since various locales might have laid claim to it. In fact, most people in the area identify the fall strictly with Fontenil. He adds with a hint of pride, "The first fragments fell here," and I look reflexively at the parlor window.

Philippe also tells me that when he was a child he read a Jules Verne story in which a cleaning lady knocks into a ray gun, accidently turning it on; its energy transforms a meteorite into pure gold. Once again, I marvel—the meteorite as treasure. Philippe pulls from his rare-book collection the first French translation of Seneca and reads aloud passages about weather, about Pliny, about the sky. With a firmness that surprises me, he says that "the sky of mythology" and "the sky of science" are always opposed to each other. Always. I'm not so sure. Both skies are full of wonder. Both seek explanations of wonder. Of course there are differences. The sky of mythology delivered fate and was a consequence of design. The sky of science delivers fact and is a consequence, I think, of random causation, of happenstance, which, for me, makes it no less beautiful. Under both skies we can tie ourselves to the universe and to one another. Philippe quotes Yuri Gagarin, the first man in space: "I did not meet God." I nod, but I still think that one doesn't need to believe in a Creator to find creation divine. The parlor's tall windows frame the night, and the lights of the château glow yellow and late. It's time to sleep.

Kathe and I go up steps of stone and steps of wood to our bedroom, which faces southeast, the direction from which the meteor arrived. We make for bed, walking down a long hall with a light timer, to use the bathroom. Once in bed, Kathe and I are in long johns under blankets, and I keep my wool hat on, delighted to feel as though we are camping in a château. We rest beneath red curtains and wallpaper flowers that catch in the mirror above the empty marble fireplace. The wooden floor and bed both squeaked as we climbed beneath the covers. I snort like a kid in exhausted disbelief and good cheer.

Who slept in this room the night after the fall? Was someone napping that afternoon of explosions and fire and stones? Who might have looked through the windows, four panes wide, seven panes tall? Kathe and I talk and cuddle in bed, amazed at this strange little adventure, a night in a château from 1544 where fragments of a rock from space fell in 1803 from a cloud visible from this very window. A string of prepositions, all relation, relation, relation.

I go to sleep wondering—a bit—about ghosts and the silhouettes of leafless trees that seem to move and move, and owls call all night long.

THE NEXT MORNING, AFTER thanking our hosts for their kindness, we take the train to Paris (this was on our way to Ensisheim, which we visited next). I think of how I watched the gray light at night through the windows, how rain had tapped at the roof and glass, and how for a minute or two I did imagine ghosts, then banished them. The calls of the owls had soothed me with their wildness. We so often think of animals as wild, but why not rocks too? I kept imagining the cloud, that accidental rectangle in the sky, placing it in my vision probably too low over the horizon, but seeing puffy arches and spikes emanating from the smoke, those red-tipped tendrils, those shooting stones.

It is still all so wonderful and improbable—meteorites? The sky raining rocks? A sixteenth-century château owned by an affable, retired geneticist? Chladni's vindication? Biot standing where we had? Those high walls and ceilings, the chill, the château, the dark, the owls?

Close your eyes in the usual darkness. Meteors can be bright enough to open them. Moments spiderweb, and the sky of mythology and the sky of science keeps raining as we go.

• • •

In a stunning bit of bad timing for critics of Chladni and his confederates, a fellow named Guillaume Deluc saw one of his own petulant antimeteorite articles appear a week after news reached Paris of the L'Aigle shower. He must have been in shock. He had said once that thinkers like Chladni "do not reflect on all the evil that they produce in the moral world." For him, random rocks in space falling randomly to the Earth meant the melancholy of a random universe. Whether there was a wide public attracted to the moral aspects of the scientific debate, I don't know.

The same year of the L'Aigle fall, Étienne-Marie Gilbert began work on the first classification schemes for meteorites, work that would become very complex within two centuries, and Joseph Izarn presented a chart in his book about sky stones showing that there was still a diversity of opinions about just how they originated, whether from lightning strikes or lunar volcanism or aerial coagulation or space.

By 1810, "reports of stones falling from the clouds," writes meteoriticist D. W. Sears, "had found a respectable place in the mineralogical and astronomical textbooks." Only in 1812 did the first history of meteoritics appear. Advances in microscope technology allowed for careful study of meteorites' compositions and structures.

Many puzzles remained, however, including the huge appearance of some meteors and the small size of the stones that seemed to drop from them. Some thought that fireballs were comets that orbited the Earth. Might electricity ignite certain gases? But it was already understood that electrical currents were unimaginably fast, faster even than the velocities of fireballs. Not till later did scientists understand fireballs as an example of a particle's passage superheating the air, creating a glowing pocket of electron-stripped ionized gases much larger than the object itself.

In America, the 1807 Weston, Connecticut, shower was the first to raise widespread questions about meteorites. It was determined that Weston was a meteorite, in part by comparing it with a sample of another meteorite—Ensisheim.

Eventually, scientists realized that the atmospheric-formation theory couldn't explain how nickel was found in meteorites or how masses of such material could form so quickly in the air. Lightning was discounted (even though the alleged connection between lightning and meteorites persists in folk beliefs; one of the Kimberly clan in Kansas claimed a lightning bolt led him to another sample of the Brenham pallasite). The lunar-volcano theory was dealt a blow when estimates showed that lunar volcanoes could deliver very little material. In the 1980s, it would become clear, however, that meteorites hitting the moon and Mars could send ejected bits to Earth.

The debate about the cosmic origins of meteorites continued until the second half of the twentieth century. In the 1950s, it was understood that meteorites originated from within the solar system. But not until 1964 did Edward Anders present a thorough explanation showing that meteorites arrived from the asteroid belt. This seems surprisingly late to me. It means the asteroid-origin model of meteorites is younger than I am by only one year.

In Paris, at the National Museum of Natural History, I get a chance to see and hold samples from L'Aigle, one of the goals of this trip. The curator, Claude Perron, even lets me hold the biggest L'Aigle sample in the world, #287, which weighs more than 15 pounds. My arms register the heft of history. Then Claude and I both admire L'Aigle #17—the numbers are painted in red on the surface—which, despite two centuries of handling, retains a fusion crust as fresh as it was on the April day it fell. It's not cracked or dulled or weathered with time. The crust could have melted minutes ago. Claude then shows me how some of the dark areas are blackened from having been burned and melted *in space,* and he moves a sample into the light of his office to show me its metal flash.

Jean-Baptiste Biot wrote that "I shall consider myself happy" if people believed his report about the fall. He thought he had "succeeded in placing beyond a doubt the most astonishing phenomenon ever observed by man."

Questions of doubt and proof, of success and happiness, would come less easily to another investigator of meteorites, one who worked at the beginning of the twentieth century, a man named Daniel Moreau Barringer. Biot had proved his point, as had others, but until Barringer few, if any, would have the audacity to suggest that meteorites might be so large as to slam into planets and make craters, huge craters, in the recent past. For such craters suggested a violence unbecoming to a planet whose history the geologists wished to keep calm, as calm as the astronomers had once wished to keep the solar system itself.

Book V

Mr. Barringer's Big Idea

Sometimes coming events cast their shadows before.
—Daniel Moreau Barringer

If the ridge could talk, it would murmur. Visible for miles in northeastern Arizona, the low rise declares little. That's what distance suggests. If Robert Peary had declared some distances "splendid," he'd have used another word for this one. "Unpromising," say.

Anyone intent on reaching that drab crinkle had to cross country shaped by paucity and violence: miserly rainfall (but dropped in deluges when it comes) and mile-high air (thin, but sometimes slashing when it gusts). The sun's heat, the night's cold, and the body ground-truths cliché: the desert is harsh. Sagebrush and cacti grow here. Junipers look like bonsai trees, and the tallest living things, walnuts and cottonwoods, cling to draws. Coyote and jackrabbit, ground squirrel and rattler, roadrunner and cactus wren all have evolved their versions of patience to cope with dearth.

It's not surprising that few people walked up the remote ridge. It looked like any other, if more jagged. Not even the seemingly out-of-place chunks of iron, like bits of broken manacle, offered much incentive. The closer to the ridge one traveled, the more metal there was, but iron meteorites were poor forage for the area's grazing sheep.

If a shepherd did push his flocks closer to the ridge, he'd see how it bent in a curve, and if he chose to walk its length at the base—through pell-mell fields of boulders and sand—he'd meet his starting point, for the ridge was a circle. And if he climbed the 150 feet to the top, he'd not be on a ridge but a rim, an edge above a breathtaking hole. Crater? Caldera? He wouldn't have known. The cavity plunges 600 feet down and spans 4,000. On its walls, rocks stubble in exposed lines. Countless gullies finger toward the floor.

At the brink of such a precipice, drama, not clemency, imbues the scene. This vertiginous hole, so wide and deep it could swallow an entire town, demands long looking. Upon seeing the place for the first time, one man exulted, "The view . . . particularly about sundown or by moonlight, is weird and impressive in the extreme. The inwardly steep and even overhanging walls, profoundly shattered, surrounding on every side a broad, deep pit, accessible only by the steepest of trails, barren of all but the scantiest of vegetable life . . . present a picture . . . never to be forgotten."

Those who knew this false ridge and its hidden hole called it "Coon Butte" and "Coon Mountain," despite it being neither a butte nor a mountain and despite a raccoon never having been collected there. It was to this misnamed place that men came to find a hidden star.

THE U.S. GEOLOGICAL SURVEY'S CHIEF GEOLOGIST, Grove Karl Gilbert, left Washington, D.C., by train on Thursday, October 22, 1891, at 4:30 P.M. Half an inch of rain had just fallen and the temperature was in the 40s, quite a contrast to the Indian summer warmth the day before, but perhaps this chill felt to Gilbert like an affirmation of his inner weather, which must have been bracing. Scholars have learned that Gilbert felt almost certain he would locate at Coon Butte a "buried star"—a meteorite—which he suspected had caused the crater's formation. The discovery would be stunning, for, at the time, no meteorite had ever been associated with such a massive depression. A meteorite "crater" was unheard of.

At forty-eight, Grove Karl Gilbert was the country's top geologist. Historian William Graves Hoyt has said that Gilbert "was, perhaps, the closest equivalent to a saint that American science has yet produced." A student of the classics and math, Gilbert had served with John Wesley Powell's Western surveys. Gilbert wrote numerous papers, devised the concept of dynamic equilibrium—the concept, as Hoyt puts it, "that landforms reflect a state of balance between the processes acting upon them and their structure and composition"—and he investigated myriad phenomena, such as erosion, and many places, including ancient Lake Bonneville, whose former levels are embodied in the "benches," or terraces, in northern Utah's Cache Valley, where I live.

It would take Gilbert and an assistant nearly five days to get to Flagstaff. The trip had come together quickly. Just two months earlier, Gilbert had listened at a conference as mineralogist Arthur Foote revealed that an iron meteorite found at Coon Butte contained tiny diamonds. In detailing the location of the find, Foote avoided an extended discussion of the origins of the "crater," a feature then associated strictly with volcanoes. The implication was, however, clear. Something other than a volcano might have formed the hole.

Desk-bound, Gilbert dispatched a staff geologist, who concluded that an underground steam explosion had formed Coon Butte. As to the meteorites, their presence was coincidence, he thought; they'd fallen after the steam blowout. Gilbert wasn't so sure.

On Sunday, November 1, 1891, Gilbert, his coworker, and their field hands arrived at what Gilbert called the "Amphitheater (= Coon Butte)." The weather was dry and pleasant. As they set to exploring the area, Gilbert was guided by three crucial suppositions: First, if the cavity had formed because of impact, the meteorite should be buried under the cavity's floor; second, if such a meteorite were iron, it would attract a compass needle because of a locally intense magnetic field; third, an investigation into the volume of material that could have filled the crater might help reveal if the origin was cosmic or subterranean. Whatever had formed the Amphitheater, Gilbert saw that "the disturbance was violent. Rocks of great size are moved to great distances & all must have been moved upward." Two things could have caused this: "1) the fall of a star. 2) The explosion of some substance beneath."

The U.S. Geological Survey team charted the crater, collected various rocks—finding no iron meteorites in the crater proper—and examined strata. Winds as unruly as the hairs of Gilbert's beard twisted across the rim and pit. Gilbert wrote equations to estimate the volume-holding capacity of the cavity and how much material could be accounted for on the surface of the rim. In a tent, the men suspended a bar magnet and checked to see if a small meteorite perturbed the magnet. It did. A big one should too.

Yet they could find no evidence of a change in the magnetic field beneath the crater's floor. The magnet assumed its normal repose, and in his blunt entry for Saturday, November 14, Gilbert wrote, "The pendulum swings just as fast on the top of the rim as in the bottom of the hole." There was no magnetic anomaly that he could find.

Working through math relating the size of the crater to the inferred size of a buried iron such that the meteorite would not be detectable by the magnet, Gilbert concluded that such a scenario was "absurd. . . . Either the hole was not made by a meteor, or the meteor went through the crust." Curiously, on Thursday, November 12, he had written, "I suspect that the needle does not afford the means to determine the depth of the star, nor its presence." He'd now set that doubt aside.

Gilbert also calculated that the "hollow" once contained 82 million cubic yards of material—exactly the volume that he estimated existed surrounding the crater. As historian Kathleen Mark explains, Gilbert thought that "if [a huge meteorite] were buried . . . an equal volume of material must . . . exist in the rim." That is, he would have found more material on the rim, displaced from the visible hole and displaced from underground by a meteorite. He found no such extra material.

After noting that the limestone of Coon Butte crumbled without difficulty, he no longer ascribed its friability to impact shock. He thought it a consequence of weathering.

"It follows . . . that the great meteor is not in this hole," Gilbert wrote. "Chief attention should be given to other explanations of the crater." His man had been right. Coon Butte and its big hole had resulted from underground magma that had heated an overlying pocket of water. Like boiling water in a teakettle popping an unsecured cap, steam blew out the crater. The star had vanished beneath Gilbert's gaze on a magnet and on his numbers penned in a journal. His men packed out toward Flagstaff, through desert, then ponderosa pines.

. . .

BUT THE IDEA OF IMPACT at Coon Butte would not go away.

The Smithsonian's head curator of geology, the respected George P. Merrill, once wrote that "those who know Mr. Gilbert thought to read in his report a strong leaning toward [the impact theory], abandoned only because not borne out, so far as he could see, by the facts." Indeed, in two talks Gilbert admitted that he had believed a meteorite formed the crater, but that the lack of magnetism and other evidence pointed toward steam even if, as Gilbert admitted, the meteorite could be buried very deeply and/or be smaller than he estimated.

The point of Gilbert's talks, scholars have suggested, had less to do with the origin of the crater and more to do with his emphasis on the deliberate nature of fieldwork and how scientists must respect what facts seem to tell them. In other words, Coon Butte mattered less as an enigma and more as a lesson. "What most people heard," Kathleen Mark writes, "rather than admonitions for scrupulous care in the search for facts, was [that] . . . Coon Butte had been created by a steam explosion."

In an ironic twist, after returning from Arizona, Gilbert began to consider the problem of lunar craters, which most scientists believed were volcanic in origin. Gilbert had reason to doubt volcanism had formed all the moon's craters; there was, for example, the peculiar fact that the moon's craters were nearly always lower than the surrounding terrain, whereas on Earth volcanic craters were nearly always higher, being, as they were, tucked in at the tops of mountains.

Like a few others before him, such as Richard Proctor in England, Gilbert now argued for meteorites as a possible origin for lunar craters. He posited an Earth that once had been ringed, not unlike Saturn, by rocks. These accreted to become the Moon, and the craters record the final collisions during that formation. (Today most scientists believe that the Earth was struck by a planet-sized object early in the Earth's formation, whacking off enough molten material to congeal into the Moon.)

Gilbert was especially interested in the problem of the roundness of craters. What was the relationship between an angle of impact and the shape it produces? According to Mark, "He dropped marbles into porridge," among other tests, in order to find out. Objects arriving from a low angle of trajectory, Gilbert found, caused elliptical craters. On the moon, round craters predominate. Though he noted that solids can become plastic with enough heat—a fact that would have enormous implications for the understanding of Coon Butte—Gilbert doubted whether rock could become so hot and so liquid that it would assume a circular shape following a low-angle impact. Later scientists would indeed find that oblique angles of impact can form round craters. For whatever reason, Gilbert stuck to the idea that circular craters could not form after low-angle impacts, despite the necessity for a very complicated model to justify a series of vertical blows. Impacts did cause lunar craters, science would learn, but Gilbert's ring-around-the-earth and vertical-trajectory model would be proven wrong.

What's fascinating here is Gilbert's failure to make a connection. Although he noted that impacts could produce heat and that impactors themselves could melt, Gilbert did not publicly reconsider the problem of Coon Butte, where these insights might have led him back to further investigations (and where, fittingly, decades later, astronauts would train for Moon landings).

He convinced no one. Volcanism remained the entrenched explanation for lunar craters, just as, for a time, Gilbert's steam blowout would for Coon Butte.

In Grove Karl Gilbert's day, geologists understood that both gradual change and violent events were part of Earth's history, the planet being ancient enough to contain both categories of phenomena. Vast waters, like civilizations, gathered, then seeped away. Sediments, like outmoded ideas, slowly deposited until something different covered them, sometimes for good. And though volcanoes spewed and fault lines shook, such violence constrained itself to local or regional scales—a few exclamation points in the long run-on sentence that was the Earth.

The principle of uniformitarianism, or gradualism, had triumphed over the catastrophist view of Earth's history that had been championed by many Christian leaders. The religious association with catastrophism had tainted it as unscientific. Many theologians argued for a Earth only a few thousand years old, but this was preposterous to Victorian scientists and to much of the public.

News of fossil discoveries and the force of Charles Darwin's theories meant the planet was far older than some bishops would have liked.

To this world change came slowly, like an elderly aunt arriving on a visit. Gradualism, as developed by the commanding nineteenth-century British geologist Charles Lyell, taught that large-scale geologic change was inexorable and, one is tempted to say, decorous.

For centuries, Christians had believed in a harmonious geocentric solar system free of debris and trauma, while arguing that God visited catastrophe on the Earth itself fairly regularly. There were, after all, sinners to be punished. Astronomers eventually displaced the geocentric model and found the solar system to be messier than they'd once supposed. That old desire for order was, in a sense, transferred to how geologists viewed the history of the Earth. The messy, massive catastrophes of the Bible had no scientific place in Earth's geology. An appropriate calm had reasserted itself, and gradualism persisted deep into the twentieth century. Further, when the gradualists—that is, most geologists—saw violent change taking place, it was due to processes confined to the planet, like, say, a mudslide. As late as the mid-1980s the notion of cosmic impacts affecting anything but small areas prompted terrific scorn.

This distaste for disaster must have its roots at least somewhat in our daily lives, lives in which, for the most part, disaster doesn't occur every second of every day. Of course when calamity strikes, it's horrifying, it's frightening. A lion attacks on the savannah: Fight or flight? The tornado bears down on your house: Hide where? In an instant, a man has grabbed your purse and shoved you onto a city sidewalk: Did that just happen? It's no news that our propensity for consuming tales of disaster—from ancient myth to cheap paperbacks—is a way of subsuming fear, of controlling it through design, of giving it and us a narrative arc. But it can be surprising to see such psychology play out in shaping a culture of science.

Part of the reason gradualists had difficulty conceiving of large-scale catastrophic events was their insistence on interpreting past processes strictly from the processes seen to be operating today. This was called actualism, and since no one had seen floods the size of states or giant iron meteorites whacking into the crust, it made no sense to invoke them as causes for anything, ever. Gradualism honed away catastrophe with the sharp blade of Occam's Razor.

Seen from this perspective, Grove Karl Gilbert's abandonment of an impact origin for Coon Butte was banal. Science is inherently conservative. After all, new ideas and observation must pass muster in a culture built on consensus, care, accuracy, and verification. When new ideas or data appear, this is when science needs to be most skeptical, even though—after a change in paradigm—such skepticism can reveal less noble traits. Scientists, like artists, can have fits of egoistic stubbornness and protective quiet. How would you feel if someone came along and proved your own working assumptions wrong and ruined a career built on now-destroyed assumptions? By the same token, how would you feel if everyone told you that your new ideas were garbage? Of course without scientists getting things wrong, they'd be less able to get things right.

The monolith of gradualism not only outlawed, in a sense, global catastrophes, it made arguing for anything other than local or regional disasters of known types—lava, flood, and quakes—pretty far-fetched. Anyone making such a case had to do so with care and be prepared for rejection, even insults. Anyone doing so needed years of training in a scientific discipline—geology, say, or physics—and a dash of charm and diplomacy wouldn't hurt. Alas, Daniel Moreau Barringer was a mining engineer, lawyer, and businessman. And he was a blowhard.

How much Daniel Barringer learned one particular evening in October 1902 is not entirely clear. A 1956 letter from his son says that Forest Service employee Samuel Holsinger told Barringer about Coon Butte while the two men talked over a campfire in southern Arizona, where Barringer had a mine and Holsinger had been on patrol against illegal logging. According to a family history and Barringer himself, Barringer was at the opera in Tucson, where he'd slipped out to smoke and found himself in Holsinger's company. They were sitting on the spacious porch of the San Xavier Hotel. It seems that Holsinger said at least this much: There was a huge crater in the desert with iron meteorites sprinkled there, and those meteorites were part of a larger one that, some said, had formed the crater. Holsinger had never been there, but he'd heard of it from F. W. Volz, a trading-post operator. (It's doubtful that Holsinger had much to say about Hopi stories of Coon Butte, since even today members of the tribe will not speak of it.)

The fact of a crater surrounded by meteorites suggested the unthinkable: that a large body from space had smashed into the Earth recently enough that terrestrial geology and weather had not erased the crater. Barringer must have known or intuited both the scientific and financial import of this possibility. Not only could prevailing assumptions about meteorites and the Earth be challenged, Barringer might also find the impactor itself—and mine it.

Astounded, Barringer let his cigar slip to the ground.

Holsinger may have filled in other historical details for Barringer. For a time, the crater was called "Franklin's Hole," after a U.S. Army scout. Then, in 1886, a sheepherder discovered a 154-pound iron meteorite and mistook it for silver. He later told a camp cook, who saw that it was iron and began to pursue mining claims. The cook sent samples to a company in New Mexico. He borrowed some cash, then vanished like a smoke trail. Another fellow filed claims but he too disappeared from the story. Samples eventually arrived for mineralogist A. E. Foote to analyze, and his lecture would interest Gilbert. As to Volz, he was uninterested in mining and had turned his attention to gathering and dealing the meteorites themselves. Soon Canyon Diablos—named for a nearby canyon—were appearing in collections around the world. From small and cheap to large and pricey, Canyon Diablos remain a popular choice among collectors.

Barringer seems to have been convinced from the moment of Holsinger's revelation that the iron meteorites were the detritus of a huge impactor that carved out the crater. He'd not even seen it, but "upon returning home," he wrote, "the matter 'would not down' in my mind."

Barringer had found a passion equal to his ambition.

BORN IN 1860, Daniel Moreau Barringer was seven, twelve, and thirteen, respectively, in the years his mother, sister, and father died. Whether among the estate's figs or beneath one of the paintings in the house or beside a pigeon coop where he and his siblings raised birds, the boy must have sought comfort to fill, if partly, those absences. After his father's death, Daniel Barringer left the family home in Raleigh, North Carolina, for the care of an older brother in Philadelphia.

Barringer was kicked out of a military school but by twenty-two had graduated as class president from the University of Pennsylvania law school. Soon bored with law, the athletic Barringer pursued Western adventures with his pals Owen Wister and Teddy Roosevelt. Within a few years, he wrote books on minerals and mining law and took geology courses at Harvard and Virginia. His affinity for geology steadied Barringer and gave him a fortune from mines.

This cigar-smoking, whiskey-drinking cannonball of a man—five-foot-nine and more than 200 pounds—fired away at doubters, and it was precisely his confidence that made it possible for Barringer to buy claims on Coon Butte sight unseen.

With partner Benjamin C. Tilghman—a friend, ballistics expert, and fellow member of the Boone and Crockett Club—he founded the Standard Iron Company in 1903. Samuel Holsinger would supervise operations at the crater.

Everyone expected the meteorite to be found right away. So heady was Barringer that his son Brandon recalled "stock was actually reserved for father's partners in a Mexican iron venture which he was sure would be adversely affected by this discovery. (Talk about confidence!)." Barringer wanted both the iron and the rarer nickel mined out of the meteorite.

Drills were sunk, workers scoffed, and clouds of silica puffed up from the holes. Then bits hit hard metal and stopped cold. Then quicksand flooded the shafts. Drill holes that were to have cost $1,500 each soon cost more than three times that.

But they revealed something important. Pieces of meteoritic iron had been found in some shafts, mixed in with sand, rock, and water. The meteorites weren't just sprinkled by chance on the surface. They were buried. They were part of the deep earth here, and this was the strongest confirmation yet of an impact origin to Coon Butte. Barringer knew he could shake geologists to their core by presenting a strong case that Coon Butte had been formed by an impact, showing the world that large meteorites hit the Earth and, at the least, affected the planet on regional scales. No more would meteorites be associated with merely punching small holes in the ground. They could be huge and deadly.

Meanwhile, in 1905 alone, the company spent $40,000 on drilling and support operations. The sixteen drills sunk in 1907 cost $24,000 total. By 1908, twenty-eight drill holes had failed to reveal a single massive meteorite, and Standard Iron was bleeding money. Barringer began to lose confidence in Holsinger and pushed him to the periphery of company operations. Preparations for a new steam-powered shaft got under way.

To compound difficulties, all the work at the crater could attract attention Barringer did not yet want. He knew that the most compelling evidence of impact would be the entire meteorite itself, which he hadn't found. But he had other evidence favorable to the impact-origin theory. So before newspaper reports published inaccuracies and outright falsehoods, which could affect future investments, Barringer and Tilghman needed to present their case for the formation of the so-called butte. And it would have a new name. Meteor Crater.

SLICE INTO SOMETHING—psyche or rock—and the hidden shows itself. Beneath external pride, say, a stratum of defensiveness. Beneath that, insecurity, a fear of ridicule. The need for approval.

Or in the case of the desert around Meteor Crater, first the youngest rock at the top, Moenkopi siltstone or sandstone, laid down in the Triassic. Then Kaibab limestone, then Coconino sandstone—more of this than anything else, about 1,000 feet of it—then, farther down, the Supai sandstone, all three of those formations having been laid down in the Permian.

This order reversed itself in the material ejected from the crater and found on the rim. There, in general, the oldest rock, not the youngest, lay on top—with the notable exception of the Supai sandstone. This layer, the drills had revealed, had not been ejected. Although fractured, the Supai had rested in place for millions of years; *nothing* had blown through it from below.

Benjamin Tilghman and Barringer each published papers in the *Proceedings of the Academy of Natural Sciences of Philadelphia* in 1906; then three years later Barringer read a paper to the National Academy of Sciences in which he forcefully argued for the evidence of impact and, very much against the fashion of the day, directly attacked Gilbert. The audience expressed dismay. ("Science and single-minded promotion of a speculative commercial enterprise . . . do not mix," writes geologist Wolfgang Elston.) The inventor Elihu Thomson, there in the crowd, told Barringer to continue, despite the negative reaction. Barringer did.

In their various treatises, the scientific outsiders arrayed several points against Gilbert. They cited the reversed strata. Ejecta could be found three miles away, including some truly massive boulders. They pointed to the uplift and tumble of commingled rocks of various ages—and meteoritic debris—along the crater rim. The fact that meteoritic iron was intermixed with the country rock argued against the purported coincidence of a meteorite fall after the so-called steam blowout. The undisturbed Supai sandstone appeared to be incontrovertible proof that no subterranean explosion had occurred there. Further, rocks showed evidence of fracture and fusion under sudden pressure and heat; quartz, for example, had been shattered microscopically. Exposure to a buildup of pressure would have revealed different patterns of stress and probably would have resulted in multiple blowout holes, not a single hollow.

Gilbert had paid little heed to massive amounts of crushed rock ("silica" or "rock flour") found at the site, but Barringer focused on this material, which could not have resulted from volcanic activity. Barringer could not contain his disdain for Gilbert's passing mention of the silica, since the stuff was everywhere. Hoyt writes, "Barringer noted particularly the massive, washcut exposure of pulverized silica on the south rim, declaring that it is 'difficult to understand how this exposure could escape the eye of any careful geologist.'"

Barringer and Tilghman also had found many pieces of what they called "iron shale," or magnetic iron oxide, which usually contained nickel, a meteoritic substance. (The term is misleading, for the pieces are not composed of shale but just look like it.) The rusty shale balls and the meteorites numbered in the thousands and, according to Barringer, formed a "compact cluster or swarm" instead of a single body. If the meteorite existed in a cluster, Tilghman said, then its many fragments would set up lines of force among themselves, making magnetic detection at the surface difficult—a neat answer to Gilbert's unmoved magnet.

Gilbert had also been wrong about how much material was on the rim. He had failed to take into account centuries of erosion and so had the impression that no extra material had been displaced by a meteorite or meteorite cluster beneath the crater floor.

And Tilghman believed that Gilbert's estimates for the size of an impactor were exaggerated, since something just 150 feet in diameter could leave a massive crater if it fell at a very high speed. Tilghman had nailed the impactor's size and was right to suspect a high-speed collision. Barringer, on the other hand, believed that the air would retard a large object just as it does with smaller ones. As Brandon Barringer muses, Tilghman "may have been brooding" on the consequences to the large meteoroid if the air had not, in fact, slowed it down.

Lacking a mathematical background, Barringer was in no position to deal with questions of mass or velocity or the energy of impact, but he made from his own limitations a categorical statement that "it is practically impossible" to make such estimates. Such rhetoric and the attacks on Gilbert needlessly hurt Barringer and Tilghman. The former's use of such phrases as "there can be no doubt" and "incontrovertible proof" have the earnest tone of an irritating high school debater.

And Gilbert's reaction? None, none at all. He never deigned to comment publicly on these various arguments.

THE SMITHSONIAN'S GEORGE P. MERRILL would be more polite in an article he published in 1908, two years after the initial Barringer and Tilghman papers and the year before Barringer's National Academy piece. In "The Meteor Crater of Canyon Diablo, Arizona, Its History, Origin, and Associated Meteoric Irons," which appeared in the *Smithsonian Miscellaneous Collections,* Merrill noted how Barringer and Tilghman had heightened awareness of the crater and how Gilbert had rejected the impact theory on "the facts then available." Merrill wrote that "the plausible suggestion that . . . [the crater formed] due to the impact of a stellar body is of so unusual a nature as to warrant the fullest investigation."

That one word—"plausible"—was a major coup for Barringer and his allies. Merrill had visited the crater with Barringer and Tilghman and was one of a handful of prominent scientists who were convinced the men were right. Among other points, Merrill had been impressed that sandstone had been heated into quartz glass, which required temperatures of nearly 4,000 degrees Fahrenheit, in excess of what exploding steam would have produced. He did say, however, that "the greatest difficulty in accepting the meteoric hypothesis lies in the absence of sufficient evidences of such extreme temperatures. There are no volatilization products and but slight evidence of slags." (Others would later find these very things.) Merrill held out the possibility that volatilization had taken place—that is, the meteorite vaporized—for no huge iron seemed to lurk in the crater.

Barringer despised this idea. Products that could be taken for the results of vaporizing—such as tiny nickel-iron grains—were, for him, parts of the meteorite cluster's comet-like debris train. Barringer called volatilization "utter absurdity." As the debates continued, Barringer never lost sight of his primary goal: to find the main mass in order to mine it.

Most of the scientific establishment could scarcely contain its disdain for Barringer, his ideas, and his financial motives. For example, after much effort, Barringer once arranged for a visiting trainload of scientists to stop at Meteor Crater for a meal and field trips. Barringer put great stock in this visit, hoping to impress the researchers by dint of his personality, the facts that he had arrayed, and, of course, the impressive crater itself. The event was a disaster. The visitors stayed only a short while and hardly trooped around the crater, and some even accused Barringer of putting out rocks in the field as a deliberate fraud to help prove his case. Barringer fumed over the show of disrespect and the allegations.

And he and Tilghman had to shut down the main shaft at Meteor Crater. They had lost tens of thousands of dollars. They'd resume again, years later, but still not find the purported massive meteorite.

Unbeknownst to Barringer—and almost everyone else in the world—on June 30, 1908, a massive fireball exploded over Siberia. Throughout the region, buildings and tents with precious stores of food and clothes disappeared in the blast. Reindeer turned to ash. Hot winds laid down whole forests. People were literally stunned for days. Rail tracks trembled like rubber bands. The explosion was 2,000 times more powerful than the Hiroshima bomb. A slight change in trajectory would have put the explosion, possibly of a comet, not over

Siberia but over the teeming city of St. Petersburg—lethal proof for Daniel Barringer's ideas.

No one in the West would know of the Tunguska impact until many years later. So from London to New York and beyond, people marveled at dawns and dusks colored beautifully by the dust of an unknown fireball's explosion and the earth it had lifted into the air. Newspapers speculated it was the northern lights. I've often wondered if Daniel Barringer looked up at those 1908 skies. Did a bit of Tunguska's dust settle onto his suit? His hair and lips? Did he breathe in the dust of impact? He must have inhaled that hidden evidence, then gone back to his reports.

IN 1908, NEARLY THIRTY HOLES DOTTED the floor of Meteor Crater. Within a few years, Daniel Barringer, ever confident, would estimate the value of the hidden meteorite at $100 million. Later, he'd raise this figure to $700 million.

Benjamin Tilghman, however, could no longer afford to participate and was increasingly uncertain that a large meteorite could be found. Barringer and Tilghman fought over drilling plans and investment strategies until Tilghman pulled out. He would die months later.

In order to cope with his crater investments, Barringer, with his wife and eight children, moved six times in fourteen years, all to find cheaper living quarters.

The crater had its hold of Barringer—"it is the ambition which is nearest my heart"—and not him alone. Elihu Thomson, the inventor who had become an ally in Barringer's fights, once wrote that "the crater has been in my thoughts at each opportunity." Thomson said later, describing a trip that took him nearby, "I can't tell you with what a longing I looked out over the dusk lit plain towards Meteor Crater as we went by in train the other evening."

After canvassing, unsuccessfully, some of America's wealthiest and most famous men—including Thomas Edison—Daniel Barringer got, in 1918, the investment break he'd been needing. Officers of the U.S. Smelting, Refining and Mining Company agreed to back the drilling.

Something else happened in 1918. Grove Karl Gilbert died. Privately, sources had told Barringer that Gilbert had been convinced by the impact arguments.

. . .

AT THE CRATER, U.S. Smelting would have no more luck than Standard Iron. Drill after drill had revealed no main mass—many shafts flooded with water and sand—though stuck drill bits tantalized. Might they have found the main impactor?

A drilling supervisor wrote in his March 26, 1922 log, "I guess there can be no doubt but that we are passing thru a series of shale-balls, or sparks from the meteor." That diction—"I *guess* there can be *no doubt*"—unwittingly embraced the two attitudes that seemed always present when the fate of the meteorite was discussed: skeptics' uncertainty, Barringer's confidence. Here too, the drill failed to establish that the main mass had been found.

U.S. Smelting backed out, having spent almost $200,000.

A mining engineer named George Colvocoresses soon stepped in. Then, after signing on to the Meteor Crater Exploration and Mining Company, Colvocoresses began to have his own doubts. He began to wonder about Barringer's old bugaboo. Had the impactor vanished? An American astronomer named Milton Updegraff told "Colvo" he thought so. Others had suggested the same. In a 1924 paper, New Zealand astronomer Algernon Gifford said that encountering resistance at high-enough speeds caused an object to explode. This was one way to explain the round-versus-oblong-crater argument. Angle of impact didn't matter, because it was the explosion that created a round hole. Barringer hated Gifford's latest version of volatilization as much as he hated Merrill's earlier proposition. Over time, though, Barringer would agree that there had been an explosion, just not big enough to destroy most of the meteorite swarm.

Colvocoresses decided the buried star must be there. Barringer and Colvocoresses had fixated on the fact that the 1,376-foot-deep drill hole under the south rim—the one whose drill bit got stuck—indicated the presence of a large meteorite. Colvocoresses wrote in his report that the minable meteorite probably weighed *some 10 million tons* if not more.

In one investment letter, to Henry Osborn of the American Museum of Natural History, Barringer said, "We may easily have in the Meteor Crater one of the greatest, if not the greatest, nickel mine, as well as the greatest platinum mine, in the world." The qualifier "may" is subsumed by the rest of the verb construction: "easily have." Not just have, but easily, and the sentence builds its earth-spanning assertion around the thrice-used adjective "greatest." Osborn declined more than once, and Barringer kept on writing him.

Despite the poor track record at the drill sites, investors coughed up funds. In 1928, drilling began anew.

Barringer had another reason for optimism. An article in *National Geographic* would be coming out soon and could only help his effort to raise funds and to convince skeptics of Meteor Crater's true origin. But, shamelessly, the piece credited Grove Karl Gilbert as being the champion of the impact-origin hypothesis. Barringer was stung at the publication and railed against the editors. The article didn't even mention Barringer's name. Again and again, Barringer had championed his ideas and prospects, only to be repaid with disrespect.

The incident would pale, however, beside a set of complex equations, numbers that Barringer confessed he could not understand even as he grasped their horrific implications.

AS NEW SHAFTS AT Meteor Crater vomited sand and water, investors grew so nervous that they asked a scientist outside the project to consider if the main iron mass could ever be found.

In 1929, Forest Ray Moulton of the University of Chicago filed the first of two reports. Barringer knew of him from his pioneering work on the formation of the solar system. Moulton had contacted Barringer that year to let him know that he wanted to include information about him and Meteor Crater in a forthcoming book and even took umbrage over the *National Geographic* article, so it's likely that Barringer expected a friendly assessment.

He didn't get it. Moulton's dense mathematics showed that the meteorite was much smaller than Barringer had believed. The object had hit at higher velocities than Barringer and his allies had assumed and, yes, most of it had exploded away. There wasn't enough left to mine. This was devastating news.

William Graves Hoyt details what must have been an arduous and shocking experience for Barringer—a meeting of the board of Meteor Crater Exploration and Mining Company on September 11, 1929. Barringer, who had lost some of his hearing, strained to listen to a summary of Moulton's conclusions, and he watched the board vote to stop all drilling at once.

A few thought Moulton wrong, including Elihu Thomson. For years, Thomson had dismissed the idea that a meteorite could hit at high speeds; he thought air would slow virtually anything down, thus preventing vaporization. But even Thomson, an old and reliable friend, grew weary of defending Barringer against Moulton's numbers.

Moulton had noted that air resistance would be—is—negligible. A large enough meteorite will not slow down, and if an entry speed of about 5 miles per second is assumed, the impact would cause much "kinetic work" to take place. That is, once the fast-flying iron (and the pressure wave in front of it) struck the ground, the energy of velocity would be transformed into heat that vaporizes the object, that sends out a shock wave, and that blows out a crater.

In his second report, Moulton assumed various angles, weights, velocities, and coefficients of retardation in some 1,300 calculations. He even offered to pay for dinner at the crater rim if his conclusion was mistaken. It wasn't. Moulton thought the meteoroid had been between 500,000 and 100,000 tons originally—far less than Colvocoresses's estimates. Scientists now know that any meteoroid heftier than about 100 tons will be destroyed on impact with the Earth.)

So the meteorite had just plain exploded and fragmented. Some geologists think that about half to a third of the original mass exists about the crater in the form of small pieces.

Despairing, the frantic Barringer turned to a friend, the famous astronomer Henry Norris Russell, who let Barringer down and drove the crisis home. When Russell wrote Barringer in agreement with Moulton, Barringer refused to share Russell's reply with the investors. Incensed, Russell so lambasted Barringer that the usually brusque businessman relented, sending Russell's letter and Barringer's own assessment to a project backer.

In the autumn of 1929, Barringer saw that he'd wasted years and a fortune on a ghost stone. According to descendants Nancy Southgate and Felicity Barringer, the endeavor had cost a cool million, more than half of which came from

Barringer himself. Adjusted for present prices and expenses, he had spent the equivalent of about $5 million.

Then the stock market crashed.

To Barringer's chagrin, sensationalist articles appeared, claiming erroneously that the iron had been found after all. For a time, the only success at the site was mining the Coconino sandstone, in which occurred bits of the asteroid belt; marketers called the abrasive cleaner "Star Dust."

Barringer was at home for Thanksgiving in 1929. On November 30, he died of a heart attack at age sixty-nine.

There seems to be so little pleasure in Barringer's quest, so much concern with reputation and argument, for credit and convincing—and, obviously, for profit—that it's clear he died, almost literally, of a broken heart. Tie contentment to just one thing, this history seems to say, and you are bound to be disappointed. In the family Bible, Daniel Barringer had written many words of advice to those who would follow, including these: that they have "the right sort of ambitions."

IN THE YEARS SINCE Daniel Barringer's death, scientists have reconstructed the event that led to the formation of Meteor Crater. This is what we know: About 50,000 years ago, the area that would become the home to Meteor Crater had no permanent human settlements—just bison, mastodons, camels, mammoths, and ground sloths that roamed hills and drank from streams. Junipers and ponderosas might have grown in the area.

When the fireball appeared, it was 10,000 times brighter than the sun, writes scientist David Kring. At 45,000 miles an hour, the meteoroid—estimated to weigh a few hundred thousand tons and spanning 150 feet—took just a half-minute to hit the ground once it had touched the upper reaches of the planet's atmosphere. The explosive force equaled 15 to 20 million tons of TNT—pretty small stuff compared with impacts that may have caused global extinctions.

Still, following the impact, earthquakes rattled the Colorado Plateau, fires raged, and the atmospheric pressure spiked to 100 times normal, which sent winds blasting outward at more than 1,200 miles an hour. Plants disintegrated.

Air was sucked out of the lungs of animals. Anything the wind could pick up, it did. According to one of the postcard sequences you can buy at the Meteor Crater gift shop, some 300 million tons of earth shot out of the impact and far into the air. One caption reads, "Clouds of melted rock and meteorite droplets rain[ed] debris on the area, destroying all plant and animal life within 100 miles." Boulders as big as trucks sailed through the air. What was left? A crater, a scarred region that life would reclaim, a sprinkling of iron meteorites. Recent research suggests that the impactor was a "tight swarm" of material—not unlike Barringer's idea—and that the age of the event may yet change, depending on further studies. Also, the trajectory of the impact remains unclear, even as scientists continue to find new kinds of impact melts. The story of Meteor Crater isn't finished.

Writers seeking to convey the scale of what remains today often note that the crater's depth would allow the Washington Monument to be set inside and not peek higher than the rim, and that 2 million people could watch 20 football games taking place on the crater floor.

Recently I looked up the word "crater" in the dictionary. The deep root extends to the Indo-European for "mix," a meaning that continues on into the Greek and Latin, and that forms the basis for our present-day "idiosyncrasy." The history of the men the crater obsessed surely seems idiosyncratic. Gilbert had been right, then wrong. Barringer had been right, then wrong. The narrative feels titanic, sometimes ludicrous, and often sad. In the end, Barringer never found the giant meteorite he was convinced lurked beneath the crater dirt, but he'd proven the crater's impact origin. For years, no one really cared.

More than once I've thought of these things as I've stood at the railings overlooking the dizzying hole that is Meteor Crater. I've been there a handful of times now. I've taken the short tour, where guides walk visitors along part of the rim and give an abbreviated geological and human history of the place. I always lag behind, having read what I've read. While the guide talks, the wind blows, and I relish the rocks beneath my boots, the sight of abandoned buildings on the crater floor, and the fallen sandstone-brick walls of quarters left on the rim.

Back inside at the Visitors' Center and its museum, I stop at a display of old Meteor Crater drilling equipment. Daniel Barringer might be pleased to know his family still owns the crater, leasing highly successful tourist and educational enterprises to another company. Barringer's failure has become today's bottom line, and the family actively helps to support meteorite researchers. Most visitors I've seen walk right past the old drill bit.

There are curious omissions in the materials produced by the museum. A booklet that provides an otherwise solid overview of the story of Meteor Crater fails to mention the University of Chicago's Forest Ray Moulton. And if you ask the tour guides about a late contemporary of Daniel Barringer—about a different man, who, six years before Barringer died, also became obsessed with meteorites—the guides fall silent. They do so because this man, once an obscure professor in Kansas, had come to Meteor Crater in the mid–twentieth century to seek data and dollars, then left under a cloud of controversy. This legacy still lingers at Meteor Crater, and the first time I heard the guides' silence, I wondered if there was more to that story than I had yet discovered. When I first went to Meteor Crater, I didn't know that this professor would become a pivotal figure in my quest to understand our entanglements with meteorites.

I think of these histories whenever I'm at Meteor Crater, but I'm always drawn outside, stepping away from the tourists and displays, leaving the stories and characters and foreshadowing behind for a time. I zip my jacket against the fall or spring chill so I can watch clouds cast black shapes across the fractured creases and crumble of the crater until it's closing time and I have to leave the place behind.

Book VI

Harvey Nininger Sees the Light

Do with your life something that has never been done,
but which you feel needs doing.
—Harvey Harlow Nininger

On a hot day in late summer 1907, a shy, short, and skinny farm boy—twenty but looking fifteen—arrived at the State Normal School in Alva, Oklahoma. Classes would begin soon at this teachers' college, but he had time to explore, so he walked by a row of young trees up to a stone building, which, with its several floors, towers, and faux battlements, lived up to its name: "The Castle."

Harvey Harlow Nininger walked inside, escaped the heat, and found the library. The heat he knew, having grown up as a farm kid on the plains. But he'd never seen a castle before, authentic or imitation. He'd never seen a library either. Harvey had read only the Bible, the Montgomery Ward catalogue, and "the Three R's"—school primers. Now he stood in a room that contained about 4,000 more books than he'd ever read. "Feverishly, I glanced at title after title," he remembered decades later.

Along one wall stood wooden cases topped with the busts of famous men. Beside the metal stacks was the twenty-drawer card catalogue, above which hung a photo of the moon.

Harvey walked gingerly on the hardwood floor—unfamiliar with such a polished surface, he'd embarrass himself in the months ahead by repeatedly slipping and falling—and he kept looking, perhaps pausing to run his finger along a spine, to tug and tip out books from the shelves. The boy who for his keenness of observation had been nicknamed "the Squirrel Dog" found himself on a different kind of hunt. Harvey pulled out Ralph Waldo Emerson's *Essays*.

In a reverie, he paged through the entire book, though many of the sentences must have escaped him. In the first paragraph of Emerson's "Self-Reliance," however, he'd have read this clear expression of courage: "Else, to-morrow a stranger will say with masterly good sense precisely what we have thought and felt all the time, and we shall be forced to take with shame our own opinion from another."

Shame had been one of the dominant emotions of Harvey's life. In his old age, he wrote of "a painful memory of . . . killing . . . sparrows"; of twanging fence wire to kill songbirds ("The memory shames me"); of how his mother made him apologize to a neighbor for stealing a strawberry ("The lesson was burned deeply into my conscience"); and of how, so small at the plow, he'd cry as he tried to keep the draft animals going in straight lines, trying to do his best. But he was also curious about the world around him. Harvey caught turtles, grabbed skunks, trapped bees, and pursued ants. He was frequently stung and stinky in his methodical attempts to investigate suspicions about nature and to master fear. Among people, however, Harvey remained bashful, teased as he was about his diminutive stature. He was so frightened of being the center of attention that during his first public speech as a student he "blacked out."

Harvey Nininger realized in grade school that he had to work harder "to be regarded as equal" and soon found himself excelling in geography, spelling bees, and arithmetic by way of compensation. All this contributed, he thought, to "the core of stubbornness . . . that probably became the governing feature of my character."

That stubbornness helped when, the year before he arrived at the State Normal, Harvey had passed a county examination that was equivalent to a high school diploma. But when his mother showed Harvey the blue ribbon he'd received for this honor—he was out "picking peas"—he "didn't take any interest in the thing at all. I think she cried a little . . . [a] thousand times I have wished I could see her again to apologize for the way I acted," he said years after the fact.

Despite such seeming indifference, Harvey soon decided to go to college in Alva. His fundamentalist parents were distraught. It was the beginning of his departure from a world of literal Biblical views and into a world of empiricism and investigation. At the State Normal, Harvey would come under the sway of a teacher who implored him to "Observe Nature." Harvey would follow that advice during his short stay in Alva, then through his undergraduate degree at McPherson College in Kansas, a master's in entomology from Pomona College, and a career in teaching and science, a career that took Harvey back to McPherson, where he became a professor.

All that would come to pass in the years ahead. On that summer day in 1907, however, on that first foray into a library, Harvey probably wore the plain black garb sanctioned by his Anabaptist faith, the Church of the Brethren. The outward conformity of his clothing notwithstanding, he felt deeply conflicted about his religious upbringing and the lure of the world at large. Did Emerson's essays allow Harvey Nininger to forget, even for a little while, his awkwardness on that strange campus and his turmoil over religion and education?

As Harvey read, his God-fearing father may have been working the fields he thought his son would inherit.

This boy, who had once seen a star in daylight not knowing what it was—Venus, a professor would later tell him—felt as though he "had discovered a new continent." Surrounded by books, Harvey Nininger was starting to understand something: That a journey is a question that travels. It carries a person like a bundle on its back.

· *Chapter VI.1* ·

Epiphany on Euclid Street

In McPherson, Kansas, at the corner of Euclid and Maxwell, two friends stood talking in the evening air. Harvey Harlow Nininger and Elmer Craik, professors at McPherson College, had just walked from a revival in the campus chapel, and the men had stopped in front of Craik's home. Friday, November 9, 1923, had been a warm day, though now, at nine P.M., the air promised a night of barely open windows and unfurled blankets. The mercury had dropped below freezing twice already in November, and the thirty-six-year-old Nininger, a biologist, knew the seasons and their markers far better than most. The kites and vultures had gone, and during the day light mellowed over the prairie. Pegasus would ride the sky near zenith, while Orion the Hunter peeked above the horizon, endlessly pressing his quarry.

Then an obscure Kansas naturalist, Harvey Nininger knew that standing with a friend on a pleasant autumn night was a gift (he would have felt that), and though the Earth's orbits around its star would go on—and how could he now begrudge any passing season?—Harvey Nininger had for weeks wondered what his future promised, wondered if he should do something new, if he should study something mysterious, a thing so remarkable and little-known that few people gave the matter much thought. On a small-town street corner, two friends talked. Perhaps such moments are enough. Perhaps not. Warm reprieves always pass.

"Suddenly a blazing stream of fire pierced the sky, lighting the landscape as though Nature had pressed a giant electric switch," Harvey once recalled. "The blade of light vanished . . . leaving a darkness seeming thicker than before."

One instant a man is standing in the dark and the next he's riding a trail of light to the end of his days. I think of Rilke: "One moment your life is a stone in you, and the next, a star." For Harvey, stone and star would be one, for, as the poet also said, "It is impossible to untangle the threads."

Three months earlier Harvey had been amazed by an article on meteorites in *Scientific Monthly* magazine. "I cannot remember ever reading anything that so completely captivated me," he said. He told his wife Addie, "I've found a new subject that I am going to make a hobby of. I don't know how I'll do it but . . . I'm so interested in this new word." By the time of the November 9 fireball, he apparently had been visiting with Kansas schoolchildren about this new word, about meteorites.

Harvey Nininger remembered his first encounter with the word—and the rocks—differently in a self-published autobiography. Visiting Chicago's Field Museum with another participant at an ornithological conference, he stopped by the meteorite display. The two men weren't sure what to make of claims that rocks come from beyond the planet. After all, Harvey remembered that "all during my childhood meteors were regarded in about the same light as ghosts and dragons." He no longer believed in those, but meteors might as well have been myth, for they were hardly the stuff to make a scientific career, especially for a man trained in biology.

As a Brethren, Harvey would have been baptized out-of-doors. "The blade of light" on November 9, 1923, was his second outdoor blessing.

The meteor presented the usual effects—enough light and noise to suggest the apocalypse, which, this time, had not arrived. The football banquet at Lloyd's Cafeteria could rest easy and finish dessert. Apparent calamity can force reconsiderations as surely as calamity itself, however. Craik could not speak. Harvey leaned over the sidewalk and marked where he stood. The fireball had shot toward the southwest and vanished in the distance, behind a pine at 1010 Euclid. In his autobiography, Harvey wrote that he told Craik he was "going to find that meteorite and that I was plotting its path from where I saw it. He laughed, but I was serious. . . ."

"Well, where do you think it landed?"

"Probably within 150 miles," I estimated.

He laughed again. "Now I know you must be kidding."

"No, I'm serious, and I'm going to hunt for it."

Craik asked, "Has such a thing ever been done?"

The answer was no. Others had searched for meteorites after they'd been seen to fall or had left craters—such as Jean-Baptiste Biot after the epochal shower at L'Aigle in 1803 or Daniel Barringer digging in the Arizona desert—but no one had tried to chase fireballs, no one had scoured the landscape to find a meteorite where none had been discovered before. It was lunacy.

The idea, the light, this new word, would lead Harvey Nininger to quit his professorship, take a low-paying position at a regional museum, and, from time to time, deliver truck chassis in towns with colleges so he could lecture about meteorites, thus becoming, during the Depression, the world's most enthusiastic full-time collector, popularizer, and researcher of meteorites.

Given this future of scientific enterprise and blue-collar labor, perhaps it's appropriate that the streets—Euclid and Maxwell—do not, in fact, directly memorialize the Greek geometer and the Scottish physicist as I had hoped. Rather, the former is named after an Ohio town and the latter for a local settler. My desire for a perfect beginning to Harvey's new ambition disappeared after some research, but the names honor pioneering, something he understood well.

Photographs of the professor show more dignified blandness than zeal, however. He was thin, with a high forehead, an angular nose, and a dimple in his chin. He wore glasses. As an adult in suit and tie, Harvey could have passed for a bank clerk. At five-foot-five and some 120 pounds, he was far from imposing, though one of his daughters, Doris Banks, says he had a "ramrod straight back," and he had the sinewy strength of a wrestler. He also sometimes showed a narrow-lipped underbite that hints at petulance, probably the result of how he'd push out his lower lip when in earnest. The photos and his published prose also belie his nervousness—an "eagerness," as one friend has called it.

Before this unassuming man went after the fallen sky, no one believed that systematic searches and public education could yield hordes of meteorites or help us understand how these stones pelt our planet. Then came Harvey Harlow Nininger, rational romantic, scientific entrepreneur, a small professor bent over a sidewalk beside streets named for the obscure.

After he made his mark—imagine being able to draw an X on the spot where you can say forever, "This is where my life changed"—Harvey said good-bye to the befuddled Elmer Craik, walked past the house above whose pine the fireball had disappeared, and arrived at his quaint home at 803 E. Euclid, with its apricot tree, front porch, and trellis. In an unpublished memoir, *It Wasn't Always Meteorites,* Harvey colors life in McPherson with Addie and the family in rose tints—a life of daisy-picking, kite-flying, winter sledding. Over the years, the three young Nininger children would play trolls under the bridge, and Doris, the middle child, recalls her white cat Fluffy. Home movies from the twenties show son Bob playing on stilts and Harvey swinging daughter Margaret about by the arms. In a word, the Niningers were comfortable. Too much so, it would turn out for Harvey.

On November 10, 1923, McPhersonites added the fireball to the gossip at the Echo Restaurant, the Maple Tree Meat Market, and the Walstrom Grocery. Harvey needed more than banter, though, so he contacted newspapers to solicit reports from eyewitnesses, those who could say where they'd seen the fireball start and end, its height and its angle of descent.

Harvey and physics professor Charles Morris soon received letters from all over Kansas and Oklahoma. One man had seen the meteor drop "straight down . . . leaving a tail of bluish vapor. . . . [It was] about the size of a football and bulb shaped . . . and was followed by at least two . . . more smaller bodies." Others saw "a very peculiar shade of green" and an "awful Red." Another man included a stand of trees in his drawing, trees behind which the meteor had vanished. Others offered exact descriptions of where they saw the meteor: "about 60 Yds. southwest of the stand pipe." Harvey needed such precision. On letterhead of the Larned Livestock and Land Company, from Driftwood, Oklahoma, manager H. O. Stockwell—who in the years ahead would be inspired by Harvey Nininger and would discover that huge mass of the Brenham pallasite in the 1940s—sent his account of the startling event. But most reports were less than helpful. A father wrote to say that his family "saw the light from [the] meteorite and we all felt the gas." Just what "the gas" was he did not say. A different eyewitness captured something of the spirit of the moment: "The meteor caused so much light that when it was gone the night seemed inky black. There is so much mystery concerning a meteor in the average person's mind, that it left me feeling very strange, and rather shaky."

Many letters made clear a problem that plagues meteorite hunters even today—equating disappearance with landing. Most people upon seeing a meteor dip below the horizon assume it's landed just over there. Usually it hasn't. Landmarks do tie the earth to the sky, but more than one report is needed to trace the disappearance of a meteor. With enough reports showing where someone stood (by this shed) and where a meteor vanished (just to the left of the dead cottonwood), Harvey could pull together the information by using a pencil, a protractor, and a ruler to make his map. He could estimate path, distance flown, and a possible strewnfield. When he interviewed witnesses, Harvey avoided tromping to where someone mistakenly believed a meteorite had fallen. He just wanted to stand where they had seen the meteor.

In his talk "My Introduction to Meteorites," Harvey says that he and Morris went through the letters on a Saturday—and he teased Morris, noting that these weren't scientists writing in. "He was," Harvey said, "thoroughly disgusted." After six letters, Morris quit.

If the fireball had dropped meteorites, Harvey surmised it did so either near Coldwater or Greensburg, Kansas, where some decades earlier Eliza Kimberly had found her own space rocks. Harvey Nininger would have to search over many miles of territory, and it would take time, money, and stamina. Still, the chance to hitch his life to something big was on his mind, especially after he'd gotten the scare of his life two summers before.

THAT SEASON IN 1921 Harvey had been hired by a private collector to gather beetle specimens, so before heading to the mountains where he'd be camping and hiking, he had applied for more insurance. The collector had been showing Harvey how one could transform a hobby into an expertise, into an idée fixe.

Harvey had his rugged summer—he loved being outdoors studying nature—but when he returned to McPherson, he learned that his insurance had been denied because of a heart problem.

More than one physician told him the same thing. "Well," Harvey remembered at ninety-five, "there were restrictions . . . for me to live . . . you've got to quit riding a bicycle, you've got [to] quit playing . . . volleyball . . . you got to quit carrying any weights, don't carry a bucket of water, don't go up any steps." One doctor told him "you have just as good [a] chance . . . correcting that lesion as I have of growing a new thumb," then held up his thumbless hand for effect. Weekday mornings, Harvey, a self-described optimist, would walk a mile to the college hearing "that ol' heart thumping, bumping." Only in his mid-thirties, he "got more frightened all the time."

Eventually, the insurance company sent a corpulent doctor to McPherson to lecture Harvey about limiting his activities. The professor grew incensed, flying into a rare rage, and the doctor, stunned and angry, turned "the color of liver." Harvey huffed across the street to see the physician he most trusted, who finally confirmed that the best thing to do was what Harvey wanted to do. Just forget it—"forget you have a heart"—and live the life you need to live.

"I told my wife that they were scaring me to death . . . and I'm going to disavow the whole group . . . but I was careful and went about my business," he said.

His business was that of a beloved professor becoming too complacent in academia as well as growing insecure about the effects of family and obscurity on his career. He taught at an out-of-the-way college. He lacked a doctorate. And the medical crisis reinforced his disquiet. Certainly he'd not forgotten his first failure to make a name for himself.

It had happened when Harvey was an undergraduate at McPherson College, which he attended after leaving the State Normal in Oklahoma. He'd made a discovery that seemed to him monumental. Textbook drawings of grasshoppers' mouths were incorrect; they didn't match actual specimens. And he had the audacity to tell his teacher. (This effort also earned him the moniker of Mr. Grasshopper.) The professor—"a man of temper"—was not impressed.

As he outlined his findings, the student, with his "natural stubbornness," pleaded, "I'm not asking you to accept me or my drawings; I'm asking you to look at nature! The book shows mouthparts better suited to a flesh-eater than a vegetarian, and the grasshopper is a vegetarian. Here, it is obvious!" Harvey displayed the text, his drawings, and a cigar box of specimens.

After a few moments, the teacher, intent on the specimens he studied with a magnifying glass, asked, "Are you sure this is the same species as in the book?"

Harvey Nininger said yes. The upstart had proven his case. His teacher exclaimed, "You must write this up . . . it will change every zoology textbook in the nation, in the world."

So Mr. Grasshopper published the paper, but according to Harvey not until 1960 did a textbook correctly illustrate grasshopper mandibles. Without any credit to him.

His heart disease and his disappointment over the grasshopper paper, which Harvey nursed his entire life, had primed him for a radical change, for something more than biological articles in small journals, something more than the grind of teaching the same classes again and again, something more than family routine in his pleasant small town.

To TRACK DOWN ANY POSSIBLE ROCKS from the fireball, Harvey Nininger called on schools and even climbed a pulpit to say a few words about meteorites before a Sunday service. This was the start of a life spent talking with migrant workers and miners, children and bankers, farmers and shopkeepers. He became the Vachel Lindsay of space rocks, lecturing in theaters, yakking on street corners, plunking meteorites on bar counters.

Though Harvey failed to find any meteorites from the November 9 fireball, the canvassing did yield other meteorites, including one brought to him by a deacon of the church he visited. This was success he could not have expected, and Harvey began to make his purchases, albeit on a tight budget and with "borrowed funds."

A year later Harvey had accumulated enough specimens of the Brenham pallasite—meteorites bought or traded for—that he could offer a 230-pound cache to the American Museum of Natural History for $690. The letter shows how quickly, and with no small amount of savvy, Harvey had transformed his pursuits into an enterprise. In 1923, Addie Nininger could buy a pork roast for

17 cents and a 48-pound sack of flour for $1.50. The sale of meteorites could not only finance Harvey's new passion but maybe pay some bills. At the time, Harvey was making about $3,000 a year.

Soon after obtaining his first two meteorites, Harvey headed to Lawrence, Kansas, where he believed geologists at the University of Kansas would help him unlock their rocky puzzles. He took the train and stayed one night in a cheap motel. Already he had pored over O. C. Farrington's *Catalogue of the Meteorites of North America,* from which he had learned the types of meteorites then known, their physical characteristics, and the historic falls and finds.

He knew better than anyone at the University of Kansas it turned out. Professors there cared little for their visitor's enthusiasms, but Harvey might have taken satisfaction in identifying a specimen in their collection about which they knew nothing—a pallasite, probably one of the Brenhams collected by the University of Kansas's own F. H. Snow.

Harvey continued reading all he could about meteorites. He'd learned that from 1803 to 1923 there were just fifty-three known U.S. falls. He believed that by delivering lectures, visiting communities, and writing articles, he could increase public awareness and, therefore, the number of finds. This felt to him like a mission, not just a business, and he wanted to be *the* person to pioneer the discovery of more meteorites and their scientific truths.

Harvey could not peddle this excitement with the Smithsonian's George Merrill—he and Farrington were America's only scientists paying much attention to meteorites. While Farrington was encouraging, Merrill was not, telling Harvey that "if we gave you all the money your program required and you spent the rest of your life doing what you propose, you might find *one* meteorite." Harvey told Merrill that the next time they'd meet, Merrill would buy a meteorite from the Kansan. When they met again, Merrill didn't. He bought two.

But some six years would pass before Harvey found another meteorite in the field. "Those were," he said, "trying years." He had trouble sleeping. He watched constellations creep across the sky, whole seasons passing in a night. Addie worried too, wondering where this obsession would take their family. When his "nerves . . . seemed to be at breaking point," he turned to birds, to "the carefree song of a meadowlark or robin."

To allow Harvey the time to pursue his new interest, McPherson College gave him a lower teaching load; perhaps also the administrators knew he was under some strain. Privately, he speculated that meteorites could become a centerpiece in scientific training, pulling together such disciplines as geology, astronomy, physics, and, if meteorites contained life or its organic constituents, even biology.

While Harvey kept "talking, talking, talking meteorites," as he put it, he continued his work in biology—publishing a guide to Kansas birds, writing articles on orioles and hybrid grosbeaks—and starting McPherson College's Rocky Mountain Summer School.

However worthy, these endeavors never brought him closer to what he most craved. Meteorites, meteorites, meteorites, and the freedom to find them.

FREEDOM'S NAME WAS "HENRY."

Henry came into the Niningers' lives because one evening in December 1924, Harvey was reading an article on Carlsbad Caverns. He sat by the stove, while Addie, pregnant with their third, took care of the two children. Harvey looked up from the magazine, saying, "We're going to see those wonderful caverns next year!" Addie wasn't sure.

"We'll take the whole year off and travel," Harvey said. "The baby can celebrate its first birthday on the road." Addie wasn't sure.

"I'll lecture in schools as we go. We can live as cheaply on the road as we can here and if I can get only ten lectures a month at $25 each it will equal my salary here," Harvey said. "And, besides, we may find salable fossils or meteorites or both. If worse comes to worse, I'll pick cotton."

Addie—by the time she gave birth to Margaret—was sure. She said she "did not want to stay at home . . . while my husband went traveling." This was a couple, after all, that had honeymooned in a tent. For Harvey, "just talking . . . about it was bringing back the courage I had been so afraid of losing."

So they bought Henry, or the Runabout, an early version of a recreational vehicle. They paid $1,800 and traded in their car for the truck, with its 16-foot-long, specially built house atop the chassis. Leaving behind $300 of debt and many unenthusiastic relatives, friends, and colleagues, the Niningers set out with about $20, some meteorites, and materials for Harvey's three talks—one on birds, one on fossils, and one, of course, on meteorites. The adventure began poorly. Henry got stuck in mud. Not long after, the brakes gave out, and Harvey nearly crashed the vehicle. The family then prayed on the floor and decided to continue.

Soon on the road, Harvey found that although he had some talks scheduled, no one would pay in advance, and he often got as little as $5 per engagement. Sometimes he would climb into Henry, turn to the desk, put his hat in the compartment above, then sit down and pull out a typewriter from among the books. He'd write letter upon letter seeking lectures in towns yet to be visited. What money they earned they kept in a pitcher, which, as they drove toward Texas, "lost weight rapidly." The day when they lacked even three cents to mail a letter "Addie for once was really downcast." The possibility of a Christmas in poverty, more than loose screens and a leaky roof, "really stung us," Harvey wrote.

So, unannounced, he visited a local school official—this was in Texas—and thunked a meteorite on the desk of a man absorbed in paperwork.

"It won't explode," Harvey explained when the man looked up.

"What is that? Who are you?" the official blustered.

Harvey ended up selling $300 worth of meteorites and scheduling some talks, and for the holidays he and Addie cut and decorated a bay tree, which they placed beside the driver's seat.

They had their adventures—hikes and sunsets and stars. They endured a sore car ride to glimpse the rare whooping crane, and they even visited Carlsbad Caverns, just as Harvey had said they would. Twice on the trip Harvey stumbled upon fossil finds, one of which would earn the name *Triavestigia niningeri*. An honor, but fame too obscure for this man. After months of travel, weary and happy, the family noted blossoming wildflowers like the dates of a calendar—paintbrush, penstemon, primrose, flax—while Addie sighed in her diary that the Niningers didn't want to give up their "little house on wheels." It was time to go home, it was time to go back to work.

BUT THEY DID NOT STAY at home for long. Just a few months later, Harvey and Addie led an innovative expedition of McPherson College students on a natural-history trip across the country, the "College on Wheels" or "Natural History Trek." Harvey needed more of "the zest and dare" he had rediscovered in himself, and Addie was game to go once more. People again scoffed. Still, the Niningers led thirteen students on a 19,000-mile journey, seeing everything from Venus in daylight while in Zion Canyon—this time Harvey knew what it was—to a Florida swamp where they hunted for ivory-billed woodpeckers. After an exhausting few months, the Niningers returned home. The likelihood of another family or college trip was small, but Harvey had the itch to stay on the open road, even if alone, which was a prospect that frightened Addie.

And that was what Harvey was planning to do. To go far away without her, to look for meteorites, to do so in a country that had just survived a revolution, a country whose poverty would make an excellent business climate and whose meteorites could give him more materials for study. He would drive overland to Mexico City. Perhaps for the first time, Harvey's obsession had become a choice between himself and family. Addie must have expressed her dismay.

Harvey had tasted excitement—two cross-country trips—and he had experienced failure: no new meteorites. The November 9, 1923, fireball must have seemed far away. If he didn't take this chance, how would he stave off the bleakness that had crowded upon him after the return to McPherson? If he didn't go, what would happen to the rest of his life?

No meteorite collectors had been in Mexico in years, and no one had ever reached Mexico City by automobile from Laredo, where Harvey planned to depart. It was a 750-mile trip. Though in debt, he again obtained a teaching leave and budgeted $200 for the trip, asking for the same amount from his student Alex Richards.

The dark-haired, handsome twenty-year-old of Spanish descent spoke the language, which Harvey did not. He was handy too. Harvey told him to make a car from spare parts, "a real husky car, a very ugly one, that nobody would want to try to steal. He succeeded better than I thought anybody could . . . [it was] a veritable scarecrow in appearance." With its 80-gallon gas tank, this pioneering sport-utility vehicle had twelve gears, five just for reverse. It had huge shocks, a large metal clip like a safety pin attaching the front axle to a skid plate, big headlights, and a strong roof.

In one photo, Alex stands beside the car in dry mountains, his arm resting on the chest-high right-door (the jalopy had no side or rear windows), his sleeves rolled up and trousers tucked in high boots. The tall, slender student looks strong and confident. Alex would be more than fix-it man and translator. He would pretend to be Harvey Nininger's property, a kind of bodyguard and servant. Alex Richards's daughter, Barbara Buskirk, says, "Prof was his *patrón*" and "people expected him to die to save the life of his *patrón*."

Taking supplies, firearms, a movie camera, and meteorites for trading, they headed to Mexico in September 1929, the same month Daniel Barringer received the devastating Moulton report. In Laredo, Harvey and Alex were told they'd not survive the trip.

The two made 10 to 100 miles a day on oxcart trails full of ruts and potholes. Harvey relayed dispatches to the McPherson College newspaper, the *Spectator,* including one report that said "it would be easier, perhaps to travel from [McPherson] to Kansas City going across open fields and taking down fences." Locals gave directions that sent the men to footpaths the car could not negotiate, and film of the jalopy shows it jerking across rough terrain.

Alex and Prof got stuck in swamps and bothered by drunks. The Americans brandished weapons to discourage anyone from thoughts of robbery. Once, three men began to make threats, but the Americans distracted them "with some trifle from our pack," broke camp, then hurried to a hacienda that required Prof and Alex to pass through four gates. After their papers had been scrutinized, the two were allowed to stay. Without warning, a machine-gun-equipped Buick dashed into the compound, and out spilled government agents on the hunt for murderers.

The agents found no killers, and Alex, of course, did not reveal his own plans for revenge.

ONE SUMMER DAY WHEN I meet Alex's daughter Barbara and his elderly widow, Dorothy, in Kansas, they tell me that Alex had a reason other than meteorites for going to Mexico. He had planned to kill the man who had married the woman he was then in love with. During the College on Wheels, Alex had served as the cook and had fallen for a young woman from Ottawa, Kansas. In the interim between the College on Wheels and the trip to Mexico she'd gotten married to a dope smuggler working in Mexico. She apparently didn't know his true occupation when they became husband and wife. Frantic, she wrote to Alex for help.

The story is that Harvey learned of Alex's plans to travel to Mexico to find the drug runner at about the same time that the professor was considering an expedition. Richards had shared his plans with his father, who, according to the women, advised him, "Look, Alex, if you want to destroy the man, you go down to Mexico first, and then you come back up north of the border, kill him, and go back south—and you will be taken care of by the people down there."

"And that's what my dad had in mind to do," Barbara says.

Did the professor know of Alex's intent? If so, did he dismiss it as youthful bravado? Or did Harvey's own darkness at the time allow the men to somehow bond? According to Barbara Buskirk, Harvey was close to or in the midst of a nervous breakdown. The grasshopper disappointment, the heart scare, the failure to find more meteorites for study and commerce, the contrast between life on the road and the habits of academia. Family pressures. Bills. Insomnia. Harvey could not see a way clear, except via overland to Mexico City. Harvey Nininger's granddaughter Peggy Schaller says that talk of a breakdown probably is "not far wrong," for her grandfather "did not care for ruts. . . . He needed to go beyond where he was." Grandson Gary Huss, himself a meteoriticist, "suspect[s] that [his grandfather's] bouts like the one in 1929 had more to do with his trying to make things happen without taking account of the realities around him. The situations that he faced were at times enough to make anyone depressed." Later Harvey "confess[ed] that our decision does seem to have been . . . risky beyond justification and like some other decisions. . . . I don't quite understand how we could have taken such chances."

One of those chances never came to be. By the time the men reached Mexico, "the narcs had gotten" Alex's rival, according to Barbara.

. . .

THERE WAS DANGER ENOUGH without attempting murder.

Needing to protect their camps and lacking a guard dog, Prof and Alex got the next best thing, a Harris's hawk. Harvey had purchased the bird for 20 centavos, and after he cooed, "Jack likes me, don't you Jack?" the bird nearly sliced open his hand.

"I mean," Harvey said, "he's *going* to like me." Alex laughed. Soon Jack was playing with their clothes and noodling like a cat. At night, tied to the car, Jack would cry alarm if startled.

One night Jack screamed. Alex stuck his head through the front tent flap and, seeing a group of rowdy men, kept the rest of his body hidden. Alex started to yell back to "different" men until Prof understood that he was to change his voice each time Alex spoke. Apparently dissuaded, the men made to leave, one extending a hand to shake. Alex gripped a rifle in his left hand (hidden in the tent) while extending his right. One of the bandits took Alex's hand and yanked him clear, only to have a barrel slam into his gut. This move was impressive enough, but so too was the gun, for the Mexicans carried only muzzle-loaders. Alex fired shots in the night and claimed it could fire a hundred before being reloaded. He knew he had six shots left. The men retreated. Alex ducked into the tent to grab a shotgun, then flew outside, where the robbers had regrouped in the back—till a blast of rock-salt buckshot sent them fleeing into the dark.

Alex must have told Prof about his strange vision as he and the robbers sized each other up. Amid the bandits, Alex thought he'd seen a vision of Addie, who told him *Take care of Harvey*. Addie would say that she had felt her husband was in danger that night.

"It's a wonder we get through any day," Barbara muses, finishing this story.

After twenty-one days, the two men reached Mexico City and met the German geologist and paleontologist Frederick Mullerried. The November 5, 1929, *Spectator* announced the arrival and said that "Mr. Nininger told the scientists at the college in Mexico City where they could find a meteorite that fell in 1878. The next day it was found." Harvey also reported "no trouble with bandits." For the next month, he and Alex settled into the tedium of museum work—cutting, polishing, identifying, labeling, and organizing meteorites. Staying at the YMCA, they seem not to have noticed the October 1929 stock-market crash. If they had access to a radio, they might have picked up stations in the States playing one of the hits of that year, Hoagy Carmichael's "Stardust."

Probably they talked of Jack the Hawk. After Jack had killed a cat belonging to the man who was storing their car, Alex had to hand over the bird as payment. "Jack must have known we were parting," Alex recalled. "He whistled mournfully, turned his head away and never looked back."

SOON HARVEY AND ALEX had plans to distract them from losing their companion. Harvey had his eye on a place called Toluca, where the Spanish had found tools made from iron meteorites, some of which had been removed over the years by villagers and visitors alike. But Mullerried was unsure about going to a place he thought was dangerous, and Alex had fallen ill, so Prof proclaimed he'd go alone. This bravado forced Mullerried to agree to go along.

To buy meteorites, Harvey cashed a $50 check and found his bulging pockets "jingled" so loudly that Mullerried "would groan" about the likelihood of theft. The going was rough, but they reached the village nestled in mountains at 9,000 feet, with its houses and huts on ridges, a central plaza, and pens for livestock. Bells rang out over stone streets and meadows. Among the men in sandals, the women with their ribbon-braided hair—toting water buckets on shoulder yokes—came the white-suited mayor, who greeted the visitors with a friendly *Buenos días,* though he wasn't quite sure what to make of them. So, being a good host, he took them to breakfast. As they ate, the mayor pondered. Then left and returned—with a 20-pound iron. Afterward, Harvey found a three-pound iron in a corn field, his first find in a very long time.

"When we returned to the . . . square, men, women and youngsters were standing about with bags and baskets containing meteorites. . . . We bought until our money was barely enough for our fare back to Mexico City and as we left . . . some were standing by the road holding up their offering," Harvey said. They had purchased 700 pounds of iron meteorites from Toluca. (The original fall site is several miles away, and this meteorite and the locality are sometimes referred to as either Xiquipilco or Jiquipilco.)

According to Barbara Buskirk and Dorothy Richards, Alex had in fact made the trip, and before he and Prof left town, they secretly sawed off part of a Toluca iron sacred to area natives. True or not, other ventures were more amusing. When Prof wanted milk at a café, the waitress hit him over the head with a tray. In mangled Spanish, he hadn't asked for milk; he had asked if the waitress was a wet nurse.

In Mexico City, the Americans sold their jalopy ("like parting with a fond, if somewhat crazy, friend," said Harvey) and left on December 3, 1929. Prof and Alex packed up meteorites at the border and shipped them home. They took a train up the Pacific coast on their way back to Kansas, and this long trip must have given the nervy professor a chance to reflect on his life, his success, his hopes. George Merrill had told Harvey that he'd be lucky to find one meteorite in his lifetime. Now he had almost a half-ton cache. As he stepped onto his porch in McPherson on Christmas Eve 1929, Harvey had arrived in more ways than one.

DESPITE THIS WILDLY SUCCESSFUL TRIP to Mexico, Harvey Nininger still had seen no new meteorites from his educational programs in Kansas. Then, a boy in Paradise who had heard one of the meteorite lectures found an odd rock while he was shucking corn. Alex knew the boy, and they relocated the rock in 1930, which was—finally—a newly discovered Kansas meteorite. Harvey's first Kansas meteorite in years, designated "Covert," took him back to the area, after which more meteorites emerged for purchase and study.

Then came the most important report of a meteorite fall Harvey Nininger ever received. At four A.M. on February 17, 1930, a blue fireball roared over Paragould, Arkansas. Wide and bright, the meteor was seen across much of the Midwest and South. Soon thereafter, farmer Ray Parkinson found an 85-pound stony meteorite in a fresh hole and sent a fragment to Harvey, who upon opening the parcel, promptly left the kids with Addie's sister next door and drove 700 miles with Addie in one day. When they arrived, Parkinson broke some bad news.

"I haven't got any meteorite," the dejected farmer told the Niningers. "They stole it."

He had loaned the meteorite to a teacher and a school principal so students could see the rock, but instead the men sold it for $300 and mailed it in a typewriter box to a buyer out of state.

Never one to just turn around, Harvey talked to witnesses, mapped a possible strewnfield, and retained a lawyer to buy Parkinson's meteorite, if it was ever returned, as well as other specimens that might be found. Harvey asked folks to keep searching where he thought bigger pieces should be. (Though it appears contrary to intuition, the heavier fragments travel farther and typically mark the far end of a strewnfield.) Before driving home in his Plymouth on muddy roads in the rain, Harvey managed to find a stone driven into the ground, a Paragould specimen the size of an apple. It was a good sign that more specimens could be found.

Not long after, a man named Joe Fletcher listened as an acquaintance explained that the huge, water-filled hole on Joe's land wasn't dug out by dogs but had been created by one of those falling stones. The men poked a stick in the water and found an *800-pound* rock from space.

Harvey Nininger drove back to Arkansas.

Borrowing money from McPherson's Citizen's Bank, Harvey outbid the Field Museum of Chicago and bought the stone for a whopping $3,600.

At about the same time as the Fletcher stone was being loaded on a truck next to the drugstore, something else fell to ground—the high-school principal, decked by Mr. Parkinson in the schoolyard. Parkinson paid his $2.50 fine for assault and soon received the $300 owed him.

The *Topeka Daily Capital* boasted that "Small Kansas College Beats Field Museum" and said that Harvey Nininger was "gloating" over his success. Apparently the professor had engaged in "some hasty telegraphing to the management of his college" in order to pay $100 more than the museum was offering. Paragould was then the largest meteorite whose fall had been witnessed, so the prize was no small matter. Harvey now had meteorites galore and national notoriety.

Once in Kansas, the huge rock was exhibited at the bank, where the public could gawk over the big gray stone. Then as now, meteorites like Paragould are intriguing not only as objects of profit and anecdote but as pieces of a larger story—as samples of what became the Sun. Perhaps some of the bank visitors, in suits and dresses, pondered that very fact.

. . .

IF YOU ARE LUCKY enough to hold an ordinary chondrite in your hand—ordinary only because they are the most common of the chondrites—then you are holding a stone that almost matches the chemical composition of the Sun, minus helium and hydrogen. Dull-gray chondrites such as Paragould are like scooping up light in a summer sky, like fetching sunrise from a potter's wheel. Chondrites are therefore, says scientist Robert Dodd, "The closest things we have to pieces of the sun." Perhaps it's right that we call them falling stars after all. The chemical match arises because chondrites are little changed from the formation of the solar system.

Despite being so venerable, chondrites are numerous, constituting about 84 percent of all the stony meteorites. According to meteorite researcher Ralph Harvey, more than 90 percent of meteorites that fall to Earth—of all kinds—are ordinary chondrites. There are three primary categories of chondrites: ordinary, which is further subdivided into classifications based on amounts of iron and metal; enstatite, which is named for its primary silicate mineral component; and carbonaceous, whose composition is the closest of all to that of the Sun and whose organic materials raise questions about the origins of life. The majority of ordinary chondrites are of two types, the H5 and L6 chondrites. The H is for high iron content, the L is for low iron content, and the numbers indicate they have a lot of chondrules (type 1's have no visible chondrules; the numbers range up to type 6's, whose chondrules blur together).

And it turns out that chondrites are like three rocks in one. They are igneous, sedimentary, and even slightly metamorphic, as Dodd points out. Chondrules are igneous products made of olivine and pyroxene, which are iron-magnesium silicates common to meteorites; they are the little glassy globes (sometimes broken up) that quenched out of the early solar nebula. A typical chondrite's matrix (the porous, friable, sandstone-like material that holds the rock together) evolved as a sedimentary rock, one deposit of material over another over another. The matrix also consists of fine grains of pyroxene and olivine, as well as nickel-iron alloys, and includes all those chondrules that settled onto and into asteroidal dust that had never been melted. And some chondrites as a whole also have been heated—not melted completely, as the achondrites have—but heated just enough that they can be considered a metamorphic product. Dodd says that chondrites are "*physically* evolved but *chemically* primitive."

Paragould is of a type—in today's jargon an "LL5 chondrite"—that possesses very little iron and whose chondrules are less distinct than in other ordinary chondrites. This nuanced classification was still decades away in the days of the Paragould fall. Only in the 1950s and 1960s did researchers settle on grouping ordinary chondrites into high-iron, low-iron, and low-low-iron categories, with those additional numbers to rank the visibility of chondrules.

Certainly by 1930 the three broad divisions of all meteorites—stony, iron, and stony-iron—were clearly recognized by scientists and they were used by Harvey Nininger himself in a popular book he wrote three years after Paragould. Today, at my rough count from one science book, there are some fifty categories of meteorites within those three divisions.

As to the origins of chondrites themselves, nineteenth-century chondrule detective Henry Sorby postulated two ideas, one that didn't last and one that did. The first was that they were fragments of the Sun itself. He meant this literally, not in a chemically metaphoric sense, for Sorby and many others believed that the Sun was actually solid. Until insights about atoms and thermonuclear reactions, the Sun's power was a real mystery. Sorby's second thought about chondrites was that they were the stuff that never became planets. As scientist Harry McSween notes, "His second hypothesis was surprisingly close to our current concepts."

Harvey Nininger commented at length on the origin of chondrites and other meteorites. In the 1930s, a few researchers still thought that meteorites originated from lunar volcanoes, but most thought as Harvey did, that meteorites came from rocky "swarms," which, when held together, were comets like the ones that sometimes grace our skies with fans of sun-warmed gas. Every once in a while gravity would disturb these accretions, and that's why, Harvey thought, meteorites land on the Earth.

In a limited sense, he was right—about meteor showers. Meteor showers—those annual events of shooting stars such as the Perseids or the Lyrids—are indeed the tiny, burning bits of old comet debris trains, not unlike a swath of sand pitched out from a child's hand on the beach. The Earth encounters such debris trains on a regular basis, and the small particles of which they are composed fall into the air and burn up from frictional heating, never reaching the ground. Harvey would come to see, as others, that most meteorites arrive instead from the asteroid belt.

At the time of Paragould and his stunning purchase of a stunning meteorite, however, Harvey wasn't thinking only about chondrites and where they come from. He was thinking about his job and where he might go.

THE PARAGOULD CHONDRITE WAS then the largest stony meteorite on the planet. (In 1976, a chondrite fell in China, breaking the record with a total known weight of about 4.5 tons, with the largest mass being almost two tons, the biggest stony meteorite as yet known.) Naturally, Harvey wanted to keep Paragould, but the $2,600 he made selling it to Marshall Field's nephew Stanley, who donated it to the Field Museum, gave him a needed financial break.

Harvey wanted more.

In 1930, Harvey Nininger was forty-three and had been teaching at McPherson College since 1920. "With each year," he wrote in his unpublished memoir, "the leash that was holding me became stronger." Harvey preferred "economic uncertainty with plenty of dare" over "security with routine." One might suspect personal myth-making here, retroactively casting his desperation and peregrinations in a more heroic light, but Harvey's propensity to move around had prompted Addie to mention it in her diary as early as 1912.

On Valentine's Day 1930, the frustrated professor wrote to the director of the Colorado Museum of Natural History, J. D. Figgins. Harvey wanted to loan his large collection to the Denver museum and wondered if he might be able to work there too. He wrote that "much as I would like to remain [at McPherson], I find the urge to research too strong to resist." Harvey already had applied for jobs at the Smithsonian and with the Audubon Society. "I am thinking of resigning my work here," he told Figgins, "and giving my entire time to research and collecting, in which case I should probably move to Denver." He didn't mention the hardships of the Depression.

Harvey had a family to care for, classes to teach, the summer school to lead, and a zoology textbook to write. He knew his prospects remained dim at a small school, and he knew that of all things he'd considered pursuing scientifically, meteorites "far outstripped anything else." Harvey says he talked with Addie—"by way of explaining my plan so as to assure her that I would proceed in such a way so as not to endanger the sustenance of the family"—and told her he'd rather do menial work such as harvesting or sweeping out offices than continue to teach. Without waiting for a written reply from Figgins, Harvey quit his job. Doris Banks remembers, somewhat ruefully now, her father's alloy of stubbornness and curiosity.

And is it accident that Harvey's decision to quit teaching and pursue meteorites came the same year his brother Roy died? Legacy and mortality were clearly on his mind. By the end of 1930, Harvey was more determined than ever before.

HARVEY NININGER SAW THE LIGHT at a place that I decide to nickname "Fireball Corner." Concrete's replaced the brick sidewalk, and where Elmer Craik's home stood, at 1200 Euclid, there is instead a 1960s ranch house. I stand at the corner of Euclid and Maxwell one spring day, probably the only person in the neighborhood who would have found an intersection poetic, transcendent even. On a small-town Saturday of errands and yard work, I stand at the corner of accident and decision while grackles chuff beneath a milky sky and the sun hangs in the southwest, where the meteor had vanished. I feel keenly the disappointment of not finding Harvey's house and chastise myself for not coming to McPherson sooner, before the campus building that housed his office had been razed. (Later, I'll realize I had seen the old wooden Nininger house; the address had been changed. The undistinguished house still stands, though it sags.)

Down the street from Fireball Corner is the McPherson Museum, a 1920s Tudor Revival mansion that Harvey walked by as he hurried home after the November 9, 1923, meteor. The museum displays period pieces and oddities, including Aleutian artifacts and an etching of the Queen of Italy from 1901. In the room where the minerals are displayed, a fine collection of meteorites is watched over by the skin of Leo, the M.G.M. lion. Most of the meteorites come from Harvey Nininger's McPherson College days.

That's when I see it. A sliced-open Toluca, from his fateful 1929 trip to

Mexico, a half-melon-sized hunk of iron with a red U-shaped magnet clinging to it. The meteorite shows greeny-brown inclusions of troilite and graphite. The etched face of success. "Not for me, ways of routine," Harvey had written. Of course, to reveal the interior one must cut a very long time, using many blades, blades that are ruined by the work and that must be discarded. Eventually, the pattern emerges—the crystalline intergrowth that makes the thing itself—and one must wash the surface in baths of acid.

· Chapter VI.2 ·

Never Done

To Denver, Harvey Nininger hauled more than family belongings in a borrowed truck. He took a confidence bordering on hubris. In his autobiography, the professor—the ex-professor—invokes the visions and achievements of such men as Hubble, Darwin, Galileo, and Copernicus to frame his pursuit of meteorites. He does so just before he reports that those who knew him felt "pity" for his family's move from the comfort and stability of McPherson College. It is a powerful narrative device—invoking greatness, tempering greatness. Forging an identity both visionary and, even today, misunderstood, Harvey Nininger is his own best hagiographer, because his ambition and his civility conspire so pleasantly that we can't resent the former when we are charmed by the latter.

Although by 1930 Harvey had scored meteoritical successes in Mexico, Kansas, and Arkansas (and was buying and selling specimens), he was not yet seen as a nationally important scientist. In fact, his collecting exploits would both seal his reputation and, in some quarters, hinder his reception as a legitimate researcher.

He was, however, "following his bliss," to use the words of Joseph Campbell. Often we judge a person's decisions and actions by their consequences—achievement? failure?—rather than honoring that person's drive whatever the result. "Bliss" is a more interesting word than "success," I think, but success is an ever-talkative critic that can spur us to do more than we think we can.

Harvey followed his bliss all the way to 1317 East 18th, Denver, where he, Addie, and their children Bob, Doris, and Margaret pulled up to a row house and started to unload. Atop the pile of belongings was a child's rocking chair. Harvey does not record the exact day in October 1930 or the reactions of Addie and the kids. Did it seem an adventure? Did it seem an ordeal? For him, at least, it must have been the former. "Arrival in a new country always gives a special kind of thrill," he believed. I wonder if Harvey's wanderlust originated that day when, as a child, he watched a house being moved off its foundation and hauled to a new location, as if by magic, between two lines of catalpas.

Earlier in the year the family had been "prepared for the cut-off of all visible income other than sales of . . . specimens and lecture fees," Harvey explained in his autobiography. But his correspondence with the Colorado Museum of Natural History did lead to a position there, albeit only part-time and secured just that autumn. It helped that the Museum's director was no stranger to meteorites, for J. D. Figgins had worked with Robert Peary in Greenland during the retrieval of the Cape York irons. A month before the move, Harvey had written out his contract with the museum on stationery from Denver's West Hotel ("You will never be treated better than at the West"). The contract gave Harvey a part-time curator's salary and some support for fieldwork, while also allowing him to pursue a few meteorite hunts on his own. Harvey would put his seven-year-old collection on display, and the museum would pay for moving the rocks. This was a complete reversal from Figgins's position in March, when he had said no both to the job application and to displaying the meteorites.

When he quit teaching, Harvey took a $2,400 cut in his pay; his new annual salary would be $600. "I guess they wondered [why] I was willing to take so little," Harvey once said, laughing.

That the two were even talking was something of a miracle. In 1924, the professor had written to the director seeking funds for a fossil-collecting trip, support that would include Harvey being paid $25 a day for four days. (For context, some North American public universities now pay traveling faculty about $28 per diem for food.) Figgins replied with dismay, "I shall state very frankly that [your letter] is a source of some wonder—first, because of the nature of your proposal and second, if you entertain the belief that I would invest the sums you mention, in view of the exceedingly scant prospect of return." The men continued to correspond—and dicker—as Harvey began to sell and exchange meteorites with the museum.

One year after the death of Daniel Moreau Barringer—America's first prominent spokesperson on behalf of meteorites (or, at least, the hole caused by one)—Harvey Nininger was positioning himself to become *the* meteorite man. These two men never met, so far as I know, but they were crucial transitional figures in the history of meteoritics. They each made their careers outside of traditional academic circles, combining research and business in ways that made many uncomfortable. Though Barringer never made the profit he sought at Meteor Crater, he had correctly argued for its impact origin. Finding, buying, and selling meteorites, Harvey was living on the margin between debt and profit, plowing money back into his meteorite program; his research and education interests were wider than Barringer's but the financial and spiritual risks were just as great. Harvey Nininger wanted most of all to study rocks that came from space, to share his enthusiasms with the public, and to be accepted by the scientific establishment. But he also had to pay the bills.

BY FALL 1930, Ward's Natural Science Establishment—the famous natural-history supply business—was including Nininger meteorites among its offerings. The deal potentially would earn $4,000 a year for Harvey. A year later, though, Ward's stopped selling his meteorites—the market had bottomed out—and letters exchanged in 1932 indicate the precarious position of each. To make matters worse, the Smithsonian also ended an agreement to buy $2,000 a year in Nininger specimens.

The recent move from Kansas looked downright disastrous. In early 1932, Harvey characterized the situation as "a crisis which threatens to seriously interrupt my researches" and "a serious and unexpected blow." To earn as much money as he could as quickly as he could, Harvey offered to sell several specimens to the Colorado Museum of Natural History at discount to prevent "a forced abandonment of my program" that would have been "a grievous disappointment." He apparently made that sale. A chance meeting on a Santa Fe bench with Smithsonian secretary and ornithologist Alexander Wetmore also led to modest support for Harvey; the Smithsonian backed Nininger's search for the long-lost (and, it seems, nonexistent) Port Orford, Oregon, meteorite. This arrangement was temporary, however.

Harvey was able to reverse course. Rather than selling off his collection, he began trading and buying at under-market prices. It was a quick, if costly, way to increase his holdings, thus allowing him to sell directly to those museums and universities still purchasing specimens.

A 1931 financial statement for "The Nininger Laboratory" showed that he valued his own time at $300 monthly, for an annual cost of $3,600. He bought $728.50 worth of meteorites and spent $1,300 running his lab and $910 traveling. He sold $2,816 worth of meteorites and earned $600 from the museum stipend. Thus, his expenses were $6,538.50, his income $3,416. He valued his additions to the collection—750 specimens—at nearly $28,000.

Eventually, the national crush of the Depression would force the Denver museum to institute its own cuts, requiring Harvey to bear all costs for meteorite operations. Figgins began writing to other institutions and naturalists seeking help for his meteorite curator.

Simply gathering payments could be frustrating. The director of the Clark Observatory opened a January 28, 1937, letter from Harvey that said a check for a meteorite already sent was "desperately needed" because "a sick brother is coming . . . to stay with us while taking treatments." Three years later, the director finally replied, apologizing that his "purse was so transparent that a dime would look like a total eclipse in it." And when a scientific journal wanted Harvey to pay for illustrations to accompany an article, he offered a specimen instead, "because I have meteorites and not money."

Harvey might have taken cold comfort from learning that McPherson College had cut salaries by more than a quarter in 1933. In his autobiography he wrote that a return to a regular teaching job was unthinkable, but the published version of events is not entirely accurate. In April 1935, during especially tight times, Harvey approached Denver University to discuss the possibility of a job, seeking $2,000 a year for teaching and offering a quarter of his finds if the school sponsored a meteorite survey. He did end up teaching at Denver University but on a class-to-class basis. (A 1936 enrollment roster shows Alex and

Dorothy Richards taking Geology 121b from Harvey.) In 1939 the school again declined to hire him as a professor of meteoritics. In 1941, he tried once more, but the university again declined. In a note to himself, Harvey appears to misremember these events, claiming that he turned a "deaf ear" to a Denver University job offer and to "all other jobs even when they were scarce and I needed them."

Letters from the mid-thirties between Harvey and Frederick Leonard, a professor at UCLA interested in meteorites, discussed the possibility of Harvey being hired at that school. This too would not come to pass.

And as early as 1936, Harvey asked Leonard not to pursue an honorary degree for him from a small California school, because Harvey would prefer one or two from larger institutions. This is Harvey Nininger at his most guileless, thinking that he can have his pick of honorary degrees and believing that someone who was, as it happened, a marginalized figure at UCLA could help him secure such awards.

"There were periods when the purchase of a pair of children's shoes was an occasion for a family conference, and when the replacement of Dad's suit was solved via patches on the old one," Harvey recalled. "The menu was dictated by economy rather than taste." Their only "luxury"? A "good car." To make ends meet, Harvey and Addie borrowed off their life insurance. Harvey also gardened, selling his carrots and onions through a local store. Doris sold vegetables door-to-door, and Bob had a newspaper route. "Once," Harvey wrote, "I brought home 500 pounds of peanuts which I had been able to buy for three cents a pound in Texas; sometimes I brought home quantities of pecans or eggs, or a carful of apples or peaches. Such provender served a dual use: Some went into our own larder; the rest was sold to the neighbors at bargain prices, with some profit to me as middle man to mend the budget."

MENDING THE BUDGET, as well as the clothes, usually fell to Addie. It was she who ran the household and tended to logistical arrangements for Harvey's trips. Harvey saved his mind for what interested him. "He could rejoin the world when he needed to and would do a fine job," grandson Gary Huss recalls. Mostly he wanted to be left alone with his rocks. As was then so often the case, a woman's steadying influence on family, household, and husband helped ensure that the man had the time and clarity to work toward success. Without Addie, it's hard to imagine Harvey's life unfolding as it did. Not that he was unappreciative. Granddaughter Peggy Schaller says that Addie and Harvey were "always respectful and loving toward each other." Harvey once said that Addie "was like no other"—his voice catches with emotion on a recording—and noted how she was his coworker, secretary, accountant, and greatest believer. "She believed in my intellectual providence," Harvey said. "She never asked that I quit."

In the evenings, with the children doing homework, playing, or sleeping, with the letters and balance sheets set aside, Addie might be reading or sewing, and either mulling over or trying to forget the family's money difficulties. Addie somehow kept the family going from bill to bill.

"The sleepless hours," Harvey wrote, "spent on [money] cannot be known; but suffice it to say many, many, many!" Whether Harvey turned to God for reassurance he did not say. His granddaughter notes that though he left the church behind, he did not abandon faith.

Harvey typically worked sixty to seventy hours a week, sleep or no sleep, but it was Addie's hair that turned gray first. Daughter Doris Banks has a photo of Harvey carrying a big suitcase as he ascends the stairs of a railcar. He's smiling. In most pictures I've seen of Addie, she is tight-lipped. "She'd just cope," Doris says. "She just put up with things and she'd put up with discomforts. . . . She put up with a lot" because Harvey had a "one-track mind."

Addie's diaries nonetheless reveal someone apparently prone to depression and nervousness. From 1912 to 1914, she mentioned both conditions several times. Addie also was struggling to free herself from her parents and from the Church of the Brethren. It was a major crisis. Harvey had gone further and faster from doctrine than Addie, who was "just beginning to realize what it may cost." Addie's love for Harvey thus began in defiance of norms and expectations, and this togetherness-in-isolation would be a theme in their lives for decades to come.

Sometimes it was too much. Doris says that "when things were too tough . . . [Addie] couldn't manage," and that prompted "the sick headaches she used to have. She'd go to bed sometimes very ill."

Still, in the midst of research, collecting, traveling, and coping with family concerns, Harvey himself managed to find some time to be active in the YMCA and Scouts. Doris recalls some of the other times he made for the family, such as attending concerts at the band shell, officiating over contests for the dasher of the ice-cream maker, watching the colored lights on the fountain in City Park in summer, and ice-skating there in winter.

That both Addie and Harvey had good senses of humor—he could be wry—may have helped cushion the children from some hardships, and in many ways, the Niningers were extraordinarily lucky, for they had experiences few could even dream about. Even a winter car trip presented a chance for the unusual: "To keep our feet warm," Doris laughs, "Dad would heat iron meteorites!" He'd wrap the meteorites in blankets and put them on the car's floorboard.

That kind of thrift also pushed the Niningers to move to less expensive housing, and, wherever they lived, Addie had to pester Harvey about money. Harvey remembered how Addie would say, "Have you noticed our bank account is near zero?" She might have told him that Margaret needed shoes, Doris needed a dress, Bob's trousers were beyond patches. "And we owe this bill and that bill," she'd add.

"It looks pretty discouraging," Harvey would reply. "But just remember . . . every week that passes without me finding a new meteorite means we're one week closer to the next find."

IN DENVER, HARVEY FOUND a new source of income and mobility as a cross-country truck driver for $4 a day. He'd deliver trucks—just the chassis, on which he mounted his Plymouth for the drive home—and set up lectures along the way. He'd put on a fresh set of clothes, then talk to professors, graduate students, and curators, never telling his hosts how he got there.

The man who made this possible was the truck company owner Dean Gillespie. One of the few individuals in the country who then seriously collected meteorites, the hefty Gillespie even used iron meteorites as bookends, right next to his Dictaphone. Gillespie would front money to Harvey for meteorite searches, knowing that failures to find specimens meant a lost investment. When Harvey did find specimens—and he did—Gillespie would get his share. Harvey even moved his meteorite-cutting saw to Gillespie's office, where he sometimes resentfully served as "window trim" for Gillespie's potential clients. Harvey's irritation at being a display was balanced to some degree by his storing meteorites in a vault Gillespie owned but to which only Harvey had the key. "He was a big help to Dad, but I think he gained more than he gave," Doris Banks says. Doris still rues that her father was too busy meeting with Gillespie to see her off to college.

By truck or train or car, Harvey crisscrossed the continent from Chihuahua City, Mexico, to Saskatoon, Saskatchewan. In the former—in 1931, after dropping off a truck in El Paso—he found the supposedly vanished Huizopa iron. On a Canadian venture, he found no meteorites but published a letter in the *Saskatoon Star* that asked farmers to examine rocks thrown out of the way of the plow. Soon enough, Harvey received in the mail a pallasite named Springwater, which later was found to contain a mineral new to science.

Harvey also continued his relationship with Mexico's Frederick Mullerried, who helped Harvey obtain permits to export meteorites, though some letters suggest that Harvey may not have always obtained such meteorites legally. A letter to Mullerried on April 3, 1941, detailed how, without an export permit, it was possible to remove a meteorite from Mexico. "I think it will be best to send it with some tourist," he said. "Explain to them that I have several times brought such material without concealment and with no difficulty. I explain that it is iron and let them do as they please about it. The officers never take any interest in it." Harvey sometimes employed middlemen too.

Harvey's success brought him attention from such renowned journalists as Ernie Pyle and from such publications as the *Saturday Evening Post*. Articles helped alert readers to meteorite identification and tied meteorites to Harvey's name. He was even featured in "Believe It or Not!" Such publicity would also tarnish his standing among scientists who disdained the press, even as Harvey worked hard to establish a standing with those very same academics. He must take some blame for this uneven reputation. When McPherson College awarded Harvey Nininger an honorary doctorate in 1937, he started using the degree on his letterhead, ads, and other materials—calling himself Dr. Nininger—which suggested that he had earned a regular doctorate. For some scientists, this was equivalent to fraud.

Harvey's otherwise laudable enthusiasms also meant he poked his nose into scientific areas about which he knew little. In 1942, F. R. Moulton—who had proven Barringer's meteorite could not have survived impact at Meteor Crater—matter-of-factly corrected Harvey's mistaken assumption that increased telescopic magnification would increase the brightness of any meteorites seen hitting the moon. Moulton pointed out that aperture, as well as the kind of telescope, not magnification, affected brightness. This is a very elementary point, one that could have been made quite snidely, but Moulton was always friendly and supportive.

The same could not be said for Harvey's rivals. In the Nininger papers, there is a file folder with the tab cut off concerning a man named A. R. Allen of Trinidad, Colorado. Over the years he was in touch with Harvey, from whom he bought specimens, but in time Harvey accused Allen of going into localities to exploit Harvey's contacts. Poaching, as it were. "I suppose there is no law against such interference," Harvey wrote to a lawyer representing Allen, "but it certainly is not a very man[l]y thing to do." He couldn't know this was just a taste of things to come.

Of course not everyone who contacted Harvey was a competitor. He continued to receive letters from people who saw fireballs—most of them landed just over there. He continued to receive letters from folks who sent him surefire meteorite samples—that turned out to be cinders of burnt hay or pieces of slag. Then there were the nut cakes. One man wrote that he was "just a plain chump of 70, born with a flair for deep thinking or dreaming," who included his unpublished paper "Sex in Celestial Objects." The paper explained that "our mysterious and erratic wanderers, the comets, seem readily analogous to . . . spermatozoa . . . [and] the sun-spots . . . may be cited as another reason for regarding the sun a female." Another letter outlined the history of Saturnians throwing meteorites at Earthlings. Harvey even heard the old tales of meteorites delivered to Earth by lightning bolts.

SATURNIANS ASIDE, there were plenty of meteorites to be found, and Harvey was finding most of them. Between the time of Harvey's first fireball and 1948, his several hundred talks—complete with displays of native terrestrial rocks and meteorites—yielded some three dozen finds in Kansas alone. The finds could produce multiple specimens; one strewnfield yielded 800 pounds and 300 stones.

Harvey also confirmed his belief in "a tendency of . . . finds to group themselves in the wake of some outstanding meteoric event or discovery." A sky flash often meant a news flash, which led to many eyewitness letters and a chance to educate more people about meteorites. Typically, just two of any given 2,000 mail-in rocks were meteorites. But if the correspondent knew of Harvey's work or had heard a talk, the numbers changed drastically. In those cases, about one out of every two dozen to four dozen mail-ins turned out to be a meteorite. And one in every six fireball chases led to the recovery of a specimen, a shockingly good rate.

Such chases and falls often made great stories, which Harvey loved to tell. He recovered stony meteorites from the August 10, 1932, Archie, Missouri, fireball that had prompted a child to exclaim, "Daddy, it's raining rocks out here!" and from the March 24, 1933, Pasamonte, New Mexico, fireball that had prompted a rancher to tell his cowboys the end was near.

Harvey happened to be in the area when the famous Pasamonte stone struck, an event that a Harvard eclipse-study team, also in the area, failed to photograph. Ranch manager Charlie Brown grabbed his camera, however, and took a spectacular snapshot of the fireball. Harvey and fourteen-year-old son Bob talked to Brown and others and were amused when locals accused Harvey of knowing the fireball would hit. How else could he explain his being there when the meteor sailed overhead? Pasamonte, Harvey wrote, "was more spectacular than even the most dramatic accounts by witnesses—the fireball was cubic miles of incandescence, accompanied on its fifteen-to-twenty second flight by a discharge of material which produced a . . . column of dust a mile in diameter and lasting ninety minutes or longer." Buildings shook nearly 100 miles away. Beneath the start of dark flight—that moment when objects have lost cosmic velocity and are traveling under the force of gravity—people complained of sore throats, presumably because of the falling cosmic dust.

The Pasamonte meteorite was not only spectacular but rare—a howardite, a jumbled, calcium-rich mix of two other forms of achondrite, the eucrites and diogenites. (Achondrites are the igneous rocks of asteroids.) Howardites are chunks of battered ancient asteroid soil, some 3.5 to 4.5 billion years old. They are the first shovelful lifted from a large asteroid by an impact. Containing traces of noble gases from the origin of the solar system and microscopic scars of impact, howardites also contain minerals that must have originated at widely variant depths; thus, these meteorites are samples of deep and reconstituted asteroid soil. Similar to lunar soil, howardites are, it now turns out, a nice spectral match for the asteroid Vesta, an object that must have had a volcanic past. Indeed, it seems to have behaved a lot like a planet, in that it separated itself into core, mantle, and crust. At one point, the asteroid had lava eruptions and even, scientist Michael J. Drake writes, "an asteroidal-scale magma ocean." Vesta features a huge impact basin at its south pole and a family of so-called "Vestoids," rocks about a half-mile wide still scattered in the asteroid belt. From the Vestoids come meteoroids that land as meteorites on Earth. Other meteorites that seem sure to have come from Vesta, the eucrites, appear to have been impact-heated, perhaps more than once. Howardites were named after the English chemist Charles Howard, who in 1802 performed the first chemical analyses of different meteorites. They are rare finds. That's because they don't fall often and are hard to identify. They look similar to terrestrial basaltic rock. Of all the achondrites, however, howardites are the most common.

A mystery of the grouping of howardites, eucrites, and diogenites (HEDs) is why they seem chemically related to some pallasites and some irons—meteorite types that suggest they came from an asteroid that's been destroyed, since pallasites and irons sample the mantle and core, respectively. But since HEDs probably came from Vesta—which has been impacted but not obliterated—how did those pallasites and irons show up from under all that now-hard igneous rock? Some researchers think that "more than one parent body" developed with the same basic chemistries as these meteorite classes show, and that the irons and pallasites came from one or more of such long-since-destroyed asteroids. Vesta may have been hit, then ejected material that fell to Earth as HED meteorites, but it outlasted related kin that proffered those deep-down pallasites and irons. We'll know much more about Vesta and the HEDs when the probe Dawn arrives at the asteroid in 2010.

When not chasing fireball reports, Harvey took to other hunting grounds, such as schools, museums, and even businesses that formerly displayed minerals.

The Depression led many institutions to discard all kinds of collections, and Harvey took advantage of this. A custodian once took him to an attic filled with a ton of neglected gems and minerals. And Canyon Diablo irons—quite cheap in the 1930s at 50 cents a pound, half the usual rate—merited so little attention that masses languished on Winslow, Arizona, sidewalks and even in gutters.

At the heart of Harvey's success were farmers, especially the elderly who had first plowed the prairie during pioneer days. Of course Harvey would have to explain his financial interest in meteorites—and start buying some—to overcome typical country stoicism.

An additional boost to Harvey's program was the 1933 publication of his first book, *Our Stone-Pelted Planet.* Favorable reviews appeared in the *New York Herald-Tribune* and the *New York Times,* which wrote, in part, "He opens before . . . amazed eyes vistas of knowledge that appeal to the imagination like a Jules Verne novel."

Part of the book dealt with the Kansas Meteorite Farm. Harvey had experienced success in Toluca, Mexico, precisely because he went to where meteorites had been found before. He continued this approach on land where Eliza Kimberly had found the Brenham pallasites. When Harvey visited the Kimberlys in 1929, Frank showed him a buffalo wallow with a rim. Knowing the wallow was, in fact, a small, eroded crater, Harvey asked permission to dig it out, but Frank refused. Harvey kept calm. He didn't want Frank to destroy the crater by hunting for more specimens; a hasty dig would ruin any chance to study its nature. After Eliza and Frank died, Harvey was able to excavate using "two teams of horses and two old-time road scrapers."

Letters between the Kimberlys and the Niningers show a human side to the story not included in published accounts. Harvey spoke of a find that he believed was on the Kimberly land, thus informing them of it and showing a probity others later questioned. Harvey and Bob, it turned out, liked the Kimberlys' fried chicken, and young Effie Kimberly appears to have had a crush on Bob.

In all, Harvey Nininger found 1,200 pounds of meteorites at the old Kimberly farm. It was another major find.

Harvey's work there had proceeded, in part, because F. R. Moulton had helped him get money from the Geological Society of America. Harvey had written to Moulton on November 25, 1936, asking for funds in order "to throw some light upon the puzzling question as to what happens to a large meteorite when it strikes the earth." Two years later, he thanked Moulton for his help and said his work on the Kansas Meteorite Farm had gone well. Harvey also waxed poetic regarding Moulton's theory of the origin of the solar system, which emphasized the slow, spinning accretion of bits of matter, calling it an "inspiration since that eventful evening in 1923 when a great meteor changed the course of my life." He also sounded a cautious note, saying he had research ideas that could yield money for him and so he could not provide details. And Harvey hoped to find research support at the staggering levels of $30,000 to $50,000 a year.

Moulton was impressed with Harvey's results at the Haviland crater—as the wallow came to be called—and said that Harvey had set a "new standard for investigations of old meteor falls" and that he had "put a life into the study of meteorites that it should have had long ago."

Not everyone agreed. When *Science* rejected a Haviland manuscript in 1933, Harvey was so irritated that he wrote to the editors almost lecturing them to run it.

Harvey's self-assuredness regarding the importance of his work is reflected in a letter he wrote to J. D. Figgins during one of the Haviland excavations. He said that "the most significant discovery . . . is that a meteorite crater can exist in a populated area which has been visited by numerous geologists. . . . No one beside[s] myself ever even suggested such a possibility. It argues very eloquently in favor of the exploratory program which has been intriguing me for several years, and surely demonstrates that meteorites may constitute a much larger factor in the earth's geology than has ever been suspected." This was Harvey's own Big Idea: that meteorites were the prime mover in all geological processes, from building planets to seeding life to triggering volcanism.

Of these ideas and others, Harvey wrote, publishing more than 160 articles, talks, and letters between 1928 and 1967 in a range of periodicals from *Popular Astronomy* and *Arizona Highways* to the more scholarly *American Journal of Science* and *Meteoritics*. He chronicled his surveys, described newly discovered meteorites, reviewed information regarding crater formation and shape due to impact and explosion, argued for the importance of meteorites, noted that fieldwork was beginning to show (correctly) that most meteorites were stones, not irons as previously thought, and speculated about the moon as the source of tektites. He even advocated using an atomic bomb on the moon to verify his tektite-origin theory. (Most scientists now believe that tektites are melted ejecta from major meteorite impacts on the Earth that fell back to the ground.) Like Grove Karl Gilbert before him, Harvey Nininger was one of the few who argued that lunar craters were formed by impacts. Harvey also visited, studied, and collected at the Odessa crater in Texas. Long before many others cared about the infall of cosmic dust into the atmosphere, Harvey was trying to gather such microscopic matter with magnets attached to alpine trees and weather balloons. He even "searched the rough bark of trees by means of a magnetic device." He stuck magnets on steeples and dragged magnets on the drained floors of fountains.

Harvey paid attention to small things but he also had his dreams. He wanted to create a National Institute of Meteoritical Research, an idea backed by many scientists, including a luminary like astronomer and physicist Moulton. The foundation that heard the proposal turned it down, and Harvey would wonder if a rival more powerful than A. R. Allen was responsible for the defeat.

By the late 1930s, collecting and studying meteorites was no longer such a queer enterprise either to scientists or the public, and to the latter Harvey sometimes stressed the profit available in finding and dealing rocks from space. He wrote that he'd "put several thousand dollars in my own pocket" from meteorites. More important, in one year alone, in 1937, he and his "staff"—Addie, Bob, Alex, and Dorothy Richards—recovered and recorded an astonishing thirty-one falls new to science. In one year, he'd found thirty more meteorites than George Merrill thought he'd find in a lifetime.

TWO YEARS LATER, Harvey and Addie helped recover the Goose Lake, California, iron. It was the biggest meteorite that he had then been involved with, as it weighed more than a ton. Hunters had found the iron on rugged, rocky Forest Service land, and Harvey had received letters about it. From a sample, he saw immediately that it was meteoritic.

Working on the Goose Lake meteorite recovery was a major operation, involving several men who had to configure a block and tackle to lift the massive iron and load it onto a wagon. Joining them was UCLA professor Frederick Leonard, a punctilious and, for some, inspiring oddball who wore sweaters in the desert heat.

Harvey writes fondly of their rendezvous with the Goose Lake iron. "Frederick Leonard had shed most of his academic dignity by this time. When we came within sight of the big iron the pudgy little professor ran on ahead, placed his hands lovingly on the great meteorite, bent and kissed it. Then he lifted his hands skyward and turned to face us.

"'This is the greatest day in meteoric astronomy!'

"Now we all joined in to fondle the prize, an almost unbelievable chunk of metal that seemed to link us, there in that wild corner of the earth, with worlds beyond."

That chunk of metal could not, however, remain linked to the hunters who found it. As it was on public land, the meteorite belonged to the U.S. government.

It's a good story, but there appears to be more to the Goose Lake saga than Harvey wrote of in his autobiography *Find a Falling Star.* According to meteorite expert O. Richard Norton, in 1999 the then elderly codiscover of Goose Lake, a man named Clarence Schmidt, shared papers that seem to show that Harvey had attempted deception in order to acquire the meteorite for himself. Harvey, Schmidt said, claimed to be working for the Smithsonian and even offered a bribe to get the meteorite. In an odd turn of events, after the meteorite had been found again (after the hunters' initial discovery) and once Harvey informed the Smithsonian's E. P. Henderson of this, Henderson hired Harvey to remove the meteorite even though he suspected Harvey of misrepresentation.

Henderson's dislike of Harvey Nininger would grow over the years. Indeed, when Harvey published his autobiography in 1972, Henderson wrote out a defiant rebuttal to Harvey's published version of the events and suggested that he had stolen samples off the main mass of Goose Lake. Norton's research found that the Henderson rebuttal was never published, and Norton himself concludes that "Nininger was guilty of deceit in his misrepresentation as a government agent [but] this was more than paid for by Nininger's assistance in the . . . recovery."

A file of "Unfinished Papers" might shed some light on the Goose Lake matter. In this file Harvey speaks of "an agreement with Mr. Henderson . . . to the effect that in case my work turned up a meteorite and the finder chose to sell it to the Smithsonian I should pay half the purchase price and we divide the meteorite." Harvey provides no date of the agreement but suggests he believed it in place for at least two meteorites, though not, it seems, for Goose Lake. Harvey also claimed that a Smithsonian official had said that Henderson wasn't authorized to secure such a deal. The account appears to have been written in Harvey's old age. Family members today tell me that Harvey felt the Goose Lake iron might just be stashed away by the Smithsonian and that he, who had worked so hard to retrieve it, would never get a chance to study it.

Harvey Nininger is not here to defend himself, but it's likely that at the least he would have told Schmidt of his for-hire associations with the Smithsonian (such as the Port Orford search). But did he exaggerate? Or did Schmidt misunderstand Harvey's Smithsonian claims?

ONE THING THAT CAN BE SAID about the Goose Lake search-and-recovery was that it was like so many of Harvey's excursions. It was challenging. Searching for meteorites often meant fixing "tires in ankle-deep mud while a raw wind numbed my fingers, or creep[ing] in low gear through mud for hours . . . only to find an ordinary rock." It meant searching "badger holes, coyote diggings, clumps of grass and clumps of weed, rocks, holes [and] ditches." Looking for fragments from the Pasamonte fall, Harvey joked that "windmills . . . seem to be in the habit of moving away from you as you pursue them."

Some of his difficulties he brought on himself. He rarely took a direct route to a report. By meandering, Harvey could interview witnesses until he had mapped a possible strewnfield. Meteorite hunters today still use these methods: talk to witnesses (or get electronic mail from them), gauge where they stood, estimate the angle of descent, determine the start and end points of a meteor, then correlate reports to map the possible strewnfield. Modern tools can help too. Video cameras capture meteors on tape, and Global Positioning System satellites allow for speedier mapping than in the days of protractors and pencil nubs.

Unless it's an exceptional fall, with a large number of witnessed meteorites, finding space rocks is usually a matter of chance and perseverance. Meteorites can be next to impossible to find in wooded terrain, especially in hills or mountains, but a desert's dry expanse offers a better visual background for picking out rocks different from the local geology. In fact, the best places on the planet for meteorite-finding are deserts such as the Sahara and Antarctica. The type of meteorite makes a difference too. Irons stand out because of their dark color, heavy weight for their size, and often fantastical shapes, which is why for years geologists assumed iron meteorites were the most common. Chondrites and achondrites are often the hardest to find, because they are so visually dull—expert eyes can discern fusion crusts, of course—and carbonaceous chondrites are so friable they may not last long enough for hunters to find them.

But against the long odds of finding a meteorite, a maniacal patience can pay off.

On a December 1933 afternoon, Harvey and his brother John, who was on his first meteorite hunt, drove into the town of Plainview, Texas, where, Harvey announced, they would look for meteorites before turning in. "John looked at me with real alarm," Harvey wrote. "Was I feeling all right? . . . Hadn't we just wild-goose-chased a half dozen reports in Mexico? . . . The food had been terrible, the water worse. Fleas, ticks, ants, scorpions, and snakes had helped to make our days and nights in Mexico miserable. We had found absolutely nothing, and here we were, three days before Christmas, returning home nearly broke." Harvey knew, however, that some samples of a stony meteorite had been found at Plainview years before. After renting a cabin, Harvey set out to talk to farmers and offered to pay a dollar a pound for any they had. He interrupted dinner at one house and soon procured an 8-pounder that the farmer had used to hold down his chicken coop. "I laid the meteorite in my brother's lap," Harvey wrote.

Two years later, in 1935, Harvey and his son Bob slept exhausted in the western Kansas town of Hugoton after having driven through dust storms. Spending the night with a former student, a Mr. Hubbard, father and son "awoke in what had been a freshly cleaned room the night before" to "bed linen . . . the color of creamed coffee, with two incongruous white spots where our heads had rested on the pillows." It had been a very bad year. Weeks of investigations and lectures amounted to squat, and dust had fairly ruined Harvey's car. Harvey and Bob wanted to get home but were convinced to visit with another former student, a Principal Riggs, who insisted on a meteorite lecture to his own students. Because the janitor didn't have time to clean up the auditorium from the latest dust storm, everyone gathered in the library.

After the lecture, a student named John D. Lynch, Jr., walked up to Harvey to convey some news. He thought his father had discarded a plowed-up rock like the ones Professor Nininger spoke of. Their farm was seven miles away. Would the Niningers like to see it?

Riggs, Harvey, Bob, Hubbard, and young John drove to the Lynch farm, where—amazingly—a meteorite was barely visible under swirling dust. Had they been a few minutes later, they might not have seen the 16-pounder. They dug it out, and Bob called out, "Look, Dad, it is only a fragment because it has fusion crust only on one side"—a sure sign that it had broken off a larger mass. Harvey was pleased his son had noticed.

He asked Lynch to take them to where his father had originally hit a mass with the plow and moved it with the lister. Lynch paused and rubbed his head.

"If we follow this corn row," he told Harvey, "it will lead us to the spot and I think I will know when we get there."

Bob and Harvey each grabbed a shovel, and they all walked the half-mile till Lynch blurted out, "It was just about here."

"Look for that kind of thing," Harvey said, tossing a fragment from the 16-pounder.

Mixed in with the dust from Kansas, sand from New Mexico, dirt from Colorado, clay from Oklahoma were fragments from space, and the fragments were everywhere.

Harvey spoke again. "If you find one that seems to be anchored, drop your shovel and dig with your hands." He probably paused for effect. "This is a very old meteorite, and we want to recover it intact."

John soon hollered, "I've found an anchored one!"

Harvey was intent, for just two feet away he'd also found an anchored sample. The meteorite could be big, very big. John and Harvey began to dig in that ocean of dirt with their hands, until Bob took over for his dad, who went, before the mass was recovered, to pay for the stone. Lynch senior refused the money until he and Harvey agreed that Lynch's son would be paid for showing the Niningers the location of the meteorite.

Few scenes from the annals of meteorite-hunting are as surreal as this one. Men and boys digging with their hands in the Dust Bowl dunes of Stevens County, Kansas, for an ugly piece of rock that had landed long, long before they were born. Their faces pancaked with grime, their clothes covered in soil, they were on their hands and knees to find the sky in choking dust, and they were ecstatic, they were ravished by their find, what was then the second-largest stony meteorite yet found, the 749-pound Hugoton mass, a brecciated ordinary chondrite, second only to Paragould. (A breccia is rock made up of various, mostly unabraded fragments.) Harvey was beside himself with glee.

He wrapped the stone in paper, plaster, and burlap and rigged pipe "to be used as handles" for moving the mass to Denver. Wind covered the tracks of the men back in the Lynch field. Years later, Harvey would write, "There is something very impressive about footprints. They seem so fragrant of life. This is because ordinarily they are so ephemeral."

Hugoton was not his only stunning find in 1935—his luck having turned after all—for Dean Gillespie had advanced Harvey funds for a recovery trip to Morland, Kansas. There he found a 600-pound chondrite, which "set my nerves tingling." The hunts were addictive.

So addictive, in fact, that his family once expressed surprise when Harvey showed up in Kansas in 1931, driving out from Denver to see his dying father, for there had been a fireball, one that he didn't know about. His kin had expected he'd rather chase a meteor than visit his father's deathbed. Of course Harvey may have felt uncomfortable around relatives who didn't understand or perhaps even respect his work. And, he said, eventually, "it was decided that I could do no more for my Father." He left to chase the fireball, and his father died a month later.

Harvey's long hours driving in open country, the passing land and its lives, and his seeing and naming all along: blurred meadowlark song, cottonwoods by a stream, the flittery call and rise and dip of a flying goldfinch, the dive of the prairie falcon, and then the meeting of strangers, their kindness and their indifference, their curiosity and their eagerness, after all the explanations of rocks that fall from heaven and how he paid for them. Harvey felt a grand and sweet connection to the enterprise of discovery, what with the universe gifting meteorites in places where ordinary days and nights went on—hanging out the wash, gathering it in, slopping pigs, walking at dusk with the local pastor you wanted to tell your secrets to, all the muted quotidian—then the stop of a trowel halfway down in the dirt, the soil flung up beside boots, then the first glimpse, a regmaglypt perhaps or a chipped, black fusion crust showing light gray beneath. Whose nerves wouldn't tingle? What better office than the land beneath your feet and the air all around you? The chance of discovery keeps one looking, with that hope for the sparkle beneath skin—elation.

WHEN WORLD WAR ARRIVED again for the United States, Harvey Nininger was already fifty-four, and his associations with the museum in Denver had loosened considerably. (The First World War had interrupted Harvey's plans for a doctorate, and instead he got government work as an insect-control specialist, traveling among farmers and staring at the hastily built coffins for victims of the flu epidemic.) The Second World War would be another time for change in Harvey's life. A new museum director had chosen not to support his fieldwork, and Harvey passed the war years as a salvage metal inspector for the government and as an oil explorer. With just a 40-hour workweek, he managed to squeeze in some meteorite-hunting.

Harvey could take satisfaction in knowing that a handful of classes in meteoritics were being offered at the university level. Much good had happened. Three craters on the continent—Meteor Crater, Haviland, and Odessa—had been identified and studied. More than 200 new meteorites had been discovered, many of them by Harvey Nininger himself. At least two textbooks included passages on meteorites. Now calling his enterprise "The American Meteorite Laboratory," Harvey owned some 5,000 meteorite specimens, the largest private collection anywhere.

As he searched abandoned mines and factories for usable salvage, as he negotiated petroleum leases, as he scoured drainage ditches and fallow fields for meteorites, Harvey Nininger again had time to take stock of his life. At an age when many would be looking ahead to retirement, Harvey knew he had no certain job once the war was over. Increasingly, he'd been thinking of leaving Colorado to start anew.

Perhaps when he was asleep in his car or on a bedroll out in the desert or in some cheap motel, Harvey had that dream again, the one he used to have as a young man. He dreamt he was leaping, nearly flying, along a road. But in his dream the road he's traveling isn't climbing to the sky. It doesn't move him closer to clouds or stars. Instead, the road is going down.

· *Chapter VI.3* ·

Strongly Spent

Strongly spent is synonymous with kept.
—Robert Frost, "The Constant Symbol"

Harvey and Addie looked ahead at the wrecked moving truck on the opposite lane of Route 66. When they saw it belonged to the company they'd hired to move their 16,000 pounds of meteorites from Colorado to Arizona, Harvey felt nauseated. He and Addie were driving to Winslow, Arizona, to get to the nearest phone. They needed to call the movers; the van was days late.

Once again Harvey Nininger had gambled. Unable to continue working for the Colorado Museum of Natural History and unwilling to work for an oil company, he had decided to move with Addie in 1946 to a stone building beside Route 66, a few miles from and on the way to Meteor Crater, the site of Daniel Barringer's dreams, a place that Harvey believed was the most moving on the planet. Not far from the crater, he would display his collection in the world's only museum devoted strictly to meteorites. Harvey's own museum.

The Niningers could only stare as they crept toward the chaos of boxes burst on the roadside like dropped eggs. Weeping, Addie exclaimed, "Why didn't we insure it?" They had, but for just a "nominal" amount, this for a collection that would later be valued at hundreds of thousands of dollars. Spilled across the road were dark—Harvey saw them—"bolts and nuts." Nuts! Bolts! Harvey and

Addie heaved with relief and drove on to make their call, though they received no information on the missing truck.

The next day they scanned the road. "About the only thing that kept us from breaking down was our knowledge [of] the great meteor shower . . . due to take place that evening," Harvey wrote. On the afternoon of October 9, 1946, the van arrived with its 189 containers of meteorites, 8 tons total—the tiny ones wrapped like china and others crated like statues. Then Harvey looked into the hold and blurted fury at the driver. No one had said that his cargo had been shifted to another truck. Now the boxes were broken and in disarray.

Addie, Harvey, friends Elmer Hokanson and his wife, the driver, and his helper unloaded for the next three hours, pausing at times under the veranda. At least no meteorites were broken or missing. The last item of business was shoving 6,000 pounds of iron meteorites out of the truck to the ground.

Soon the women set a picnic on blankets, and the two couples watched the Draconid shower, which, despite moonlight, blossomed into a storm. At points they saw as many as 100 meteors a minute. In England, the Jodrell Bank Observatory used radar to track meteors for the first time. Harvey called the storm "a fitting prelude to the opening of our museum. And it could be said truthfully that some three tons of meteorites fell within [our] view."

AFTER DAYS OF UNPACKING, hauling and cleaning, the American Meteorite Museum opened—without electricity but with hefty Canyon Diablos loaned from Meteor Crater as the display-room centerpiece. The sixty visitors who came the first day could see a derrick waiting to move one of Harvey's great prizes, the 700-pound Hugoton chondrite. Peering into cases that the Denver museum had given Harvey, the tourists listened as he gave a three-to-five-minute lecture. They saw the Braunau meteorite "that crashed through a bedroom where two children were asleep," others that "almost broke up a burial service near Denver . . . in 1924," and a Toluca iron worked into a crowbar "by a Mexican blacksmith." A later brochure told visitors to "think of holding in your hand a stone or lump of iron which, 3 hours before it landed on the earth, was farther . . . than the moon!" Tourists could wander among displays ranging from moon impacts to meteorite types. In twelve months, the Niningers recorded 33,000 visitors.

But what was good for the checkbook—and for Harvey's goal of public education—was not always good for the psyche. Harvey, Addie, and their employees George and Ruth Thompson found customers at times "rude" and offended at having to pay a fee. Tourists were wary because the building, nicknamed "The Observatory" for its squat tower, had once been a tourist trap. Controversy still attends the tower, with some claiming that visitors were charged to barely glimpse the crater's rim from that vantage point, while Doris Banks says her father rarely allowed anyone up the stairs.

Regardless, the visitors came at such a pace that Harvey wondered if the museum could become the major research center he yearned to operate.

In between ticket-selling and dreaming, Harvey and George tended the sixteen-year-old rental. Plumbing failed, rain sent streaks of ochre down the interior white walls, wind sent dust everywhere. Eventually, the Niningers used meteorites to brace roofs and doors against storm, just as many farmers had done in Kansas. Scorpions and snakes found their way inside.

The Niningers ate and read by lantern light in a back room divided by bookcases, the kitchen on one side, the bedroom on the other. Addie did wash by hand, put up new drapes, and if the couple was hungry for news or other voices, for they had no neighbors, they'd listen to the car radio. On the way back in, they might water some Chinese elms they'd planted. Supplies were a drive away in Winslow, and Doris paid an early visit to the museum, helping with the work. The other children did not make the move to the remote locale.

Mornings could bring the calls of coyotes and cactus wrens as Harvey and his dog Blondie went out for exercise. Some mornings when they caught their breath, Harvey must have felt the desert as he most wanted to feel it, as fountainhead, as the place for "great poets, philosophers, artists and prophets." And he may have thought of the first time they saw Meteor Crater, on the "Henry" or Runabout trip, as sunset colored virga above that epic ground.

He had been to the crater during his Denver years. In 1939, with funds from the American Philosophical Society and Dean Gillespie, Harvey and son Bob had driven around the crater with a rake that bit into the soil. The rake and its six large magnets attached to the rear end of a Studebaker netted 43 pounds of tiny nickel-iron bits—and the complaints of ranchers who didn't like the ground getting torn up. Harvey stopped. He still wanted to map the distribution of specimens—even very tiny ones—all around the crater and, of course, he was obtaining the specimens too.

Now, living along Route 66 and near the dirt road that led to the crater in the distance, Harvey knew the "risks were large. . . . We had not heard of a museum of meteorites anywhere, much less one located on a lonely highway." Nor had his friends, who again were less than enthusiastic. One warned Harvey, "Now, don't you let those people down there hurt you."

SURROUNDING METEOR CRATER—STILL OWNED by the Barringer family—was the Bar T Bar. In 1941, Burton Tremaine, owner of the ranch, bought grazing and tourist rights to the Meteor Crater land itself, while the Barringers of course retained their mining rights.

The war had interfered with tourist traffic, but Harvey Nininger had always thought ahead, even before moving to Arizona. Back in 1940, he'd written to Brandon Barringer, Daniel Moreau's son, to say he'd just visited "the grand old crater" and spoken with the Tremaines, telling them that the Barringers might be interested in a lease that would allow the ranch to put up a tourist facility. Harvey said he was offering this idea because he "would like to see that crater more appreciated and your father's work better recognized." Harvey also said that he'd "slip out of the picture." He did anything but.

A few months later he wrote to Brandon again with the off-the-wall suggestion that the Barringers finance a museum building for $50,000 in which Harvey could display his meteorites. He could lead tours at the crater rim. Brandon politely declined. So Harvey switched his focus, telling Brandon of his talks with the Tremaines to search for meteorites on some kind of royalty basis; Brandon responded with a friendly letter giving Harvey the go-ahead to work out such an arrangement.

But Harvey continued to try to insinuate himself into the tourist operations at the Meteor Crater rim, which were then contained in a meager stone building. (Harvey spoke later of broken windows and "a very unsociable and blunt young chap with his aged mother" who sold tickets to see the crater.) In one letter he even said he had a lead on another location where he could place his collection, which might set up competition, not cooperation, something he was not keen to see. He must have been thinking of the Route 66 building, where he ended up. At times, Harvey's pleas to the Barringers and Tremaines were utterly tone-deaf: "I want to make it emphatic that we cannot await your convenience, much as we would like to. Our situation is an emergency." That was but one of many such entreaties, arising out of financial urgency and poor judgment.

Not that the Barringers had ever wished him ill. On the contrary, less than a month after Harvey opened the American Meteorite Museum, Brandon wrote in a November 6, 1946, letter, "I am delighted to hear that your venture has gotten off to such a good start and only sorry that arrangements have not yet been made to let you sell tickets to the Crater or to put a caretaker physically on the property." Such letters from Brandon surely encouraged Harvey in his attempts to control the crater's tourism operation. For the time being, though, anyone who wanted to see the crater had to drive by the American Meteorite Museum first, a fact that quickly led to tension.

In early 1947, Harvey complained that the blunt young chap at the rim—Theodore Johnson, who worked for the Tremaines' ranch manager, Boss Chilson—wouldn't tell tourists at the rim to visit the American Meteorite Museum. On the other hand, Harvey said he always promoted the crater to his tourists and clarified that admission to the museum didn't cover admission to the crater itself. He also had heard from a scientist who had gone to the crater "without my knowledge" that Johnson was slandering Harvey as "unfair, crooked [and] uncooperative [and] purely commercial." Harvey was outraged, calling these accusations "absolutely false."

Chilson, not surprisingly, sided with Johnson. He also told Harvey to stop collecting Canyon Diablos because he had brought no meteorites to Chilson for them to apportion between each other nor had Harvey offered any details about the locations of his finds.

Despite a more conciliatory tone in response, Harvey gave little ground, saying that his staff had "defended your right to charge in hundreds, if not thousands of instances." As to the meteorite-hunting, Harvey said he'd found only one meteorite "larger than an ounce" on the ranch in years, though he did now have a six-pounder for which he needed to pay Chilson. Most of his collecting had been on state property, he claimed.

A "shocked" Brandon Barringer came to Harvey Nininger's defense, saying that "it is obvious to me that [Johnson] has been fed some of this stuff by your professional rivals. . . . I certainly hope that you will give him . . . a chance to mend his ways. The Tremaine enterprise and yours are . . . naturally complementary rather than competitive." Brandon even took the step of writing Johnson himself, calling Harvey Nininger "the leading exponent of the science of meteoritics" after Brandon's father's death and "a personal friend of long standing. . . . I know Nininger has enemies among the professionals, just as my father had, but I have never known him to misstate a fact."

As to the collecting, Harvey had written the Barringers before, wanting to make sure that the "latest shipment of meteorites [had] reached" them and noting that his records of where specimens were found seemed to threaten those who, until his arrival, had stolen meteorites with impunity. Indeed, a series of 1947 letters between Harvey and the Barringers spelled out the terms for meteorite-collecting on the Barringer land, where the Tremaines grazed their cattle. Harvey accepted a renewable six-month agreement, which specified portions of finds the Barringers wished for themselves and which required a map of his finds.

With these various flaps settled, the Niningers might have felt a growing sense of routine. Visitors paid their admission fee and put their hands on astral metal. They listened to the former professor lecture about how, though scientists knew meteorites came from space, no one had yet calculated the orbit of a single one. His pointer touched fusion crust. His lenses magnified chondrules. The cash register bell rang. Someone from, say, Big Stone City, South Dakota, put a five-cent postcard in a purse or vest pocket. Teenaged girls giggled and cooed when they held pendants made from the sliced and polished pith of Canyon Diablos. Their fingers touched the intersecting lines of minerals. Perhaps they quieted to think of such force, but probably they kept whispering, pointing to the sullen boys in plaid standing for a photograph.

After the business day was over, Addie unrolled the register's tape and held a pencil, beginning to add. In shirtsleeves, Harvey sat lit by candlelight, with a copy of *The Autocrat at the Breakfast Table* on his lap. Blondie settled at his feet. The planet tilted, from summer to autumn to winter. Looped like crimped Christmas ribbon, the sales tape draped down. Addie watched her step on the wavy linoleum of the kitchen and maybe thought of movies she saw in Denver years ago, a random flash of longing for some place other than this high and empty desert. She taught employee Ruth Thompson how to knit; they made pot holders. She went to wash her hair, talking to Blondie as if the dog were one of her children, and to their children they wrote, asking for *The New Republic* and books and hose and underwear for gifts. Sometimes Harvey worked in the front room, writing pamphlets or display labels. Don and Ruth walked to their trailer at day's end, and Harvey and Addie watched another sunset sharded and fat with pink, with orange. Now Harvey wore a wool coat and watched stars from beside the elms. Maybe he imagined he could see the dark things there, the ones that move through rivers of attraction, downstream, toward Earth.

SPEWING FIRE, a huge red meteor geysered through the sky and blew up. Barber and Normandy veteran Tony Sebaugh fell to the floor, then rose to continue snipping his customer's hair. Horses ran pell-mell into ditches and died in a panic. Harold and Glenda Hahn were remodeling their house when they heard the boom. "Oh my goodness," Glenda said, "a propane tank exploded." Six-year-old son Richard shimmied down a tree, yelling, "Mom, Mom, the sky is on fire." Dale Leidig, then a thirteen-year-old, stood with other children beside a school bus stuck on a wet back road. They heard a blast that "scared the *de-vil* out of us" and looked up to see a "*tre-men-dous* cloud of smoke and dust." A B-29 Superfortress crew witnessed the long-tailed, cherry-red fireball, then watched a cloud form after the bolide's burst, a cloud that grew to a half-mile wide and lingered for an hour. Somewhere near the northwest Kansas town of Norton, on Wednesday, February 18, 1948, at 4:56 P.M., in a clear, unseasonably warm sky, the universe had redistributed some mass.

Before exploding near the Kansas–Nebraska line, the streaking light had been seen as far away as Texas and New Mexico. Had it been a V-2? A nuclear weapon? Whatever it was, the explosion had been heard more than 50 miles away. Twenty miles from Norton, buildings shook so hard that dust puffed from walls and roofs, as though houses, offices, and barns were letting go of a long-held breath. Stone walls trembled. Windows clattered like busy telegraphs.

M. R. Krehbiel, editor of the *Weekly Norton County News,* said it had "sounded like the gas station on the next block had exploded. . . . After that first blast there was a terrific roar for about 10 seconds like a blast furnace getting ready to blow its top. . . . That roar was the most terrifying part of the whole thing." Krehbiel saw "a long streak of blue-grey smoke . . . [that] curled into jelly rolls, then went straight again, and then jelly-rolled once more . . . like the vapor trail of a plane out of control."

The *Norton Daily Telegram,* in its February 19, 1948, front-page story "Sky Blast Here Was a Meteorite?," correctly reported that smoke trails came from fragments dropping off the main mass of the bolide. Curiously, one man who would have wanted such fragments remained far from the scene. The newspaper said that Lincoln LaPaz, a mathematician, astronomer, and director of the recently established Institute of Meteoritics at the University of New Mexico, "will come to Norton for a personal investigation as soon as it can be reported that fragments have been found." The newspaper also said that Oscar Monnig, a Texas meteorite hunter and friend of the Niningers, believed the meteorites "may be scattered over [an] area or line 5 to 10 miles long about [the] same distance beyond east of where [the] dust cloud was exactly overhead." Monnig was apparently the first to suggest in print where the strewnfield might be.

After being alerted to the fireball reports by the *Denver Post,* Harvey and Addie Nininger headed to Kansas on February 19, interviewing witnesses along the way. They also asked one of the local Norton County newspaper editors if other researchers, such as LaPaz, had arrived in Norton. The word was no.

On February 20, the *Norton Daily Telegram* continued its coverage with a note from Dr. LaPaz asking witnesses to write to him with "compass direction of meteor-burst and estimate of time between flash and sound of meteor explosion." The newspaper reported that one person saw "objects falling . . . looking to be about the size of snowbirds."

Harvey arrived on Saturday, February 21, and probably read that day's paper, with its postexplosion vows of sobriety and mention of the "real money" the meteorites could be worth. On Monday morning Harvey—the newspaper called him "Dr. Nininger"—told a lecture audience that, based on his interviews, they should search, in the paper's words, "along a line starting at [nearby] Norcatur and running northeasterly to a point on highway US-283 about eight miles north of Norton." Such an estimate placed the upper end of the possible strewnfield below the Kansas–Nebraska border, an initial reckoning that would prove, in one crucial instance, to be off by about three miles. The paper also noted Harvey's emphasis of both the meteorite's importance to science and the "ready market" for specimens, even though the largest specimen he then thought possible would be, perhaps, as large as a trash can. Later, he'd say that he usually "concentrate[d] on the search for small pieces . . . to let these . . . guide the later search for large fragments." It was not always a good strategy.

The weather soon turned from false spring to late winter. A snowstorm forced Harvey and a local pilot back to the airport after an attempt to survey for craters, and the Niningers left for Arizona at noon on Tuesday, February 24. Perhaps they heard Red Skelton's joke on his radio show that the meteor in Kansas had caused an earthquake in California.

Harvey's "preliminary investigation" suggested that the bolide exploded nearly five miles "east-northeast of Norcatur at an altitude of about 15 miles." Harvey acknowledged that the "momentum would carry fragments for a considerable distance" and gave a high school principal a map of the strewnfield. This map no longer seems to exist.

Meanwhile, Lincoln LaPaz continued to receive reports and letters and calculated his version of the strewnfield. On March 3, he had made his first map. (In a 1961 book, LaPaz calls it "a first determination of the probable area of the fall," a claim that seems inaccurate.) Decades later, LaPaz would say that he mapped the strewnfield on February 20, though I found no evidence to substantiate this assertion. In any case—and this would be important—the LaPaz estimate extended the strewnfield farther north and east than Harvey's description.

Why did LaPaz stay away? He believed it was more efficient to let reports arrive by mail than to ground-truth accounts right after a fireball, which could be a waste of time if no meteorites were located. It also couldn't have helped that LaPaz was teaching classes at the time. A month after Harvey had done his preliminary work, a team from the University of New Mexico finally arrived in the Norton area and talked to residents, but they found no meteorites.

Harvey knew none of this, and March continued oddly silent back home at the museum in Arizona. No calls from Kansas or Nebraska, no letters or samples.

In Norton, those eating dinner at The Steak House come mid-April may have brought along the *Saturday Evening Post* with its profile of Harvey Nininger. But frankly, after the meteor, the big news was the grand opening of the Silvaire Roller Rink on Highway 36.

AT LEAST THERE WERE GOOD TIDINGS closer to home. Harvey could hunt meteorites at Meteor Crater without interference from anyone. Brandon Barringer told Harvey, however, to no longer send Canyon Diablo finds to him. Instead, he was to give the Barringers' due portion to the testy Theodore Johnson.

Once again, the situation worsened. According to Harvey, Johnson shadowed him, finally telling him that he and his employees couldn't search for meteorites at all. On April 27, 1948, Harvey wrote Brandon with this news—in a letter marked "STRICTLY PERSONAL AND CONFIDENTIAL"—and said that "we have never been treated so shabbily as we are now being treated by the ranch and by the people on the rim." Harvey claimed he'd been careful "to avoid any slight over-stepping of my rights ever since we moved here. Yet, I've been accused of everything, even theft. It is evident that they are trying to drive me out of here, simply because *my coming has made it more difficult for certain people to help themselves to meteorites without reporting them*" (emphasis his). Harvey also accused Brandon of apparently being "completely under the domination of men who insist on being my enemies," men who saw no value in studying meteorites or their distribution in the area.

Brandon responded in measured tones. "If we treated all your letter as strictly personal and confidential . . . how could we have told them to stop ordering your men off the property without revealing that you had told us it had happened?" Brandon also brushed off these insinuations that either Chilson or

Johnson would steal meteorites, chalked the whole situation up to different personalities, and told Harvey he could work "anywhere you please."

This was not the first time that Harvey and Brandon had discussed the poaching of meteorites on Barringer property. They had corresponded about the matter in January 1948, when Brandon revealed that he had learned, in his words, that the Chilsons and Johnsons had "a business . . . selling meteorites which have come off both the Tremaine property and our property," adding, though, that "they had made a regular accounting."

Concerned that rancher Burton Tremaine was trying to drive him out of business, Harvey would hint, in the months ahead, that someone was "responsible for the misinformation which started the rift between us and the Tremaines." Harvey probably was talking about Lincoln LaPaz, who, over the years, had become his most serious rival.

Harvey clarified that he didn't think that Boss Chilson had been stealing meteorites, but that "some of the people whom he has favored much more than he has us" were taking irons. He didn't name names. In addition, he couldn't understand Chilson's negative reaction to the American Meteorite Museum workers "studying the distribution of oxides and metallic fragments of minute size" on ranch lands, though they stopped doing so after Chilson objected. He wrote Brandon, "My staff have been told so often by the Johnsons and by others that you had surrendered all of your rights to the meteorites. . . . These people even [said] that if you had given us a letter you must have forgotten your agreement with [the] Tremaines. . . . I still wonder whether there might be a misunderstanding . . . between you and the Tremaines."

Brandon's response was crisply supportive: "The Tremaines have no rights in connection with the meteorites and you have all the authority you need to carry on your studies."

But perhaps Harvey's desire to see his collection at the crater itself—one of his most important dreams—kept him from telling Brandon of a conversation he'd been having with his friend Harlow Shapley, the well-known Harvard astronomer. After receiving a request for help from the National Parks Association, which wanted to see Meteor Crater turned into a public park, Harvey wrote to Shapley, saying he had the "utmost respect" for the Barringers' position, which was to keep Meteor Crater in their family. Harvey nonetheless agreed with the goal of making the property public. "Some one should lead a movement" to make it so, he wrote, clearly hoping that Shapley would take up the cause.

Three months earlier, Harvey had told Brandon that an official with the State of Arizona had questioned him about the possibility of acquiring the land. Brandon replied that he couldn't criticize Harvey for speaking with the official, even adding that the family would "be interested to hear from" the state. Harvey must have taken this as a softening in the Barringer position, but it appears in retrospect that Brandon's expression of openness was mere politeness.

In June 1948, Harvey dropped a bombshell. "Now here is something that I did, to which I trust you will not take an unfriendly attitude," he told Brandon. "Under pressure from many imprtant [*sic*] people who have been urging it during the life of the museum, and particularly stron [*sic*] pressure was brought just as I was leaving for the meeting and after my arrival there. So, I submitted a resolution, a copy of which is enclosed." Presented at a scientific conference, the resolution stated that Meteor Crater should be obtained by the government and become a public park. This was a stunning act, given the Barringer family's long-held insistence against government ownership and Harvey's own position relative to the Barringers' good graces.

"I made clear . . . that the property would have to be acquired by purchase, if at all," Harvey hastened to add. Tellingly, he concealed his communications with Shapley. "But please believe me, that there has apparently developed a rather firm determination on the part of several very influential people to see this natural feature nationalized." Almost pleading, Harvey argued that he wanted to protect the crater and the good name of Brandon's father—as well as stop interference with his surveys. He closed by saying, "If you want the most important reason of all, you will have to wait until we can get together." They never did. The reason may have been Harvey's distress over silica mining on the property, which he thought threatened future research.

When Harvey read the salutation "Dear Sir" on Brandon's July 6, 1948, response, he may have experienced the same nausea he'd felt at seeing the wrecked moving truck. Brandon was "shocked beyond measure that you should have introduced the 'resolution urging the creating of a public park at the Great Arizona Meteorite Crater' . . . and that you stated that you believed we 'would be willing to consider a fair proposition.'" He called his move "completely unprovoked" and "hostile," telling Harvey that when the collecting agreement ended in October, it was over for good. Harvey Nininger would be banned from Meteor Crater. Period. All that remained was for him to return the loaned display meteorites back to Theodore Johnson, hand over the Barringers' share of recent finds, and give them a "full report" of his research, which was to include a map of his find locations. An angry and crestfallen Brandon Barringer closed his letter, "I am sorry our relationship had to end this way."

Amazingly, Harvey could not see his position as "in any sense a betrayal," blaming "the conniving and plotting that have been going on at the crater headquarters" as reason enough to support the resolution. Rashly, he told Brandon that "your reaction . . . does seem to place you in full sympathy with these doings."

Brandon Barringer privately called Harvey's act "treachery," even as he said that he and his family had for years "considered ourselves good friends of Nininger and were, and in fact still are, among the greatest admirers of his achievements." To Lincoln LaPaz, Brandon wrote, "The publicity has completely gone to his head and . . . he literally thinks he owns or should own every meteorite that has ever fallen as well as all that may fall."

Letters indicate a careful if prickly accounting between the American Meteorite Museum and the Barringers involving the final delivery of specimens owed to the family. (A letter from another family member in 1977, however, questioned whether Harvey had in fact supplied all the specimens he should have.) And for the rest of his life, Harvey Nininger refused repeated requests for a map showing the distribution of Canyon Diablo specimens he had discovered. He argued that without find information from other parties, such a map would be pointless scientifically. The Barringers would always respond that the agreement was for a map of Harvey's finds only. This refusal to supply such a map has been a source of rumors that continue to this day, rumors that Harvey took more than his agreed-to share of specimens and that he wandered off state lands (where he had permits for a time) in order to poach specimens from Meteor Crater property.

At least he and Theodore Johnson had another chance to butt heads. They argued over when and how Johnson could come by to pick up the loaned Canyon Diablo specimens at Harvey's museum. After Johnson loaded his truck with the irons, Harvey would have seen dust rising as the young man drove back to the crater, the crater that Harvey Nininger loved and from which he'd been banished.

MONTHS EARLIER, on a late April evening, the Niningers had heard Lincoln LaPaz announce on the radio that his team had found hundreds of fragments from the February 18 Norton bolide.

The first such fragment, found in a Kansas clover field on April 6 by farmer G. W. Tansill, came to the LaPaz team as soon as they arrived in Beaver City, Nebraska, on April 27. He had waited three weeks before revealing his find because he was afraid of being ridiculed for producing a meteorwrong—something that looked like a meteorite but wasn't. The meteorite had sparkled, he said, "like a million diamonds."

In his several unpublished manuscripts on Norton, Harvey Nininger said nothing about any worries on their long drive to the prairie that April. Instead, he boasts that he and Addie were "confident that the populace we had alerted and instructed would rally to our support." He could not have been more wrong. According to Harvey, one of the local newspapermen, the son of a friend (this was probably M. R. Krehbiel), warned that the New Mexico team had turned the community against him. Alleging some military value of research on meteorites, LaPaz and his workers had told the residents, at least according to Harvey, that Nininger was interested in making money—and jewelry—from the meteorite, nothing more.

Anyone familiar with meteorites even then would have known that no one, including Harvey Nininger, would make a necklace or ring out of the Norton County stone. The jewelry sold at his museum was made from commonly available irons, such as Canyon Diablos. Norton County was not an iron.

Fragile, friable, and quite rare, Norton County is in a class of achondrites called aubrites. Achondrites are those igneous or volcanic-like meteorites that have been melted through and thus changed from their original chondritic state. And today we know that aubrites are "unlike any other group of volcanic rocks *anywhere*," says meteoriticist Ralph Harvey, of Case Western Reserve University.

He explains that aubrites, having been melted, are still chemically related to enstatite chondrites, which have *never* been melted. The oxygen isotopes in both aubrites and enstatite chondrites match exactly—which is baffling.

"Somehow we've got both the original dust of the solar nebula *and* the highly processed igneous rocks that were made from them by a planet; but this seems quite startling to a geologist because planetary processes like igneous activity should be the result of a whole planet process . . . and not leave any preserved, unprocessed material behind. . . . We've got nothing like this for Earth or any other planet," Ralph Harvey says.

Another oddity about aubrites is that if they represent a sample of an asteroid's "settled crystals" (the mantle), then we should have a lot more specimens of the igneous material (the basalt) that occurs *above* the mantle. "Yet the sum total of aubrite parent body basalts is zero," Ralph Harvey states. Why and how aubrites have been blown out of an asteroid mantle without related surface specimens is puzzling.

There's yet another mystery. The aubrites and enstatite chondrites also share the same type and ratio of oxygen isotopes with the Earth and its moon. This is curious, because the chondritic parent body would have been far smaller than the Earth and would have formed farther from the proto-sun, or at least have been pushed farther out at some point. So the oxygen link among aubrites, enstatite chondrites, and the Earth-Moon system is usually explained away as a fluke.

"The alternative—that the Earth/Moon, aubrites and [enstatite] chondrites are all siblings—is too wild of a theory," Ralph Harvey explains, who compares aubrites to the Rosetta Stone. "These rocks somehow 'explain' the distinction between bodies that behave like planets and bodies that remain primitive remnants of the earliest solar system. And they may speak not only about the aubrite parent body, they may say something about the Earth's erased early history, but if we're trying to read that book . . . we've got just a few pages out [of] . . . hundreds."

IN 1948, LINCOLN LAPAZ was trying to find a few of those pages. So too Harvey Nininger, once he arrived in Norton that spring. But when he did, no one would listen. University of New Mexico students working for Lincoln

LaPaz had been in the area on March 25 and 26 and they found no specimens, according to LaPaz. But they may have begun to tell people that Harvey was just a businessman with little or no interest in real research. According to Harvey, people literally turned their backs on him when he returned in April. And "one mother was vehement," he wrote in an unpublished account of the episode. "That man said he represented the American armed forces and that you were merely a peddler, and that if I were to let you have even a tiny scrap of that meteorite it might cost some poor American soldier his life."

"That man" had to have been LaPaz, who had worked for the military and who, in a newspaper article published in Norton, emphasized the scientific importance of the meteorite specimens. The front page of the May 4, 1948, *Norton Daily Telegram* carried the headline "Norton Meteorite of Military Value." In the press, LaPaz said that study of the meteorite could have an influence on rocketry, which seems odd, given that aubrites are mostly stone. This supposed interest of the military seems to have been a ruse by LaPaz to outfox his competitor.

LaPaz also said that researchers could test the meteorite for exposure to cosmic radiation—a way of knowing how long it had been in space. There was a clear and inaccurate implication that Harvey didn't share this interest.

Indeed, researchers would learn that the Norton aubrite had been in space for 230 million years—a far greater span than, say, chondrites, which have an upper cosmic-ray exposure age of 60 million years. This suggests that the parent body of aubrites—that mysterious Rosetta Stone—is tucked away in a part of the asteroid belt that doesn't regularly interact with Jupiter's gravity.

"Some collectors of meteorites purchase them for resale to museums or to firms and individuals who destroy them for manufacturing jewelry and novelties and the specimen becomes lost forever to science," the newspaper said. This was gross oversimplification and a clear reference to Harvey. (LaPaz's criticism of meteorite jewelry did not prevent him from making a necklace out of a Canyon Diablo to give to his mother's housekeeper.)

The only stone that Harvey Nininger would find in April 1948 was a metaphoric one: an ostrakon.

After leaving Norton, he and Addie drove north to Furnas County, Nebraska, a place not included on his primary strewnfield map published later that year (but which was included as a suspected locality) and a place, it seems, where no negative gossip had arrived, at least according to Glenda Hahn. Glenda had seen the fireball with her husband and son. Harvey struck up a conversation with Glenda's husband Harold, who farmed on land rented from a Helen Whitney, an absentee landlord living in California. Harvey could have cajoled Hahn into walking at least part of the field at the time in order to search for stones, but he didn't. "The wheat in those fields was rank and thick," Harvey recalled in an unpublished manuscript, "and I felt that my walking those fields would be resented and rightly so." Harvey told Harold "to be on the lookout when he harvested."

Throughout this second trip to Kansas, Harvey suspected that LaPaz had somehow taken his information to help the New Mexico team locate specimens. Harvey apparently had left that strewnfield map at the high school in February. And he'd made his pronouncements in the press and at his public lecture, but I could find no evidence that LaPaz relied on Nininger's materials.

Strangely, in fact, Harvey's first published piece about Norton in a science magazine appeared in the October 1948 issue of *Sky & Telescope*. It is unclear why Harvey did not immediately write a report on the Norton fall for the national scientific community in late February or early March. Doing so would have been equivalent to making a claim on the territory for future searches.

The University of New Mexico team that had arrived on April 27 stayed in the area through early May, and the highlight of this—their second foray into the strewnfield after the quick, fruitless search in March—came from fourteen-year-old Ralph DeWester. On May 1 the teen found a 130-pound specimen, much to the delight of Lincoln LaPaz. (The McKinley stone purchased by LaPaz was one he acquired for his own collection. The University of New Mexico would buy LaPaz's personal collection years later.) The find occurred on the farm of the McKinley family, though not everyone there was excited. "Little Jimmy McKinley considers the fragments 'only a piece of rock.' He was playing with a chunk Sunday half the size of a man's fist, saw a strange dog, and flung it. It struck a cement sidewalk and shattered into small bits," the *Daily Telegram* reported. LaPaz paid $500 for the 130-pound meteorite, according to a handwritten bill of sale, and on May 5, he gave a lecture at the high school.

Harvey Nininger left Furnas County for Arizona hopeful he might hear from Harold Hahn and deeply upset at how Norton residents had reacted to him. Back home, Harvey wrote to LaPaz alleging that the latter had been the cause of the community's negative reaction. Harvey wanted an apology and the "semblance of a fair settlement regarding the material which you acquired so unethically from the territory in which I had, according to recognized custom, established prior rights." There appears to have been no reply.

So how to sort out the competing interests of Harvey Nininger and Lincoln LaPaz? While Harvey was the first in Norton, the first to lecture there, and possibly the first to map a strewnfield, the crucial question, as LaPaz would see it, was whether the Nininger map included the area where the Tansill and McKinley specimens were found. LaPaz would claim that Harvey altered his map for publication in October 1948 to include areas that he had not originally believed should be searched.

The Norton community had no prior experience with the acrimonious relationship between these two men, but their mutual history and eventual mutual hatred stretched back for years. It began, though, with a cordial invitation for lunch from an ambitious professor to an ambitious scientific outsider.

"IT WOULD BE A PLEASURE," Lincoln LaPaz had written to Harvey Nininger, "to discuss some of our common interests." He wanted to have lunch the day of Harvey's October 1938 lecture at the Ohio State University, where LaPaz taught before moving to New Mexico. LaPaz knew of Harvey's work and was widening his own interests to include meteorites. Harvey might have been wary, if many of his unpublished writings are accurate, though it appears he himself had contacted LaPaz to set up the talk.

According to LaPaz family members, most of Lincoln's papers no longer exist, so it's difficult to assess the accuracy of the myriad claims and counterclaims each made on many issues. One thing is clear, however: in 1944, at the University of New Mexico, LaPaz founded an Institute of Meteoritics, the nation's first. Harvey would suspect that LaPaz had privately shot down Harvey's own meteorite institute idea to clear the way for the center in New Mexico.

One public difficulty between the two men took place within the context of the fledgling Society for Research on Meteorites, which Harvey and UCLA professor Frederick Leonard had helped to found. (The SRM later became the thriving Meteoritical Society, a professional scientific organization.) LaPaz and Harvey began to argue in the early 1940s over something that LaPaz called "contraterrene meteorites." LaPaz presciently recognized the need to explain several apparent impact craters that seemed to lack meteorites near them. LaPaz invoked what were, in essence, antimatter meteorites that destroy themselves completely when they encounter the terrestrial environment. Harvey scoffed, saying this was no better than resorting to the occult.

LaPaz took aim at Harvey for including bits of rusted Canyon Diablos in booklets he sold to the public. Harvey called the material a meteorite, to which Leonard and LaPaz both objected. The question is both semantic and chemical. At what point does a meteorite on the Earth's surface cease being a meteorite and become, through erosion or oxidation, a terrestrial stone? Brandon Barringer came to Harvey's defense, noting that "an egg, fresh or rotten, is still an egg." LaPaz pushed the SRM to censure Harvey, who subsequently changed the booklet's specimen label to "an oxidized fragment of nickel-iron meteorite."

LaPaz was also "scandalized," geologist Ursula Marvin writes, "when Ninninger brought specimens to Society meetings and offered them for sale." Never mind that the University of New Mexico, like many institutions, had purchased meteorites from Harvey.

Despite these and other conflicts, LaPaz offered—and Harvey apparently accepted—a research assistantship in the 1940s with the New Mexico institute. It's not clear if Harvey actually did any work for hire. It seems doubtful, since he was concerned that the sole reason for LaPaz's gesture was to obtain Harvey's collection. Harvey even professed to have been concerned at times for his family's physical safety.

"[LaPaz] has always completely lacked a spirit of cooperation, attempting to appropriate whatever he could of credit in the field of meteorites," Harvey once wrote to his supporter Dean Gillespie, but LaPaz was also "a very bright man and very capable in the field of mathematics and to some extent in astronomy. . . . I do not propose to do anything to hinder. On the other hand, I want to cooperate with this man's efforts, but I definitely do not plan to come under his jurisdiction."

Frederick Leonard did. Harvey's old friend soon spurned Harvey and embraced LaPaz. A former student of Leonard's, O. Richard Norton, remembers Leonard as "meticulous to the point of being pedantic." Certainly Leonard's prose betrays a pomposity both exuberant and formal, not unlike LaPaz's own writing at times. LaPaz could deploy, sometimes with humor and sometimes without irony, a word like "cerebration." Such diction suggests the men shared a sense of grand gesture.

Bill Cassidy, a student of LaPaz's and the founder of U.S. meteorite-collection efforts in Antarctica, calls LaPaz "probably the most gifted teacher" he'd ever had. Of the rivalry between him and Harvey Nininger, Cassidy blames the differences in the two men's situations as the root of their enmity—LaPaz a professor, Harvey working privately.

Today most people in the meteorite community know only that there was bad blood between these two men and are content to leave it at that. When I ask Wolfgang Elston, a former professor at the University of New Mexico, about his take on the feuds, he shares with me his sense that "LaPaz's reputation stood high in those days, even if some of his ideas were decidedly unconventional." Then he adds, carefully, that LaPaz was "known to be unpredictable."

BY MID-JULY 1948, HARVEY Nininger's world was falling apart. On two trips, he'd failed to obtain fragments of the Norton fall and had been stung by the community's reaction to him during his April visit. His exploration rights at Meteor Crater would come to an end that autumn. At least Harvey and Addie finally had electricity at their lonely stone building on Route 66.

The couple probably found some solace from a June 3, 1948, letter from editor M. R. Krehbiel: "Believe that folks have changed their minds a little around here about LaPaz. Ran the Capper's Farmer story about you (condensation of Sat Eve Post story) and it seemed to have quite an effect. LaPaz has never returned the fragments he agreed to return to museum here and . . . folks are beginning to wonder."

Unbeknownst to Harvey or LaPaz, in July a man driving a combine for Harold Hahn nearly crashed into a large hole—in the very wheat field Harvey Nininger had told Hahn to comb for meteorites. Thinking it an old coyote den, Hahn told his worker to keep threshing. Then Harold gave the matter some more thought and called Harvey. "My name is Hahn," he said. He told Harvey he had "uncovered a stone 39 inches across at the bottom of a six-foot hole."

On August 16 or 17, 1948, Harvey and Addie drove to Norton—after Theodore Johnson had taken away the last loaned Canyon Diablo—and the drive must have felt a bit like the trip to Mexico back in 1929. A long-shot chance.

But—and this seems incomprehensible—Harvey insisted on interviewing more eyewitnesses along the way even though he *knew* a huge specimen lurked in Hahn's field. There was more going on back at the Hahn property than Harvey suspected at the time. In an unpublished manuscript devoted to Harvey's conflicts with Lincoln LaPaz, Harvey asserted that Hahn had not called LaPaz because of how Harvey had been treated, but the thresher operator who first found the hole, a man named O. E. Gill, had called the New Mexico professor. Years later one newspaper article said that LaPaz had "recruited" Gill in case of any finds.

The Niningers arrived at 10:30 P.M. Tuesday, August 17, to stay with the Hahns.

Lincoln LaPaz would question whether Harold Hahn discovered the meteorite and whether Harvey had spoken to him about looking for possible specimens. "Hahn's story to me was entirely different from Nininger's fabrication," LaPaz wrote, "indicating that Hahn himself had had no earlier contact with Harvey, although one of his neighbors . . . had."

The fact that Glenda Hahn remembers the Niningers staying with them strongly suggests prior contact, though she could not recall her husband mentioning any conversation with Harvey from the spring. The closest house to where the specimen had fallen was owned by a Horace Collins Sr., who had rented the house to another man, probably the neighbor LaPaz mentions.

The *Norton County Champion* on its front page for Thursday, August 19, 1948, carried this headline: "Huge Meteor Fragment Found by Harold Fahn [*sic*]." The story reminded readers of Harvey's "original survey" in February, adding that Hahn wanted the meteorite saved for science. Meanwhile, the *Daily Telegram* had reported that Lincoln LaPaz had calculated the possible fall site and noted that LaPaz arrived at 11:30 A.M. on Wednesday, August 18.

The *Omaha Morning World-Herald* on August 20, 1948, stated that the meteorite measured 9 feet in circumference in a hole about 4 feet deep and showed a photo of Harvey with a magnifying lens studying fragments while Hahn and others looked on. Harvey, the paper explained, "said that no effort would be made to get the piece out of the ground until word had been received from the [absentee] owner of the farm." The story reported that Hahn had called Harvey. LaPaz wasn't mentioned.

The hole was ringed by a mound of dirt and surrounded by wheat stubble on a side hill that had been terraced and from which one could see, to the west, a line of trees along Sappa Creek. Close by stood a small earthen dam with its pond.

After the one-ton stone was discovered, Harold called Helen Whitney, the landowner, who told the farmer to call her lawyer, Frank Butler. Then Harold called Harvey, who, Glenda says, got to the small crater before Lincoln LaPaz. "We really liked him," she said of Harvey. "He explained and told everything to you. . . . He was a very private man. . . . He was seriously minded about that meteorite." Glenda didn't go with the men to the meteorite hole the morning after Harvey's arrival but vividly recalls Harvey being "so anxious" to get there that he skipped breakfast. Addie, meanwhile, took notes and photographs.

Harvey and Harold squeezed down between the cavity and a meteorite that was "almost white" with a "straw-color" fusion crust. In his white shirt and dark trousers, Harvey squinted at a specimen. Harold watched, dressed in overalls and a straw hat.

"I don't know who is going to own the meteorite," Harvey recalled saying to Hahn, "but whoever does will appreciate the preservation of those small fragments. Let's you and I get down there and wrap individual fragments in Kleenex."

An hour later Lincoln LaPaz and Frederick Leonard arrived and soon so did a team from the University of Nebraska allied with LaPaz. According to Harvey, LaPaz began "giving orders for us to get out of that hole, proclaiming the meteorite was not our property. This I well knew." Harvey then explained why he felt wrapping the small specimens was important. "After this I could not understand what was said but . . . some arguing was being indulged in." Harold's father told his son that the two men should get out of the meteorite hole or the police were going to be called.

Climbing out and then standing by a farm truck, Harvey watched—he claimed—LaPaz unwrap specimens, pretend inspection, huff, then toss each aside.

Harvey said that LaPaz "stormed about emphasizing his authority as a representative of the armed forces, telling the small group of farmers who had gathered that Nininger was a peddler . . . who would commercialize this great opportunity . . . and that he [LaPaz] represented pure science." Harvey asserted that LaPaz had a photographer take a picture of him "as he proceeded to descend upon that fragile specimen as a prize-hungry cowboy would descend upon a roped steer!" LaPaz crushed many fragments underfoot, Harvey alleged.

The fate of these small specimens is not quite what Harvey Nininger implied, however. "I wouldn't trust what Nininger says as much as what I've seen with my own eyes," Smithsonian meteorite curator Tim McCoy says. "Namely the presence of thousands. . . . of individual fragments . . . that were collected at the find site and are part of the collection of the University of New Mexico. . . . Included is one fragment still in the dirt in which it impacted. . . . So, the truth is that both [men] collected the small fragments and . . . almost all of the research has been done on these smaller pieces rather than samples removed later from the main mass."

Glenda Hahn doesn't recall Harold talking about LaPaz tossing any specimens aside nor rumors in the weeks before about Harvey's reputation, but she remembers Harold saying that LaPaz ordered him and Harvey out of the hole.

Soon enough, lawyer Frank Butler wrote to Harvey and the LaPaz/Nebraska team to "propose to take bids from all parties interested." On the back of Butler's undated missive, Harvey wrote down some notes: "Bidding includes big stone alone or with all fragments? Any strings?" and "Has any one been in hole today? Anything been taken away?"

LaPaz, Leonard, and the Nebraska team met in Butler's office with Harvey, Addie, and H. O. Stockwell, who had joined with the Niningers. There don't seem to be any firsthand recollections in LaPaz's hand, so I've relied on Harvey's memories, which, despite an understandable bias, surely are right in their characterization of "the atmosphere of the little room" as "growing more and more tense as the alternate bids written out by the competitors were announced by the legal assistant." Perhaps Harvey thought of his success years ago in outbidding the Field Museum for the huge Paragould meteorite. He hoped to be so lucky.

The final bid was $3,515, and it belonged to the University of New Mexico, the University of Nebraska, and Frederick Leonard, who put in $2,415, $850, and $250, respectively. A letter from LaPaz to C. Bertrand Schultz, director of the Nebraska State Museum, claims that "Leonard's intervention probably saved the meteorite for the universities." The final, winning bid against Harvey had been made by his former friend.

A photo shows Lincoln LaPaz smiling under his dapper mustache with his hand on the stone. In another, LaPaz has his left arm outstretched, his fingers touching a pail that could be lifted and lowered from the deep excavation. He holds a long-handled trowel in his right hand, the scoop crimped like the edges of a pie crust. With a sweatband across his forehead, LaPaz has his head turned to the side and slightly up, his expression serious, as if he disliked the interruption. He probably did.

After he won, LaPaz "got a little bossy," Glenda Hahn remembers. When Harold accidentally kicked some dirt into the meteorite hole, LaPaz "threatened to have him kicked off the land." After that, "Nobody dared get close to the hole. Oh Harold was mad." LaPaz, she says, "got nasty with everybody"—even with her young son Richard, from whom LaPaz simply took a bag of small meteorites. "He just wasn't a likeable person after he bought it." LaPaz told Schultz that "Harold Hahn has in his possession 4 pounds or so . . . purchased by our universities. I am perfectly willing that he should retain this material if he is on the side of the universities, as he definitely was when I last talked with him; but not if (by his silence) he gives support to Nininger's fabrication."

The Hahns did keep some small fragments of the stone and donated one to the University of Nebraska. Glenda Hahn keeps the family meteorites in a safe deposit box, but on a summer day when I visit Norton, she retrieves one to show me. It's pale and fractured, with a partial fusion crust that looks like a charcoal smudge. With my hand lens, I can see every bump and edge. It could fit in the palm of my hand but I don't even ask to touch it. She folds her hands in her lap as I take a picture of the meteorite that caused so much heartache, resting there on tissue paper, like a gift, atop a folding table in her modest home near downtown Norton.

After Harvey left town, the excavation continued. It was like, Glenda says, "a county fair. Even tourists came out." Most left disappointed, as LaPaz had limited access to the small crater. The meteorite was trucked to New Mexico after the excavation was complete.

Harvey, then sixty-one, fumed. "As things stand now, absolutely no legal consideration is given to the suveyor [*sic*]. When it came to the sale . . . the land owner who was 1500 miles away was the sole owner, although she would never have known that it landed within a hundred miles if the scientist hadn't dug into his pocket and into his years of experience . . . to make the survey." Harvey called his journey home a "long drive." He and Addie discussed whether to call it quits, whether to sell the collection and be done. But even that would require a catalogue and money up front—money they had spent driving to and from Norton three times, $1,000 lost with nothing to show for it. Harvey admitted that "the man who gambles with her [Nature] doesn't always win." In the end, he couldn't quit the meteorite business, and even Addie had said as much before: "It gets in your blood, I guess, when you get to finding them you can't stop."

Sentiments in the Norton area soon began to shift. The director of the local museum, R. D. Bower—a man whom editor Krehbiel had characterized as once "strictly LaPaz" —wrote a friendly letter to Harvey on August 31, 1948, and said, "We feel that in spite of adverse statements made against you . . . your prestige has risen in this community, and that in spite of the fact that you were unsuccessful in obtaining the large fall . . . it is extremely possible that your chances are definitely improved if any other specimens are obtained." In September, Krehbiel wrote again to Harvey, saying that he'd heard that landowner Helen Whitney was "pretty much stewed up about the deal—she hasn't been paid yet. . . . [Bower's] not too pleased with LaPaz, either, as he had been promised some specimens . . . for his museum and nothing's come forth yet."

The next month Harvey submitted a polite request to the Institute of Meteoritics asking for samples of Norton in order to conduct studies of its surface features. But he refused to fill out a form that asked if the requester had ever charged fees or made jewelry. LaPaz privately made light of Harvey's request. He never responded to Harvey's application.

After the Norton episode, an old friend of Harvey Nininger's—Frank Cross, a former journalist—began a campaign to discredit LaPaz. By Cross's account, he received support from a number of scientists, and Harvey encouraged the efforts, letting loose a couple of epithets for his two enemies— "old Doc Splurge and Jabber." He even journaled about LaPaz on Christmas Day 1948. Over time, Cross's campaign fizzled, which was probably for the better.

Harvey would go on to resign from the Meteoritical Society or be ousted, depending on whom you believe. He'd not rejoin until 1963.

Some acrimony was left behind in Kansas as well. During the one night the mass was unguarded, some portion of it was broken off. LaPaz and Schultz would suspect Harvey Nininger. And despite getting about $300 for his trouble, Harold Hahn was sick of it. A friend told Harvey that "the Hahn family got a great plenty of LaPaz to do them for a long time." Meanwhile, according to the Nebraska State Museum director, who was no ally of Harvey, "The farmers out there certainly do not trust him [Nininger]." LaPaz even gathered statements from people who agreed that LaPaz did not try to bar Harvey from seeking specimens. He didn't get a statement from Hahn.

LaPaz still believed that Harvey had seen his announcement about the Norton fall in *Science* on May 21, 1948—thus giving Harvey location information that could have helped him in a later search. In fact, Harvey had admitted that some of the fall specimens were not in the area he, Nininger, originally suggested for searching.

Amazingly, the Norton controversies flared up in the late 1970s and early 1980s after Arizona State University published the collected papers and articles of Harvey Nininger. The book included the text of a leaflet about Norton that Harvey had privately printed and distributed not long after the whole affair, and when LaPaz saw it reprinted in the anthology, he exploded, saying it was full of "monstrous falsehoods and innuendos." In a long critique, LaPaz raised many issues, but focused especially on the strewnfield maps, noting that Harvey's October 1948 strewnfield map in *Sky & Telescope* was divided into 25 sections, not the typical 36 for townships. (Harvey himself said the different division in the magazine was a necessity for illustration.) LaPaz suggested that Harvey Nininger changed the township divisions in order to include the 100-pound McKinley mass within Harvey's strewnfield map. He brought up his suspicion that Harvey had used LaPaz's notice in *Science* to serve his interests. And there was much more, including a denial that LaPaz had told the local newspapers he'd wait till fragments were discovered before coming to Norton. The critique is tortured and pedantic, carrying more than a whiff of paranoia. It has moments of odd phrasing, as when the sound of the fireball was called "thoroughly demoralizing."

Decades didn't soften Harvey's feelings either. Even into his nineties, a question about Norton could provoke anger. In a taped discussion about Norton, Harvey recalled his dismay when he read a reference work listing LaPaz as the discover of the Whitney/Hahn stone. "It's the only discovery of a meteorite ever made with a couple of men sitting on or at the edge of a meteorite," he said. "I—I don't know how you happen to have the nerve to write that."

While Harvey had lost the meteorite in the summer of 1948, LaPaz faced loss of another kind; he had been in Wichita in early August for his mother's funeral. That summer also turned tragic for Frederick Leonard, when his wife contracted polio.

TODAY THE AREA WHERE the Norton main mass was found is terraced fields and cedar shelterbelts, cow dung and old junk, snow-on-the-mountain and what one resident calls "five-fingered funny weed!" —wild marijuana left over from the days of hemp production during the Second World War. That resident is Horace Collins, Jr., the son of the farmer who lived near the excavation and who now owns the old Whitney property. On the same trip when I visit with Glenda Hahn, after driving in from Utah, I walk with Horace, who points out the little, now-dry pond near where the meteorite was found.

"Here, now here, the end of this terrace, this is about as close as you could get," he says. We're near the dirt dam, a few paces east of an old livestock tank. Or—maybe it was more in the middle between the currants and the terraces. Horace mulls.

He does recall LaPaz's daughter taking notes and wanting to look at the rental house on the hill to see if there were meteorite holes in the roof. He remembers how the weather during the excavation was "awfully hot," how his father brought cold beer to LaPaz and Leonard, and how the Niningers brought watermelon to "ingratiate themselves . . . people didn't much take to it."

The next day I drive back out after getting more directions in an e-mail from

George Corner, a curator at the Nebraska State Museum. He had extrapolated from a tiny map that LaPaz had published to a larger topographic map, guessing that the site was about 300 feet due northwest of that dam.

Cottonwoods rustle, kingbirds swoop, grasshoppers hop, and I sit on a log by the dam feeling sweet with time, a bit as I did on Meteorite Island beside Peary's road. I look again at the cedar shelterbelts and terraces to the east and the hedge of currants to the west. But if Corner's right, then the site is just beyond or close to a sturdy barbed-wire fence that separates this pasture from a different field. Was the hole on this side of the fence or that side? Cicadas buzz. At the fence, I find a sandy patch about 4 feet wide, nothing there but an anthill surrounded by spent yarrow. "This could be it," I think. But nearby there's also a much larger circular feature rimmed with grass. Old meteorite site? Old buffalo wallow? Who can say? Fixity wavers in the wind.

On other back roads I arrive at the former Hahn household, where Harvey and Addie once stayed. The light is late-in-the-day honey glow, the house trashed out. Flaked red trim and green siding—Christmas colors. A view through a window reveals beer cans, mattresses, ratty blankets. At the far end of one room, a lone chair in dusty light. What Harvey must have said to Addie in those rooms. . . . And, here, beneath a big spreading maple, is the step where Lincoln LaPaz sat years ago, sorting specimens.

I sit there myself, far too late to wish them peace or even something like it.

A YEAR AFTER NORTON, Harvey suffered a nervous breakdown. But in the months that followed, he recovered, then resumed his detailed studies of all the soil he'd collected around Meteor Crater. Harvey suspected that the metallic sand he kept finding was explosive melt from the vaporized impactor. He spent "day and night . . . sifting and sorting for hours at a time."

Armed with magnets wrapped in cellophane and Butter-Nut coffee cans, Harvey, like others before him, had explored anthills made of those tiny metal globes. These weren't chondrules but what Harvey called metallic spheroids, hard drops of a metal rain that fell back to Earth after the meteorite had vaporized into a huge cloud. To sort the pellets, Harvey used "toy cake pans," grocery store boxes, a magnet, and sieves. There is still uncertainty as to whether the spheroids were metallic rain—condensation—or material shocked in situ by impact or even pellets formed on the backside of the impactor. Harvey's work is still being cited in studies today.

He also found that what scores of geologists and others had considered to be volcanic "lava bombs" actually contained cosmic nickel; they weren't volcanic at all. In fact, the "lava bombs" thinned out the closer one came to volcanoes. Harvey named this slaggy, fused rock "impactite." Science historian William Graves Hoyt says that Harvey's discovery of impactite "demonstrated that there had been more fusion and melting . . . than anyone had previously thought." Harvey concluded that Moulton's calculations were correct and said his own discovery of impactite was the "greatest thrill in 30 years of field work."

In a book on Meteor Crater, Harvey mentioned the then-recent laboratory discovery of coesite. Coesite is a high-pressure quartz that Harvey and others recognized might also be found in nature, formed by large impacts. In short order, two researchers named Eugene Shoemaker and Edward Chao published a paper confirming the discovery of coesite at Meteor Crater. It was a major find. For Shoemaker, it was the start of an important career studying impacts.

Meanwhile, a new highway had bypassed the American Meteorite Museum on Route 66. Not that Harvey and Addie wished to stay, but they had been stuck with a bad lease, were borrowing money, and were generally worrying away once again.

In 1953, things began to look up. Harvey's Meteor Crater book was out, and with help from friends and the sale of some of oil leases, he and Addie moved into a new, clean-lined house and museum in the town of Sedona. From there, they opened up shop again and sent out advertising, including radio scripts that advised, "Don't let the world of tomorrow catch you unprepared."

Harvey and Addie didn't. In 1958 and 1959 they sold most of their collection to the British Museum and Arizona State University for $140,000 and $275,000, respectively. With one banker, who was startled to see two checks totaling $275,000, the Niningers "laughed and joked . . . until I think Addie and I both were more relaxed than we had been in years."

In 1959, they moved into another house in Sedona—on Meteor Drive—where daughter Doris still lives. From Sedona, they still left on adventures. At seventy-seven Harvey continued to hunt meteorites.

In his old age, Harvey Nininger could look back with satisfaction. He had found about half of all the meteorites found in the United States during the preceding forty years, including many new to science. He'd revolutionized the search for meteorites, proving them far more common than geologists had believed. He'd understood that little-discussed aspects of meteoritics could become important fields, such as the collection of cosmic dust and the relationship between impacts and mass extinctions. Scientists Edward Anders and Michael Lipschutz of Fermi Lab confirmed Harvey's discovery of extremely tiny diamonds in some Canyon Diablos, showing that he'd been right when he'd theorized about how they formed. For some, though, sorting out Harvey's contributions isn't easy. As one meteoriticist put it, "It is . . . hard to separate [the] salesmanship from his science." But in 1967 the Meteoritical Society awarded Harvey its highest honor—the Leonard Medal, named for his old friend and one-time foe—for all his accomplishments. The scientific journal *Geochimica et Cosmochimica Acta* dedicated a special issue to him, though Harvey may not have understood some of the papers among the nearly 600 pages, for over the years the field had become highly technical.

As early as 1937, when Victor Goldschmidt presented work on the spectra of stars and compared them with chemical descriptions of meteorites, meteoritics was beginning its rapid push toward the esoteric. It was exactly the kind of highly specialized work that Harvey Nininger was not prepared to do, and in the future Goldschmidt's cosmic family tree would lead to the work of such meteoritical hotshots as Brian Mason and Stuart Ross Taylor. Ten years later chemist Harrison Brown wrote a paper called "Elements in Meteorites and the Earth's Origin," which won a major prize from the American Association for the Advancement of Science. Brown and Harold Urey began to utilize nuclear chemistry in the study of meteorites and to teach courses on meteorites and cosmochemistry. Others followed suit. Students began to emerge, writes Ursula Marvin, with expertise in such areas as "elemental abundances, isotope dating, nebular temperatures and pressures, and rates of formation of planets and satellites." Complicated tools such as the mass spectrometer and the electron microprobe allowed scientists to date rocks with certainty for the first time and to explore their complex suites of minerals. About the time when Harvey sold his collection, scientists were arguing about the size of parent bodies from which meteorites came. In 1960, it was still a radical idea to suggest that most meteorites had originated not on a large planet but on smaller bodies, on, that is, asteroids.

The well-known and respected meteoriticist Gary Huss—grandson of Harvey and son of Glenn Huss, who, with his wife Margaret Nininger, took over Harvey's operations as the American Meteorite Laboratory—clarifies aspects of his grandfather's legacy and understanding. Though Harvey "was not familiar with these tools" —things like electron microprobes— "that did not mean that [he] was unable to understand the major scientific questions of the field." Gary recalls wide-ranging conversations about many issues that his grandfather had been thinking about years, even decades, before others. And he remembers "watching the first Mars pictures return from the Mariner spacecrafts" when "scientists on the mission were expressing great surprise to see the craters. My grandfather was expressing great surprise that they were surprised. 'What did they expect to see?' he asked." From Gary Huss's unique vantage, it wasn't till the late 1980s and early 1990s that "mainstream meteoritics was dealing with all of the major unsolved questions that Nininger, my Dad and I had discussed when I was growing up." While Harvey "saw the problems," the lab scientists that followed, especially in the 1960s, were unable, he thought, to integrate data with actual, physical samples and the largest questions of the field. The disconnect between these researchers and what Harvey had done—which Gary Huss says they "did not really understand" —led to the impression that Harvey was "left behind." Gary says his grandfather didn't lose his sense of the field's technicalities until much later.

For a time there were three generations of Niningers—Harvey, Glenn, and Gary—making presentations on meteorites at scientific conferences.

"I can read meteorites; they can read formulae. We do not speak a common language," Harvey wrote of other scientists, at least those not related to him. Yet he knew he'd been the start of something unique. Harvey Nininger was right to say that perhaps his "greatest contribution to meteoritics has been the creation of proper interest in the subject."

. . .

"NATURE IS GOOD TO us and we try to reciprocate. As I often say this is the most interesting world I've ever lived in," Harvey wrote in his Christmas greetings in 1971. Even in his seventies and eighties, he'd rise at six A.M., then work in his garden. He'd have a glass of wine late in the day.

On one of the many manila envelopes that magazines arrived in, which Harvey would use as notepaper, he wrote, "3 a.m. August 12, 1980"—an entry about getting up to watch the Perseids. Then he saw a picture of Frederick Leonard and "almost wept." The two men had begun a rapprochement, but Leonard died before he saw Harvey again.

Harvey Nininger died on March 1, 1986, at age ninety-nine, from complications following a fall in which he injured his head. "My only quarrel with Nature has been that my life on this planet cannot continue for a thousand years," Harvey once wrote. The family tells me that his ashes were scattered on a hill near his home in Sedona's Red Rock country, not, as popular talk has it, over the rim, walls, and floor of Meteor Crater.

HARVEY NININGER IS VENERATED as a hero by those who find, sell, and collect meteorites. Like the most interesting heroes, though, Harvey is a complicated one, a study in contradictions. He left an important legacy for science, even if some think his fervor led him at times to skirt ethical edges. Harvey Nininger lived a life of high adventure, great risk, and sometimes terrible defeats. If anyone epitomized wonder, and the costs of wonder, it was Harvey Nininger. He was a pivotal figure in the study of meteorites, helping to create that science and helping to create a community of meteorite hunters, dealers, and collectors. He was singularly important.

In Sedona, in the house he and Addie had built, a house designed by a student of Frank Lloyd Wright, with its fireplace, slanting ceilings, and canyon views, I visit with daughter Doris Banks, who, spry and silvery, seems herself to oscillate in her reactions to her father's work. She recalls with fondness climbing through open windows at the old museum in Denver to reach her father's office, and she sounds protective when speaking of her father's scientific contributions. But as she speaks of his failure to see her off to college, Doris is stoic.

And Addie? The family tells me the story of how, after having to be moved to a nursing home, she always asked, when she saw a man wearing a hat, if that was her dear Harvey.

Another drive, this one on a cold autumn day, brings me to the old American Meteorite Museum on Route 66, near Meteor Crater. The stone building has fallen in, though the tower still stands. A raven croaks protest at my arrival, and I pick about the ruin. The elm that Harvey and Addie watered is long gone, but the creases of desert where they made their home still go out to the distances, including toward the rise that marks the upturned rock of Meteor Crater. Wind buffets me, and I think back to what Harvey wrote in 1937 while he camped near cottonwoods at dusk in Nebraska, listening to killdeer and meadowlarks: "What a great boon, this hour alone. Normalcy has been restored. . . . The coals at my feet are gloriously red."

I think, "They still are." I stoop to pick up a broken fragment of sandstone brick from Harvey's old museum wall, and dragging a magnet through the dirt, I see it glean a fur of particles to its head so that, despite all the wreckage, I gasp in delight.

Book VII

A Serious Case of the I Wants: Passions of the Dealers

"About here," Gary Curtiss says from the backseat, "I'd be so excited."

Gary is giving directions to his friend and fellow meteorite connoisseur Matt Morgan, who steers his orange Mercedes on Denver streets dusted with a light early-April snow. We're along a busy thoroughfare, about to turn into a neighborhood.

"I'd have thirty dollars in my pocket from washing dishes and would think, 'What could I get?'" Gary says, recalling his college days when he began to obsess about meteorites—and about owning them.

Matt and Gary are friends, and when not on the clock as geologists for the State of Colorado, they're active meteorite collectors and dealers who each own hundreds of specimens. We've just finished a Sunday breakfast, and over the eggs, juice, and coffee, they've told me about their passion for meteorites and their perspectives on the contemporary meteorite dealing-and-collecting community, which has mushroomed in size since Harvey Nininger's days. A longtime birder, I know a smart, savvy, and fiery subculture when I see one. This is one such. Today's meteorite collectors range from the budget-minded to the very rich, but regardless of financial status, most pay assiduous attention to available meteorites, new falls, prices, and scams, as well as the history and science of the field. Over the Internet and at exhibitions, they trade gossip, specimens, exclamations, and sometimes accusations. And I'm finding that collectors are keen on owning not just rocks from space but artifacts—historical trinkets—related to a search, a fall, a find, a place, a person. My impulse to grab some wood from Greenland seems apropos.

Matt Morgan, for example, is the proud owner of broken toilet pieces from the Niningers' Route 66 meteorite museum.

Although Matt saw an ad for meteorites in *Astronomy* magazine when he was ten, it wasn't until after college that he started pursuing them. That year he attended a mineral show, bought some meteorites, went home, and promptly sold them for a profit on the Internet.

Gary's fascination began with the song "Catch a Falling Star," sung by Perry Como. Growing up, Gary loved fossil and mineral displays at museums and dreamed of owning objects from the long, prehuman history of the world. A shop owner showed Curtiss, then in college, a Canyon Diablo iron and told him to talk meteorites with Glenn Huss, Harvey Nininger's son-in-law. Gary even met Harvey himself.

"From there it was all downhill," Gary told me, laughing.

We're coming up to a curb by a brick ranch house nearly hidden in the trees—the former home of Huss and his wife Margaret Nininger, the last site of the American Meteorite Laboratory, which Harvey had founded all those decades before back in McPherson, Kansas.

At breakfast, I told Gary and Matt about my own pursuits, how I'd been tracing stories and clippings, houses and craters, how, for me, meteorites were inseparable from the people who hunted them, who cherished them, who lost sleep and more over them. What I didn't say was that I felt myself hitched to history—how, in transcribing and translating some annals of the fallen sky, I could believe my own life added up to a story, to something more than just being here, then being gone.

We pushed plates aside. I had visited so many of Nininger's old haunts that, back again in Denver on other business, it felt fitting to drive by 7891 Osceola Street in suburban Westminster. I wanted to see where Glenn Huss had carried on the Nininger legacy and where Harvey spent his last days before dying in a hospital. With Glenn deceased and Margaret hospitalized—she'll pass away a few years after my Denver trip—the family is selling the property. The AML exists only as an empty house and a useless phone book listing. Still, we decided to drive by and see what we could.

We're not expecting much, because the place is locked up—a stroll around the yard, a peek in the windows. Then Luck pulls up to the curb. A realtor emerges from her car with a couple. Matt, Gary, and I stare at each other. *What the hell?*

Gary hurries to the door to explain our interest, and we're in! Matt heads to the basement office of the American Meteorite Laboratory while I stand in the living room looking at the translucent glass blocks in one wall. They glow with the diffuse light of a spring becoming less tentative. This is no ruin like the old Nininger museum, but my affinity for the spaces where lives have been keeps me there, alone, for a few moments. I shake my head and go to the basement, where Gary is whispering to himself in the dusty air.

Being here is like a homecoming for Gary, who had visited the house so often in years past, when he was eager to spend his dishwashing money on one of Glenn Huss's latest meteorite acquisitions. For Matt, for myself, getting into the Huss house is simply a rare chance to see a place few other meteorite fanatics will have seen. And unspoken but surely on all our minds is this: What might we find?

"Makes me sad," Gary says, looking around a basement divided into several small rooms. "Oh," he interrupts himself. "He'd do nickel tests right here!" We walk into a concrete block room with a metal door and a huge, vaguely Lovecraftian hook used for weighing specimens.

All around are pegboards and tired cabinets and shelves and, here, a green workbench.

Joining us, Matt reaches up to take two pegs.

Gary calls from another room. "Here's the desk still! I'd love to buy this desk!"

I graze my hand along paneling and dingy, butter-yellow walls. I walk on linoleum, imagining nitric acid dripping from a beaker, the floor burning away like linen under cigarette ash.

"Where were the meteorite saws?" I ask. Gary shows me, here by the washer, dryer, and sink. I love this juxtaposition of domesticity (imagine clothes sitting on the washer years ago) with outer space (imagine Glenn Huss hefting a newly cut iron in his hand).

Nostalgia pivots soon enough to . . . what? Curiosity? The hunt?

We're opening doors now, peering at the backs of shelves and pulling out drawers. I wander to a small room with a low desk built along a wall. A child's room or—*Harvey Nininger's last home.*

My God, I think, he must have been dying right here.

Months later, the night before I transcribe these words into a chapter, I will dream that I find several fusion-crusted, split-open Norton County aubrites sunk into the lawns of people I don't know, and I will take them, I will take them all.

I find a ruler, scissors, and tie-on tags, the kind you knot onto specimens and on which can be written a tiny provenance in pencil. I find a gray metal canister with a yellow screw-top lid—for film, I guess—though now it holds blue and red thumbtacks. Not long after taking this canister, I'll use one of Harvey's tacks—redder than a bolide—to post on my bulletin board a Chinese fortune I've kept for years: "Many possibilities are open to you—work a little harder."

Then, in one of Harvey's desk drawers, is the greatest treasure of all. It's a pop-up metal phone list, the kind where you slide a pointer to a letter of the alphabet, press a button, and presto, the lid flies open, revealing names. Like Harold Urey, the Nobel laureate. Like Eugene Shoemaker, codiscoverer of coesite at Meteor Crater. I recognize the handwriting. This was Harvey Nininger's directory. I actually stroke the names.

For a second, I'm unsure what to do. I feel like pocketing it, but I also want to show Matt and Gary. For a second, I imagine Harvey calling Urey or Shoemaker or just phoning the Oak Creek Store in Sedona to see if they have fresh celery for a meat loaf recipe he often made.

Matt's in the hallway and, without thinking, I motion him over, and he's so thrilled that he holds on to the phone listings and I think, *Wait, I found that.* I'm about to say something but instead I grab a handful of the tie-on specimen tags and the thumbtack container. *These are mine.*

Upstairs, Gary answers questions—no, meteorites are not radioactive; yes, some come from other planets—while Matt retrieves specimens from his car to show to the visiting couple and realtor. I move through an empty upstairs bedroom, all vacancy, the atoms of ordinary matter constructed mainly from void, as if what we know exists does so because it's mostly not there to begin with. Back to the kitchen, I write down for the potential buyers a family tree that explains the American Meteorite Laboratory's history. Then we thank them and take our leave. Melting snow drops in big happy splotches on the lawn and walkway as we get into the car.

"How sweet is that?" Matt says, turning his head toward Gary in the back seat, as we show him the directory.

"Well," Gary says to me, "you got more than you bargained for." He pauses. "I'd trade all of this to get Glenn back."

And I'm silently wondering about what we've just done. Is this preservation? Theft? I think again of what I took from Greenland. And the sandstone brick from the Route 66 museum in Arizona. But isn't it true that had we not taken the directory and other items, they'd be thrown away? If Margaret's daughter Peggy Schaller wants the directory back, it's hers, Gary says—but first, Matt adds, he'll copy its pages.

Not that hunting and hoarding is the only thing meteorite hunters do. Far from it. At breakfast, Gary and Matt talked about their donations of meteorites to schools. They're especially gratified when professors use meteorites to introduce students to geology and other fields.

"You can teach every aspect of science with meteorites," Gary had said.

"Physics, geology, chemistry," Matt responded.

"Biology!" Gary exclaimed.

I waited a beat and added, "Psychology!" And we all laughed.

PSYCHOLOGY.

For years, I had thought of "my" characters as, quite simply, eccentrics, as obsessives. They might have fallen for stamps or orchids or show dogs or View-Masters, and mere chance led them—and me—to meteorites. I'd always felt a bit uncomfortable with the term "obsession," though, because it suggested a sickness, an out-of-control pursuit that damages everything. I'd pondered the word on Meteorite Island, where it seemed to fit Robert Peary's life, but I also had sensed it might not always apply to others in the meteorite-hunting world. Nininger had harnessed his obsession, done astonishing work, and made a name for himself, even, one might say, an entire science. And though it sometimes hurt people near him—him too—this "obsession" did not destroy. In fact, it saved his life. What else, how else, to name these complexities? "Eccentricity"? I suppose. But that word seems tweedy and quaint, even feckless.

Exuberance, psychologist Kay Jamison writes, "denotes a mood or temperament of joyfulness, ebullience, and high spirits, a state of overflowing energy and delight." Jamison believes that human and animal exuberance arises from nature's fecundity, which is, at its roots, a sexual energy. "Exuberance," she says, encourages "the exploration of the universe. . . . It fuels anticipation . . . intensifies the joy once the exploration is done; and sharply increases the desire to recapture the joy." "Exuberance" and "exuberants," these are terms that finally make perfect sense to me.

Psychologist John D. Gartner uses the clinical term "hypomania." Gartner says hypomanics "are brimming with infectious energy, irrational confidence, and really big ideas. They think, talk, move, and make decisions quickly. Anyone who slows them down with questions 'just doesn't get it.' Hypomanics are not crazy, but 'normal' is not the first word that comes to mind when describing them." They take risks, they're defensive, they're frequently insomniacs, and they have a sense of self-genius and "mission." They're often oversexed. Hypomanics sometimes have experienced a crucial epiphany. I think of Harvey Nininger's Fireball Corner. I think of Daniel Barringer's dropped cigar. In fact, both men had another trait that is characteristic of hypomanics—their ability not so much to "create original ideas," but rather to "*grasp the significance* of an idea . . . and *leap on it*," Gartner writes. As many note, such traits often lead to astonishing productivity.

It's not just the exuberants, the hypomanics, that crave heightened or, at least, meaningful experiences. We're all wired to respond to the new and in ways that can motivate us to keep seeking it out. Basically, it's all about dopamine. Anticipation and consummation of new experiences (not always pleasant ones) increase dopamine production in the brain. Strongly associated with food, sex, and, for some, a range of exuberant behaviors, dopamine increases energy and, well, it just plain feels good. Dopamine is also a decision chemical. "Motivation and commitment are two facets of the same process, with dopamine acting as the catalyst," researcher Gregory Berns writes. "The release of dopamine in response to novel information is the essence of a satisfying experience and kick-starts the motivational system." Discover a meteorite in the woods, in a field, in the snow. Discover a crater that could earn you a fortune. Discover a mineral alien to the Earth. Or find the spot where someone began a journey that changed science and history. Then write the secrets down. Yes, it does feel good, but you may require more such thrills, again and again, and perhaps scarier each time.

Of course we crave both predictability and "novelty." In what measure depends on personality and circumstances. But, as Berns writes, "The sense of satisfaction after you've successfully handled unexpected tasks or sought out unfamiliar, physically and emotionally demanding activities is your brain's signal that you're doing what nature designed you to do."

Having to work at something to gain reward introduces something else to this biochemical-philosophical equation: stress, which produces cortisol, which, in turn, writes Berns, "elevates mood, increases concentration, and [can] even improve . . . memory."

In combination, dopamine and cortisol not only make us feel good, they can conjure up a sensation of "transcendence." Such welcome emotions are, of course, enhanced if one "pass[es] through the terrain of discomfort," as Berns calls it. The more skilled we become in such terrain, the rougher the country we seek. Harvey Nininger understood this. "But I'm one of those addicts," he wrote. "I never quit."

And if you hitch ardor to something bigger than yourself—while not letting go of the world and while accepting the passage of time—you can do more than produce some selfish chemicals. You can beget meaning. As Gregory Berns writes, "Satisfaction and purpose have become the same thing."

IT WOULD SEEM SO FOR GARY CURTISS and Matt Morgan. They've had plenty of dopamine and cortisol cascading in their neurons like a meteor storm.

"I think of the ones that got away," Gary Curtiss told me at our Sunday breakfast. As in Plainview, Texas, the place where Nininger had collected with his brother. And where Gary heard a sound from his metal detector that indicated a find. But he couldn't stay; he had to be home the next day for a birthday. Later he lost his field notes. He wonders what might have been.

But not always. Years ago, Gary was on his hands and knees on state lands around Meteor Crater before collecting there was banned. He would have looked like the proverbial wanderer in the desert, crawling toward some mirage, dying of thirst. He ran his magnet over the dirt, back and forth, and found a few fragments of Canyon Diablo. "I was *in heaven*," he said.

More recently Gary and Matt went to Holbrook, Arizona, to pick at the worked-over strewnfield from a 1912 fall. They expected little but danced their "meteorite canes"—sticks with magnets—over the earth. "I was beside myself," Matt recalled. "There was still some here!" On the same trip they put pennies on railroad tracks then watched the trains come bearing down to crush the coins.

"Ah." Gary smiled. "We're still kids."

It's not all fun and games. Dick Pugh, an Oregon meteorite expert, has been shot at while hunting for specimens. "It was an area of pot farmers, poison oak, and rattlers," he told me years ago. His license plate reads "Meteor," not for vanity but for safety.

Nobody took aim at Gary Curtiss when he went hunting for meteorites in the safer climes of Park Forest, Illinois, following a spectacular nighttime fall on March 26, 2003, one so loud that residents thought terrorists were attacking. The stone is just another chondrite, but a witnessed fall over a major city is a big deal. This was the first in modern times. The Chicago burb was peppered with meteorites that hit roofs, cars, sidewalks, you name it.

In a matter of hours Park Forest was teeming with some three dozen meteorite collectors in what one newspaper columnist called, a year later, "a whirlwind of greed for some, salvation for others and an education for all."

Matt Morgan couldn't go, but Gary left in a Colorado blizzard and arrived at a neighborhood lit at nine P.M. by flashlights and television cameras. "It was like the *Night of the Living Dead*," Gary told me, "with people looking down."

The keen competition for specimens drove up prices and frayed tempers. It didn't take long for someone to tell Gary—who prides himself on polite restraint and respectful behavior—"We're really sick of you." Meaning all of them, all the dealers, especially those Gary says were "pushy and arrogant," not naming names. Conversely, a few dealers accused some scientists—who were also vying for specimens by purchase or donation—of misleading the public into believing they had to turn over meteorites that they'd found on private property.

Just how crazy was this scene? Gary bid against another dealer for a meteorite and the computer it had hit. The other dealer ponied up: $12,000. Dealers love "hammers"—meteorites that hit things (computers, drywall, blinds, anything) and the things themselves, if they can get 'em. Hammers make for great stories and can dress up your average-looking meteorite in a display.

So while residents may have tired of the searching and bickering, the ordinary chondrite was indeed what Matt called a "a windfall," not just for savvy dealers and collectors but for financially strapped families in the blue-collar area.

"We haven't even broke even," Matt Morgan claimed. "We sold $12,000 worth yesterday but spent $20,000."

Matt had shown me what they were selling. At our breakfast get-together, he'd pulled out of plastic bubble wrap two Park Forests that could fit in my hands, one that he and Gary had cut open and another with a two-tone fusion crust. There were melt lines subtle as cat whiskers, an exposed gray-white interior like a chunk of old dry detergent, and a bright yellow smudge from where it hit a fire hydrant. A hammer! I stared at the rock, wishing I had been there, wishing I hadn't missed the "Night of the Living Dead," wishing for a dopamine high.

WHEN I FIRST ENCOUNTERED meteorite hunters in person, at a show in Tucson, I was years away from meeting the affable Matt and Gary, years from reading about exuberance and satisfaction, but I knew enough about psychology to know I was in a bad way. It was early 2001, a month after I had visited New York City, when I saw the Willamette meteorite at the beginning of my journeys for this book. I was still married and I'd been flirting with another woman, with Kathe, and in the predawn hours before I left Manhattan, Kansas, for my flight, I checked my e-mail to find a note from her. Kathe was in graduate school then and married. Once, before we'd ever really talked, we had passed each other on a snowy sidewalk. She looked at me and said, "There's a snowflake on my scarf." She might as well have said, "*Carpe diem.*" I was speechless.

"Heed glides into heedlessness," Kay Jamison writes, "as effortlessly as the silk chemise drops to the floor."

In Tucson, I rented a car and drove to a hotel, where I met O. Richard Norton, a well-known meteorite author and collector. He had agreed to show me around to some of the dealers. Not only is there the main Gem and Mineral Show at the Convention Center, there are dealers of all kinds who set up shop in their motel rooms, so one can park at, say, a Ramada Inn, then wander up and down the corridors where doors are open and signs are posted.

When I arrived at a hotel lobby swarming with fossil dealers and shoppers, I overheard a man remark that a women he'd just seen, another dealer, was herself "an oriented specimen." Apparently the entire world had become sexual. The air itself—February in the desert—seemed playful. The sun a sprite. There was snow up high, rain here and there on the lower elevations, scudding clouds, scumbling mist, then the clearing that heightened every outline of everything. Beside broiling parking lots, white Christmas lights wrapped around the trunks of palm trees, and cactus wrens flitted among hedges and signs that said "More Great Dealers This Way." Sliced-open amethysts the size of cows glittered on the backs of flatbed trucks. There are some 10,000 "serious private collectors" of meteorites, according to *Smithsonian* magazine, and a few of them were right here in Tucson, along with the dino freaks and the gem polishers. I stared at amber, at trilobites, at $3 penguin bones. All across Tucson mockingbirds chattered away, while a gray, puffy, affluent demographic overburdened with turquoise and cell phones descended toward $35 dinosaur vertebrae and $10,000 meteorites peddled by dealers from Australia, Pakistan, Russia, America. Giant ammonites dwarfed us. Dinosaur bones loomed. And, anyway, we could purchase caps and T-shirts that cost far less.

Richard Norton spoke with a deep voice, presented a Roman nose and gray hair, and in his blue jacket and brown pants, seemed the epitome of the casual-dignified naturalist. Prolix on all things technical, he lost me at times—I wasn't ready for a discourse on ferric chloride reactions—but his enthusiasm and good cheer felt companionable. The guy was genuine.

"Oh," he'd say, "these chondrules are like BBs!"

Passion—unaffected, glittery, real. If the world of meteorite dealing is at times shot through with envies, misunderstandings, deceptions, and rivalries—and it is—the core is without artifice. I'd seen this passion myself in birders, but something changed me in Tucson those years ago. I was unhappy, but when I plunged into a world where people weren't afraid to show a very real part of themselves, at a gem show, of all places, I felt . . . envy. Exuberance, it turns out, can be fetching. It can also save your life.

"Oh, Nantan!" one man exclaimed, invoking a Chinese meteorite notorious for rusting.

"He throws himself into everything—meteorites, his children."

"My wife is leaving me because of this stuff. Told me two weeks ago."

Exuberance can also ruin your life.

"Do you have any meteorites for ten bucks? Oh yay!"

Dick Norton took me to the room of Blaine and Blake Reed, Colorado meteorite dealers, and while he spoke of the crystalline intergrowth and inclusions in irons, I stared. My God, here was a slice of *Cape York*. Three grams of Greenland history. *Ho-ly shit.* At the time, I had not been there yet but was making my plans and I hadn't expected an encounter with a piece of one of the world's most important meteorites, here in a beige-and-floral-generic-bedspread-and-gold-framed-prints-on-the-wall motel room. Where were the icebergs? The Inuit? I gave slow thought to handing over a credit card and racking up $300. I hemmed and hawed over the rock and put off purchase. Would I ever buy a meteorite? Or would they be more alluring if I kept them afar?

I wasn't alone in being stunned by such a historic meteorite specimen. Norton himself has bemoaned their availability as portions of old, famous stones come out of museum collections in exchange for newer material. "Certain meteorites were so important," he had told me on the phone before my arrival in Tucson. "They were Hope Diamonds! Then you walk into a motel room and there's Ensisheim—Ensisheim! —on sale." For Dick Norton, some meteorites ought to be beyond commerce. That concern extends to other practices too. He has been somewhat critical of turning meteorites into jewelry, as well as literally hammering a meteorite to shatter it into smaller, more salable pieces. This has been done with meteorites from Mars.

Dick pulled me over and handed me his magnifying glass and urged close inspection of a chondrite's interior. I saw a half-sphere, a cavity where a chondrule had *fallen out.* Meteorites can lose their chondrules the way we lose our teeth! This stunned me. Where had it gone? Did it separate after freeze-thaw action for years in a desert? Did a careless collector ding it off somehow? Was it stuck in the carpet beneath our feet? Should I fall to my knees and search?

Blaine Reed told a story. He was a boy when he purchased his first meteorite by mail order and waited, impatiently, for six weeks before it arrived. His parents opened the package and they were rolling the rock—an Odessa, Texas, iron—around on a table when he came home. "They must have thought," Blaine said, laconically, "'Our—boy's—ordered—a—rock.'" He went on to make a necklace out of this meteorite, which he wore constantly and which his wife hated. For her birthday he made a faux version of the necklace so she could rip it off his neck. He still has the original. The Reed brothers kept busy firing a potato gun while I looked around, and Blaine opened a bottle of beer with an iron meteorite. I thought, *I like these guys.*

The next day the Reeds answered questions from all comers and watched for the hands of potential shoplifters, especially among the teenagers on a field trip with a science teacher named Bob, who was taking serious shit from his class. After all, here was a three-inch slice of the moon going for $65,000. "Fossils, meteorites," Blaine said, looking past me, "they're more subtle, even ugly. It takes knowledge to appreciate them." I agreed, shook hands, and headed to the room of dealer Al Langheinrich—he goes by "Lang" too.

"Allan, $10,000."

"You gotta go up."

"Get off the pot."

"Allan, you keep thinking of yourself, your collection."

"It's the stone, the complete stone."

"Listen to me . . . "

"I'm trying . . . "

I'd interrupted as earnest a conversation as I'd ever heard and watched Al, with his longish hair, intense blue eyes, and jean jacket. He casually welcomed me. Yes, I could look around. Here was a cheap shaving of the moon, from some lunar meteorite or another, for just $100. I was drawn to a slew of lovely Brenham pallasites but I didn't note their cost because the light caught in olivine was too pretty to price. Al told me about one of the rocks he's most proud of owning—a cow-killing meteorite from Venezuela. Ugly stone, cool story.

Next I stopped at the hotel room of Darryl Pitt, who runs the Macovich Auction, which specializes in "aesthetic irons." Broad-chested and blue-eyed, Darryl is a former news photographer. Just before I arrived, he had closed a deal on a giant iron meteorite big enough to be a table, complete with a deep cavity, an inset bowl. Pitt had kept candy in it. Pitt's "aesthetic irons" are popular with the high-end crowd. When I told him that another dealer had said of the meteorite clan "We're all screwballs," Darryl smiled and laughed, saying, "That's being mild. We're all sociopaths." At first guarded, Darryl became very friendly, lunging forward to make a point, then leaning back. "Oh, oh," he said, in response to my telling him I was tracking down the stories of some meteorite hunters. "Amazing, amazing stories." Later, he asked me not to portray all meteorite folks as "eccentrics."

At the room of two French dealers, Bruno Fectay was smoking when I explained my project and he said in that noncommittal but ironic and maybe pointed French way of sort-of-opening-a-conversation-but-maybe-not, "Ah, about the meteorite world." The French, I learned, are the top collectors out of Northwest Africa. He showed me his booty of rare chondrites, a diogenite, and the world's largest ureilite, the size of a loaf of bread. Ureilites are achondrites that mysteriously mix ancient, unprocessed elements with evidence of impact melting. At another French dealer's room, photos—with some faces blacked out—showed a team's work in the Sahara.

So down the colonnade I walked, mockingbirds singing in the orange trees, to where a one-ton Campo del Cielo iron stood for $95,000 by the door of Southwest Meteorite Laboratory. Inside, under his cowboy hat, was Marvin Killgore, a former prospector turned meteorite hunter and a dead ringer for baseballer Randy Johnson. He was talking on the phone about the 3,000-mag on an electron microprobe. At first his wife was cautious. "And what do *you* know about meteorites?" she asked me coldly. Foolishly, I hadn't yet introduced myself. I said something about being there precisely to learn more about meteorites, looked about, quietly exited, and let out a sigh.

My last day in Tucson I wandered through the cavernous Tucson Convention Center, which was filled with gem purveyors. I gawked at hunks of color and bought two stones—stones of this world, desert rose, a type of selenite with vertical fins, flanges crisscrossing one another in a lovely jumble, a confusion of directions. One was for my wife, one was for the woman who would become my lover. I took notes to forget the cruel absurdity of my dual purchase.

WHEN I RETURN TO Tucson two years later, I'm divorced, Kathe is too, and we've moved to Utah. Life as sudden as a bolide. I find arriving in Arizona again oddly comforting. I can't shake the feeling that meeting so many ardent meteorphiles the first time had helped convince me that appetite didn't have to be bleak. Tucson feels like pilgrimage. I'm happy and happy to be back.

I've come to the annual meteorite dinner, which is billed as "The Return of the Magnificent Meteor Mayhem Birthday Bash." It's been organized by dealers Geoff "Colonel Carbo" Notkin and Steve "Tambo Quemado" Arnold. Some one hundred people, including top meteorite dealers, crowd into the La Fuente restaurant under the gaze of tribal masks on the wall. Ceiling fans turn languorously as I stand beside salsa-stained strangers in the overly hot camaraderie of the slightly tipsy.

The man next to me strikes up a conversation. We're standing beside cakes in colors that seem radioactive. He's been collecting for just a few months but really knows his stuff. As the saying goes, "One meteorite, you're a collector. Two of the same, you're a dealer." He's frank. "It appeals to nerds like me," he admits, further noting that "like cars" the meteorite scene is very white-male.

Meanwhile, Geoff Notkin hands out awards—laminated "Harveys." One goes to a man dressed in alien garb who says, "On behalf of the Klingon Empire, I accept." Darryl Pitt accepts the "Tough As They Come" Harvey for the recently hospitalized meteorite hunter Steve Schoner, who, as a young man, knew Harvey Nininger. The "Longest Suffering Meteorite Wife Harvey" goes to "the lovely Iris Lang," Al's wife. Everyone is smiling, talking, laughing, and drinking. Geoff Notkin jokes that this is "the only night . . . we all get along."

The next day, after seeing a meteorite slice from Northwest Africa that some think could be a piece of the planet Mercury, I realize I'm driving on a street that Kathe had lived on years ago, long before I knew her. Then I pass a street called Magic that has "No Outlet" and I think, *If you were going to stay put, it ought to be on Magic you can't escape from.*

OF COURSE SOME FORMS of magic can be bought, and the forms of magic that can be purchased depend on your interests. The psychology of exuberance may shed some insight on the historical and contemporary hunters of space rocks. As one collector says, "When I hold a meteorite . . . I dream of travels to other planets, asteroids and comets. It's a magical feeling, holding a meteorite." But which meteorite? Which to buy and display?

Some collectors simply want as wide a range as possible from the myriad categories of space rocks. Some may focus on meteorites with peculiar stories attached to them. Another collector may want all the meteorites ever found in Italy or Texas or Canada. Others may limit specimens based on price, because some meteorites can be had for very little, while others require a second mortgage. But all that doesn't explain the underlying question: Why do we collect anything at all?

Freud thought it was all about poo, and it's true that some meteorites (shrapnel from Sikhote-Alin) bear more than a passing resemblance to it. "The need to keep and the refusal to give is what adults do as a substitute for the childhood habit of retaining feces," writes Geoff Nicholson, invoking Vienna's great weirdo in a book on collectors (of erotica).

Be that as it may, some collectors just enjoy the notice they receive. Some collect less out of love for the objects than for the human connections—and the right to boast—within a subculture. Nicholson goes so far as to say that the organizing a collection requires is another "way of keeping chaos and death at bay." It hardly needs mentioning that artists and writers are collectors of a kind too. All that makes more sense than poo.

The very act of collecting, many say, is sexual. One either redirects libido from people to things or, as must be the case with hypomanics or the exuberant, there's enough energy to encompass both lovers and collections. "Every collector," Freud wrote, "is a substitute for a Don Juan." The comparison seems pretty apt, since collecting—meteorites, stories, mountaintops, Chippendales, all the above—is less about the fulfillment that comes with attainment and more about the never-sated desire to attain. Recall Melvin Konner's observation that dissatisfaction is at the core of being alive. That never-stated desire is built in, of course. It can seem pathetic, but without it, we'd shrivel up and die. Curiosity is another name for hunger.

I've also always thought the urge to collect must be rooted in our evolutionary past. After all, we evolved as gatherers on the plains. It seems inevitable that when we see something that interests us, we literally reach for it. Psychologically, spiritually, emotionally, such grasping has, however, connotations of desperation and sickness, of possessiveness. On the one hand, I think of Ellis Hughes. On the other, I think of Harvey Nininger in his best moments. Sometimes the grasping-that-is-collecting transcends possession and places the holder in relation to the wider world, the cosmos, our knowledge of it and our ignorance. There comes a sense of connective calm, of affinity with creation, and that's when collecting gets bound up in another Freudian category, the "oceanic." The oceanic is that feeling of being swept up in time and space and things, of belonging to The Big Picture, however you frame it. Geologists understand the oceanic even if they don't know the term, because for them it's impossible to look at present landscapes without seeing the forces that shaped them, forces that put geologists in touch with deep time. Collecting pieces of creation—meteorites—does the same.

Personally, I find myself attracted to "aesthetic irons," the shapes that dealer Darryl Pitt compares to the work of artists. Not that I could afford his wares. One iron in his catalogue looks like "a Picasso-inspired Minotaur." No price. And I find myself attracted to whole stones with fusion crusts, especially crusts that have melt or flow lines, because they're the static evidence of downfall. I understand fusion crusts in ways that I don't understand the insanely complicated chemical differences among various chondrites and achondrites. And it's hard not to admire pallasites, with their olivine glowing yellow-green, like storm light caught in metal sky.

Education professor Martin Horejsi, now at the University of Montana, was teaching high school when he was "floored" to learn that one could buy and own meteorites. "Ever since then," he says, "I have been captured by their science, stories, history." For Horejsi, whose collection begins with a slice of the 1492 Ensisheim fall, there are three scientific issues that contribute to the "excitement" of meteorites. First "is astrobiological. . . . Meteorites have many interesting biotic traits such as containing water, salt, amino acids." Second "is identifying the specific body in space from which the meteorite came." Third "is the origin of chondrules." Horejsi calls chondrules "the most basic humanly visible building blocks of the solar system." A wonderful thought.

This passion for the science of meteorites is a huge part of the collecting community, and for me, it goes a long way to helping me accept the fact that meteorites are also commodities bought and sold. Nearly every collector I've met—from small-scale to world-class—knows a lot about meteorites. They appreciate the gee-whiz factor. But they also take the time to learn the history of how we've come to understand what meteorites are. They follow debates among the professionals studying meteorites right now. Birders keep lists of birds they've seen. Rock hounds keep collections of rocks. Both groups—the two nature-nerd movements I know best—are, at their best, expressions of curiosity. That money gets involved seems both inevitable and, perhaps in the end, almost beside the point.

Australian dealer Jeff Kuyken tells me that he pays close attention to the "unprecedented quantum leap" in meteoritics, which is helping inform amateurs about the latest issues and theories. "It's amazing that in the last five years collectors have gone from '*Wow,* what a nice L6 chondrite!' to 'Not another L6! Do you have an olivine diogenite or lunar feldspathic regolith breccia instead?'" This savvy, along with some good prices—tiny bits of the Moon and Mars can now be had for $40—has helped drive a collecting craze that Richard Norton once described as "a meteorite feeding frenzy."

THERE'S ANOTHER REASON FOR the frenzy: A self-described "long-haired rock and roller" with a mailbox shaped like a rocket and a persona as subtle as a fireball. Robert Haag—The Meteorite Man—has been the most influential meteorite dealer since Harvey Nininger. The dude looks like Sammy Hagar, and his address number is 5150—"just like the Van Halen album," he says.

In fact, when he drove into Portales, New Mexico, in the summer of 1998, he had Van Halen cranking through the speakers. He showed the locals pictures of meteorites, and the locals found meteorites from the just-witnessed fall. Bob Haag paid $15,000 for them. He was so giddy he hugged strangers.

In one of his colorful catalogues, there's a picture of Haag leaping in the foreground, his head situated in a line with the faces in the background. Mount Rushmore + Haag. In another, the foreground shows a beautiful black meteorite, the background a blurry Rhode Island Red—a chicken. The caption reads, "Chicken Little was absolutely right. The sky *is* falling . . ."

Haag began collecting and selling meteorites in 1979. Before that, he "stood around shopping malls in a tight silver space suit, hitting on matrons for small change." He was selling "space passports." "I nearly starved," Haag writes in his catalogue. "This is definitely better."

This is selling specimens of all sizes, types, and prices. This is selling watches with etched iron meteorite faces. This is selling an "asteroid belt"—a leather belt decorated with an iron meteorite, Gibeon. This is surviving a plane crash in Baja, California, while chasing shooting stars. This is "near misses with rattlesnakes, dead hits from dagger plants that ram two inches of hard spike into your shins, and tumbles down steep slopes. . . . Did you say there was beer back at the camp?" This is using "everybody I can to get information—the library, NASA, the police . . . I can't do one square mile in a year. But lots of people can. . . . The reward system works."

Mentored by the wealthy collector Jim DuPont and meteorite aficionado Richard Norton, Haag first touched meteorites in Norton's presence. "The air was electric. I handed him a chondrite," Norton has written, "then a pallasite. Here was a man possessed." After selling some meteorites for Norton at the 1982 Tucson show, Haag seized on the business opportunity. With relatively little meteorite dealing going on at the time, Haag searched the area around Meteor Crater before the lands were cut off from dealers and collectors. He and a partner also found more Brenham pallasites, the sales of which financed other searches. The Spanish-speaking Haag, years after the 1969 Allende fall in Mexico of a rare carbonaceous chondrite, found more specimens of that precious stone. Then he placed an ad in *Scientific American* that cost about $4,000, all that he had. The ad reaped $100,000 in sales. "I couldn't even talk to anyone unless it was about meteorites. I was absolutely obsessed," he says. Haag organized mailing lists and contacted the handful of other established if lower-key dealers, including Glenn Huss.

Haag was in business. And he transformed the business he was in. But then, when I meet him, The Meteorite Man is contemplating getting out of the business. I've driven to his suburban house in Tucson on a warm, sunny October day, my third trip to the now-familiar city. When he greets me, the tanned, handsome, and barefoot Haag is wearing a short-sleeve shirt and shorts. His white teeth gleam, his blue eyes shine. He has shoulder-length blond hair. I would have hated this guy in high school, I think, and Haag begins talking immediately. We're beside photo-covered walls, shelves of books, cases of meteorites. We're downstairs in his basement office with its bank vault door, mural of Saturn, black filing cabinet, track lights, black walls—one with a dent. He jokes, "It was hit by a meteorite." I ask, "How much?"

Pedestals and meteorites line the walls. There's Gibeon, "the best iron in the world." He has a big slice of Cape York beside the wall inside his "meteorite cave." The historic specimen is joined by three moon rocks, a half-dozen Martian meteorites, and, it seems, every other meteorite known to humankind.

He's traded this exotica with "Moscow, the Smithsonian, campesinos in Mexico . . . I've traded for trucks, tools, trailers, kisses, gold," he says. But it's not just commerce, and Haag is riffing now. "They're religious, they've been put on altars, they're a mystery sent from heaven, they've been anointed with oils, they've been crushed up and eaten—sent from God. I kind of worship them myself."

He's also eaten them. Moon dust too. "It's not like cheese but it has a tingle to it."

Yes, he's riffing. "Because I have to sell these to live, I have to get people interested in them. My spin? They're treasure. Each one has a huge story, who found it, its space history." He holds a meteorite in his hand and says, "This is my ticket to outer space." Yours too, if you buy one. Haag prints about 10,000 copies of his catalogue, or "field guide," at a time, an indication of a pretty hefty customer base. Nonetheless, his sales average a modest $150 per transaction.

"I've gotten shit for being a capitalist," he says. He's even shown up in a French comic book called *Chasseurs d'Étoiles*. "Look at me, I look like a monkey in this!" The plot has something to do with stealing a meteorite and getting stabbed. In another volume of this series, the character Maag sometimes blurts out in English "Fuck the . . . " and "My balls!" and, inexplicably, "Potted meats!" In the story, a meteorite is sliced open to reveal a mirror.

When I ask him if he could have predicted how it all turned out, he laughs, his blonde curls shaking. "God no. I'm glad I found my niche. I've got twenty-five years, millions of dollars, tens of thousands of hours invested in this."

I notice some gray hairs in among the blonde.

Haag's not grinning now. "It's too big." He's looking at the scads of meteorites arrayed around us. Robert Haag wants to sell the whole collection, preferably to a single institution, where people can see it. "It's not fair to keep it to myself."

But he adds, "I will *always* go to hunt meteorites."

Haag, who has talked virtually nonstop since I arrived an hour or so ago, pauses as he looks at his collection. So I decide to throw him off balance. I ask him what his favorite rock-'n'-roll band is. He looks at me, mulls. *He's stopped talking.* Then, tentatively, lists Bowie, the Rolling Stones, Aerosmith. I ask about Van Halen. "Sure, Van Halen works." Then—"The Cult! Ah, The Cult, fuck they were good." Then he's riffing again: He's made guitar picks out of meteorites and given them to rock stars.

But he's met no bigger star than the original Meteorite Man, Harvey Nininger, whom he saw at a conference. At Nininger's talk, Haag was in the first row. He had read every Nininger book and he thought, "Nininger did it, he did it, he made it!" Harvey couldn't hear the young man when "I told him how he was my inspiration. But I said it to myself."

A call comes in. A customer wants to come by and see Haag, who, even when giving directions, does so with a slightly manic edge. "Then you go uphill *barrrrrrom*."

Some people can't stand Robert Haag. I like him. His wife Heidi will later tell me of the time when he replaced a fried-chicken meal with a live chick from home, pulling back into a fast-food drive-through to complain that the food wasn't done enough then revealing the chirping package. The employee screamed. "Bob," Heidi will say, "has a fondness for chicks" and will laugh at her own joke.

HE ALSO HAS A fondness for a rock he found many years ago in a box of rocks. In among a bunch of run-of-the-mill meteorites sent to him from Australia, Haag was stopped cold by a stone with an unusually bubbly fusion crust. He had learned well. He was looking at a rock that had been blown off the Moon's surface. An article about the stone, with Haag's name in the credits, was published in *Nature.* He sold some of the stone, donated some, and traded a bit as well.

The find was not without controversy. A story in *New Scientist* noted that "Haag says he bought the specimen from a dealer. But meteorites found in Western Australia are the property of the Crown and can be exported only for scientific purposes, with a special permit. No such permit was issued." In an interview Haag fired back. "If the Western Australia Museum thinks the rock came from there . . . ," he snapped, "why don't they go out and find one themselves.'" The official who looked into the matter said that the government "can't prove Haag's meteorite landed in Australia and we can't prove when it was exported." The case was dropped.

Dealer Jeff Kuyken later tells me that "Australia has laws which are very confusing at the best of times." Kuyken, who has an export permit, is able to sell some of his material overseas.

Haag also has persuaded a museum in Nigeria to trade for several of his specimens. What did the museum give? Thirteen pounds of something called the "Zagami shergottite," a Martian meteorite. It's unclear what the museum official knew about the values of Martian meteorites or if the deal involved more than a trade. In his book *Rocks from Space,* Norton does describe Haag's leaving money on a table at the customs office and that the money vanished when he turned his back. Haag's thirteen pounds of Mars was worth a cool million. Haag was also once arrested in 1990 after an Argentine minerals trader apparently tricked the American. Haag was told a rancher would sell him a 37-ton Campo del Cielo iron, but after Haag had the big meteorite on a truck authorities took him into custody, according to Norton. He didn't know nor had he been told that meteorites belonged to the government of that particular province in Argentina. Haag got out of jail, didn't get the meteorite, but reaped the benefit of increased sales brought on by all the press.

While some have objected to Haag, nearly everyone I've spoken with respect his energy and his place in the field. The same cannot be said for Ronald Farrell. The *New Haven Register* reported on July 1, 1997, that Farrell had been imprisoned in Brazil after authorities there alleged that he swapped out worthless stones for rare meteorites in an official collection. That September, the paper said, the dealer was let out of prison "after pleading the Brazilian equivalent of 'no contest' to charges that he stole priceless meteorites from a national museum." He and a partner were both fined several thousand dollars. Lawyers for Farrell contended that he'd been set up by a museum curator who had "stashed the meteorites in Farrell's luggage so she could 'catch' him and further her career," in the words of the New Haven reporter. Farrell has also been accused by Yale researchers, according to sources used by the *New Haven Register,* of "posing as a Yale professor in 1991" to allegedly dupe Egyptian officials into loaning him two Martian meteorites, one of which was sold and one of which was recovered. Farrell wouldn't speak to me. Government agencies could provide me with no files on him or other dealers.

After the Brazil incident, scientist John Wasson was quoted saying, "These dealers are everywhere now. They are ready to bribe the curators of small museums and have largely depleted some museums in Africa." When I asked Wasson about this, he says it "was speculation. . . . On the other hand, it seems clear that it was not trades among peers that resulted in [Zagami] leaving Nigeria. I also know that dealers could get rare meteorites out of the Geological Survey of S. Africa . . . when I could not get a reply to a request for small samples."

As one meteorite insider told me, "A lot of chicanery goes on—that's part of the story too." Electronic discussion groups devoted to meteorites can flare with e-mail accusations ranging from underpaying Third World villagers for recently fallen meteorites to whether dealer competition is out of hand. Criminals even stole some hefty Gibeon iron meteorites that had been on public display in a Namibian town, though the thieves were apprehended.

One way the meteorite community guards against chicanery is through the

International Meteorite Collectors Association. On its website, the group notes that perfectly credible dealers may choose not to belong, but those that do pledge to "adhere to the highest standards of meteorite identification and proper labeling practices." IMCA is an excellent resource for collectors.

If science helps to drive the interests of collectors, scientists themselves are sometimes ambivalent about the dealers who supply them with meteorites. Apart from the National Science Foundation's annual meteorite expeditions to Antarctica, most material today comes from private dealers and field searchers. Scientists are just too busy in labs to search in the field.

Alan Rubin, of UCLA, works with a few dealers, sometimes trading specimens, sometimes purchasing. He notes that "some curators are very antagonistic toward dealers in general and some in particular." He wouldn't name those held in dim light by the research community but did mention several with whom he has good relations, including Haag, Reed, and Killgore.

Conflict revolves around three issues: cutting up specimens before scientists can study an entire sample, withholding of location data, and donating valuable specimens. In the first instance, Jeff Grossman, of the United States Geological Survey, compares the slicing of specimens before whole rocks have been studied to smashing an Anasazi pot, then selling the fragments. Second, withholding precise information about where dealers have found meteorites makes it difficult for scientists to pair up recent finds with prior ones. Is this a new meteorite or a piece of one already known? Is there a strewnfield out there scientists can't map because dealers won't say where the meteorites are coming from? Failure to report precise locality information may protect dealers from competition but it can frustrate scientists. Finally, the issue of donating material for research has caused friction. If someone wishes to publish research in the Meteoritical Society's journals, the meteorites being studied must have official names from the group's Nomenclature Committee. To have a meteorite named, the dealer must give 20 grams or 20 percent of the specimen's mass, whichever is less, to an institution. But with lunar or Martian meteorites, the price per gram can run up to $1,000 or more, and some dealers just can't stand to part with such valuable stuff.

Difficulties with the meteorites emerging pell-mell from northwest Africa even prompted a workshop in Morocco at which academics and bureaucrats grappled with such matters as the illegality of meteorite exports from nations with unsecured borders. Balancing these concerns are economic ones; how should scientists and goverments reckon with "those, from nomads on up the chain, who make their livelihoods collecting and selling the meteorites?" wonders researcher Tim Swindle.

Some researchers would like meteorite commerce banned entirely. Enforcement, however, would be almost impossible. Grossman says he appreciates that "meteorites [can] come from dangerous places, which scientists do not have the resources to . . . search." Jeff Kuyken agrees. Most scientists "have come around to embrace or at least tolerate the collector/dealer community," he says. "Personally, I am more than happy to help out any meteoriticist that asks. . . . It's very difficult for busy meteoriticists to know what is out in the market."

That last point cuts to a final source of tension, which ultimately is cultural. One researcher, who asked not to be identified, says that there is a "fundamental disconnect" between dealers and scientists because of the fast tempo of the market and the slow pace at which institutions make purchases. Further, dealers with new specimens sometimes expect scientists to classify a meteorite quickly, but the academic community does not value classification as much as original research. Then there's the tension between the dealers' natural desire to make specimens seem "as unusual as possible" and the scientific tendency "to make things fit into existing groups."

Scientists aren't the only folks who sometimes have issues with dealers. In early 2002 and in late 2007 dealer Darryl Pitt was attacked by the press and the Confederated Tribes of Grand Ronde for selling pieces of the Willamette Meteorite—Tomanowos. The American Museum of Natural History had allowed some cutting of the meteorite in the late 1990s, a not untypical practice whereby museums can barter specimens with other institutions and with dealers. This was prior to the agreement with the tribe. Matt Morgan bought one of the two pieces Pitt sold in 2002 for $11,000 and planned on dicing it up to keep some back while selling the rest. An Oregon chiropractor bought a smaller piece for $3,000 and donated it to the Grand Ronde.

Despite occasional conflict, a passion for meteorites unites amateur collectors, professional dealers, and scientific researchers. That point was driven home in meteoriticist Alan Rubin's review of *The Robert Haag Collection of Meteorites.*

Rubin's positive review, which appeared in a scientific journal, called the nearly 300 photos of Haag's specimens "impressive and of significant scientific value." His description of that value indicates the level of precision in today's meteoritical studies. For example, he notes that one photograph shows "the inhomogeneous distribution of vesicles in the Ibitira eucrite." Who knew? Rubin suggests readers "approach Haag's book the way teenage boys approach *Playboy* magazine. Don't read it!"

On the cover of the book is Haag's "Venus stone," the Adamana, Arizona, chondrite that landed on the earth ablated and oriented in the shape of a rocket nose cone. Or, as Haag's nickname suggests, a woman's breast.

Book VIII

Church of the Sky

And her fragile bones: how simply, how completely,
Janis would disappear. And all of us.
—Robert Olen Butler, "Doomsday Meteor Is Coming"

So G'sell so.
—"All's well," the nightly message of Nördlingen's town crier

Press a hand against the walls of St. George's Church in Nördlingen, Germany, on a mid-March day and stone's cold seeps into skin. Look up at the church's gray, black-flecked tower—the Daniel—and you'll see its roof tip against the blue some 300 feet later, like the nib of a pen about to write on colored paper. Like all such buildings, this one aspires to heaven. Unlike most, it's built of the sky itself.

Or, rather, built by explosive consequences dating to 15 million years ago, when a stony meteorite more than half a mile wide hurtled down from the sky at 43,000 miles an hour. The impactor's force penetrated more than 3,000 feet into the earth, blasting out rocks and destroying every living thing within 60 miles. A regional event, the impact nonetheless was equal to the Hiroshima bomb—times 250,000. The resulting crater was 15 miles in diameter—15 times larger than Meteor Crater in Arizona. This crater is now called the Ries Basin.

A satellite photograph of southern Germany reveals, east of Stuttgart, a smooth circle set in the mountains of the Swabian Alb. This is the Ries, which, after the impact, became a lake—recolonized by life—and which over time filled in with sediment, then disappeared. Uplift and Ice Age sculpting millions of years later resurrected the crater's shape. Tucked inside the crater, tucked inside the Ries Basin, is the winsome town of Nördlingen—it's been there for centuries—with its curious church made from the earth-and-sky remains of a meteorite impact. St. George's, the Daniel, and many other buildings are made of suevite, a whitish-gray breccia—a mixture of fragments—dotted with "flädle," a black, glassy melt. Until the mid-twentieth century geologists mistook suevite for tuff, a deposited volcanic ash similar in appearance. This misidentification suggested a volcanic origin for the basin.

Long before anyone knew the origins of suevite, let alone the entire Ries Basin, construction began on St. George's Church. It dates to the early fifteenth century, and the church honors a saint who battled creatures of fire and sky—dragons.

In 1960, geologist Eugene Shoemaker, his mother, and his wife Carolyn drove into the Ries Basin in their VW Microbus. Shoemaker also had with him a pick, a hammer, and a hunch. He was stopping in the basin before driving to Copenhagen, where he'd deliver a paper titled "Penetration Mechanics of High Velocity Meteorites, Illustrated by Meteor Crater, Arizona." Carolyn says, "We saw more quarries than we ever knew existed and fewer 'tourist' sites than anyone else who ever went to Germany for the first time." It was also a time for grief. Gene's father had died just two weeks before.

The 1950s and 1960s saw much ferment in geology. In David Levy's overview of this period, he notes that Shoemaker, "almost alone among geologists," thought the geology of the Moon and the origin of its craters "a fertile subject" for scientists. Even some paleontologists looked skyward—though not at the Moon—to explain extinctions on Planet Earth; they suggested meteorites or supernovae might be suspects. Harvey Nininger had speculated in 1942 that massive impacts could lead to global extinctions, though no one paid attention at the time. In addition to arguments over impacts and the origin of the Moon and its craters, new voices would soon argue for continental drift, or plate tectonics, an idea first espoused to derision in the early twentieth century by German meteorologist Alfred Wegener.

The excitement over plate tectonics would overshadow arguments over the frequency and effects of impacts, at least until the early 1980s. That's when scientists announced that they'd found iridium—an element associated with cosmic impactors—scoring the boundary between the Cretaceous and the Tertiary, the demarcation between the Age of Dinosaurs and the Age of Mammals. The firestorm that erupted when Walter and Luis Alvarez argued that the extinction of the dinosaurs had been caused by a cosmic impact may seem distant now, but it was epic in its moment, with most paleontologists vociferously favoring gradual climatic change. Eventually, geologists found the iridium signature across the globe, then the dino-killing crater itself, which is located underwater off Yucatán's coast.

At the time of his 1960 visit to Germany, the mercurial, ebullient, workaholic Shoemaker had studied salt structures as well as uranium in the Rocky Mountain West. But it was his work at Meteor Crater that had sealed his reputation. After reading a paper by a petroleum geologist critical of Harvey Nininger's pro-impact evidence at Meteor Crater, Shoemaker took a simple step. He had a crater sample analyzed by spectrograph. The result? Fused quartz, which only melts at temperatures higher than those produced by volcanism.

There was more. Shoemaker, who had visited craters created by underground nuclear tests, immediately saw their similarity to Meteor Crater. He noted that "violently eruptive volcanoes . . . do not produce deformation of the crater walls remotely resembling that produced by impact." He compared such volcanoes to popping open a bottle of champagne; the force decompresses, such that "the walls of the volcanic vents collapse *inward.*" Such force simply could not carry huge boulders, which require an explosion outward.

Where Nininger had laboriously collected, sorted, and studied tiny metallic pellets from around Meteor Crater—using magnets on a stick and toy cake pans—Shoemaker, with his advanced geological training and institutional ties, could more quickly bring together the field evidence showing, without any further doubt, that Meteor Crater was formed by impact and that the impactor had vaporized. The lingering theories, from steam blowout to collapsed salt dome, could finally be put to rest.

And U.S. Geological Survey scientist Edward Chao was thinking—like Shoemaker—of coesite, which had been discovered in the lab just prior to

Nininger's 1956 book on Meteor Crater. (Loring Coes had compressed quartz in a lab with such force that he'd created a new mineral, which now bore his name.) On June 20, 1960, Chao found coesite in rocks from Meteor Crater. This was—to use an almost perfectly appropriate cliché—the smoking gun. Chao, Shoemaker, and Beth Madsen made the announcement in *Science,* though the piece didn't mention Nininger.

A month after this discovery, Shoemaker was driving in a storm along the bottom of the huge Ries Basin. The thirty-two-year-old geologist arrived with his wife and mother at the Otting quarry near Nördlingen at sunset on July 27, 1960. The rain lifted. "He left us on the [quarry] rim watching," Carolyn recalls. "He was enormously excited . . . [and] sure that he had found the samples he needed." Found them he had, at a small exposure of what anyone but a geologist would have considered dull gray rock. "By the light of the setting Sun," Shoemaker recalled, "we looked at the shock-formed rocks that were thought to be volcanic. I took one look at these rocks with a hand lens: No question these were impact rocks!" He saw the blebby black glass caught within the finely brecciated suevite, like drops of tar captured in concrete. In his field notes, Shoemaker wrote, "Everything seen in the quarry supports the hypothesis that the suevite represents fallout of high trajectory impact ejecta." Carolyn says that Gene wanted to send the samples to Chao back in the States and worried about finding the post office in town.

The next day Shoemaker located the post office—with its eight-sided suevite columns at the entrance and German eagle above the door—and mailed the package to Chao. Under sunny skies, the couple walked the stone streets of Nördlingen, admiring the old town wall and charming flowerpots beneath window after window. They came soon to St. George's. Carolyn vividly remembers Gene's reaction: "'Look at that!!!!' . . . He pulled out his hand lens that he always wore about his neck and examined the wall closely. Then he was sure. The church was built out of suevite and probably contained a lot of coesite within it. Gene often walked carrying his geologist's pick and this time was no exception. He probably wouldn't have used it to sample the cathedral but I assured him that he had better not," Carolyn says. Of this epiphanic moment Shoemaker wrote, "The blocks of suevite in the wall had beautiful pieces of glass quenched from shock-melted rocks and also shocked granite. I remember thinking, 'Here is a church whose very walls must contain abundant coesite, our proof for the impact origin of the Ries,' and, at that moment, I was the first person to realize it."

He was right. Chao soon found coesite in the Otting quarry samples. As Shoemaker friend, astronomer, and writer David Levy puts it, "This was a major find." Confirmation that the Ries Basin was a huge crater suggested that impacts were far more frequent than geologists had realized, and these events occurred on massive scales that altered regional landforms. Further, this particular impact had taken place long after the initial bombardment of the Earth early in its history. Cosmic impacts—big ones—extended into the geologically recent past. This was a discovery that helped to overturn or, at least, complicate about two centuries' worth of placid uniformitarianism, the principle that geological processes worked slowly, that violent changes such as floods and eruptions were confined to small scales, and that present geological actions are the only way to understand past geological actions. If Daniel Barringer, Harvey Nininger, and others had pointed to the fact that impacts occurred after the formation of the Earth, Shoemaker extended that insight with his finding, because the Ries was a large crater, not a relatively small hole in the ground such as Meteor Crater. Discovery that the Ries was an impact site sent scientists scrambling to look for other craters, large and small. By the early twenty-first century, nearly 200 impact craters were confirmed worldwide. After the Ries findings, the sky would never seem as safe, and Shoemaker would be established as one of the most important figures in twentieth-century science.

Not long after Shoemaker visited the Ries, another scientist discovered what came to be called PDFs, or planar deformation features, which are sets of parallel planes of shock-formed glass. These are now among the most important visual evidence for impact events.

Eugene Shoemaker would go on to hold key positions in the Lunar Ranger and Surveyor robot missions and help train Apollo astronauts, even as his dream of becoming one was cut short because of Addison's disease. He'd criticize NASA for not doing enough science on the Moon. One can only imagine how some of Shoemaker's private conversations about Apollo must have gone, for this was a man who once shut a car door with enough strength and anger that the window shattered. He founded the U.S. Geological Survey's astrogeology unit in Flagstaff. With wife Carolyn and friend David Levy, Shoemaker organized telescopic searches for objects that might threaten the Earth. Their most spectacular find was Comet Shoemaker-Levy 9, a series of comet fragments that slammed into Jupiter in 1994. No one doubted the ongoing reality of impacts after that. Eugene Shoemaker died in a car accident 1997 in the Australian Outback while he and Carolyn searched for craters. Two years later planetary scientist Carolyn Porco arranged to have a small vial of his ashes put onboard Lunar Prospector, a craft that was deliberately crashed into the Moon.

AFTER CLIMBING 350 STEPS up a narrow, winding chamber inside St. George's Church, after passing the huge gearwork for four bells and the wooden walls inside the stone stairwell tower, after paying for tickets in a room high in the tower, after seeing the "witch" hanging over the watchmen's stove (to guard against evil spirits), and after catching our breath, Kathe and I duck out into the sudden, hard south wind, squeezing against other visitors on the narrow Daniel "balcony" no wider than two size-9 shoes, as vertiginous and tight as a razorback trail.

One week in mid-March, Kathe and I have come to the tower of St. George's and the Ries Basin. We're here to shadow the footsteps of Eugene Shoemaker and to fathom or try the events that led to a crater now clotted with postcard villages and farms. The Ries Basin is the perfect place to mull over the dangers of the sky. It's practically Chicken Little's backyard.

Meteor Crater is a feature, albeit stunning, of the landscape. The Ries crater is so huge it *is* the landscape. This you can see from atop the Daniel. The Ries Basin is about 25 kilometers, or some 15 miles, in diameter. From the Daniel, on a clear day, you can see the 99 villages of the valley. Disaster seems distant unless you understand that the treed hills to the south, beyond the town's wall, mark part of the crater's outer rim. The rim rises about 500 feet from the basin floor.

Now I'm facing Löpsinger Tor, one of the gate towers of medieval Nördlingen, as well as the church, fountain, Market Square, and Hotel-Sonne below. The Hotel-Sonne, where Kathe and I ate lunch yesterday, boasts having hosted the great German poet and naturalist Goethe and, centuries later, the Apollo

Moon-mission crews. While we ate lunch there I thought of those astronauts and recalled how, starting in about sixth grade, I'd tape-record broadcasts of the lunar flights and play them back in my darkened bedroom, dreaming of space. In high school, Miss Hawk kicked me out of advanced algebra—for what, I can never recall. Doing poorly? Goofing off? I stopped recording space missions after that.

I'm used to heights on mountains, but it's disconcerting to see a packed jumble of angled red roofs below me. At places, the town's ancient wall is almost lost behind so many roofs. Beyond the wall, more houses, warehouses, and fields. Sheep dot gentle hills. We move on to the northwest side, look into the hazy-sky sun, look straight down at a kestrel, spire-perched and preening. In Nördlingen, a man is swinging a boy by the arms, a bicyclist in red crosses the square, Peugeots and Renaults negotiate tight curves, light and shadow graze the roofs. At the Hotel-Sonne outdoor café people tarry and pigeons flutter. I hold on to the stone balustrade and look down chimneys and try to imagine a stony plunge from space.

Inside the outer crater rim lies the crystalline ring, a U-shaped line of hills that opens to the north. Nördlingen is set upon the crystalline ring and sediments deposited since the impact. Composed of granitic basement rock—the oldest igneous and metamorphic rock of a given area, which occurs beneath sedimentary deposits—the crystalline ring was forced upward during the impact.

Last night we met Gisela Pösges, a geologist attached to the local Rieskrater-Museum, who will tomorrow take us into the field. She'll show Kathe and me more nuances of this meteorite-derived landscape. Tall (she's a former high school basketball player), blonde, with a wide smile, Gisela took us to dinner at a tavern where she seemed to know everyone. I drank Rieskrater Hefeweizen, setting my glass down on a coaster that showed an upside-down beer falling over Nördlingen. The first impact on the Earth of a giant brew from space.

Before the actual impact, the subtropical Ries contained moist valleys and dry uplands thriving with snails, elephants, rhinos, and huge, lumbering tortoises. The land lay atop limestones of the Triassic and the Jurassic when the area was awash with the waters of the Tethys Sea.

Then it came: A bow wave of superheated air grew like a shield in front of the falling rock, laterally shaving off a layer of earth to the sides, jetting out what became tektites. Ejected high above the planet, then falling like rain, countless tektites landed more than 300 miles away. These particular tektites are called moldavites, and they're as green and feathery as summer algae, pretty enough to be sold today as pendants by enterprising meteorite dealers. Within one to two hundredths of a second, the half-mile-wide meteorite had thrust down into the basement rock, with the meteorite and terrestrial rock compressed to just 25 percent of their original size. The temperature increased to some 20,000 degrees Fahrenheit—almost twice the temperature of the hottest portion of the Sun's photosphere—and the area experienced pressure equivalent to 4 million Earth atmospheres. The meteorite and surrounding rock vaporized. The crater had formed within seconds, and supersonic shock waves continued to bounce throughout the rock. For two minutes, ejecta exploded up and out, sending limestone blocks—tons of them—about 40 miles away. The melted ground itself rebounded like water rippling in a pond, with the granitic basement decompressing and swinging back into what would become the crystalline ring. The superheated mixture of basement rock, limestone, glassy melt, and dust—a flying cauldron—reconstituted into suevite, some blazing outward beyond the crater rim, some falling back into the roaring center, whose depth was reaching 2.5 miles. A mushroom cloud ascended, a glowering, glowing, thundering rumble, shot through with fires and lightning, towering 20 miles, almost into space.

In a matter of seconds the inner portion of the crater filled with hot rocky material that was pouring back in. The rebounding shock waves pushed the crater floor up, as other material slid back, reducing the crater's depth from 2.5 miles deep to about a half-mile. In a quarter of an hour, it was over. Some 93 cubic miles of material had been ejected. More than 600 cubic miles of rock had been moved in rolling and gliding waves along the ground. Fires burned. Dust suffused the sky. Winds plucked trees like a comb pulling hair from a scalp.

A robin calls. People chuckle as they press against us on the Daniel wall, all jackets and sunlight. I touch this pockmarked gray-white stone with its glassy black melt. "Suevite," I whisper, and all across the stone grows yellow-green lichen.

Visiting the crater museum, which is housed in a renovated sixteenth-century barn, amid displays of meteorites and exhibits on the solar system, I had been stopped by a demure vitrine with two chunks of blackish stuff, one walnut-sized, the other a larger square. Charred and cracked with tiny fissures, these once had been trees. They were burned at distance from the impact. They were alive then, 15 million years ago. More than the rocks and the labels, this burnt-wood evidence of catastrophe held my attention. What kind of trees were they? I just stared and wondered. From them, what leaves? What buds? What song? Fossil trees cracked like the lichen on this suevite wall.

Cold, Kathe and I decide we've seen enough vista and walk back down from the church of the sky. We emerge on the street and, after coffee, we walk the wall, the perimeter of the old town, which I realize is the size of the meteorite itself, and there being no meteors in the sky that we can see to count, we name birds instead. Turtledove, greenfinch, mistle thrush. We hold hands in the shadows of leafless trees and pass buildings decorated with plastic Easter eggs. We stop and name what we know, what we're learning, and we lean in silence against each other as the light goes gray, and I relish how we array each moment as a beautiful, useless stay against time.

NOT THAT THERE IS a significant chance that Kathe and I will be killed on this trip, whether by car collision or meteorite-beaning. I've learned enough to know the former is far more likely than the latter. But statistical truth loses a bit of its power while we walk about a village set in a former scalding, rock-hurling maelstrom. Still, not a single human death has been undeniably attributed to meteorites in historic times.

There is some juicy gossip, though. Philip M. Bagnall, author of *The Meteorite & Tektite Collector's Handbook,* notes that approximately 70 people may have been dispatched by errant space rocks, including a monk in Milan in 1650 and two seventeenth-century Swedish sailors. In 1929, a guest at a wedding in Yugoslavia was reputedly killed by a meteorite. A few years ago, a mathematician from New Zealand unearthed a narrative about an English shepherd who in 1725 died along with five sheep when they had been struck, it seems, by meteorites.

Accounts of injuries also abound. In 1827, a man in India was allegedly hit in the arm by a rock falling from space. The afternoon of August 4, 1835, a meteor boomed, fragmented, and fell over Aldsworth, Gloucestershire, sprinkling black stones that playful children, extending their hands to catch them, believed to be beetles. On July 14, 1847, the Braunau meteorite—weighing about 40 pounds—crashed into a bedroom where three children were sleeping; the children were not injured. (Braunau was a rare meteorite in that, hours after landing, it was said to be too hot to handle. Most meteorites are not warm to the touch upon landing.) In 1946, an English boy may have crashed his bicycle after being hit by a meteorite. In 1992 stones hailed down on Uganda, and one boy was struck but not seriously hurt.

Only one instance of injury has been thoroughly documented: the famous case of Mrs. Hewlett Hodges of Sylacauga, Alabama, who, on November 30, 1954, sustained injuries to her abdomen and hip from a meteorite weighing 10 pounds. The wife of an arborist, Mrs. Hodges was sleeping on her couch when the stone ripped through the pitched roof beside their wraparound porch, blasting a hole in a corner of her living room ceiling, whizzing past her lace curtains and floral wallpaper, hitting a radio, smacking her hand and hip, ricocheting off her body and the couch, then falling to the floor. Had the meteorite hit her outside—not slowed by the roof and radio—she probably would have died. Though many people in the area saw the meteor, the only witnesses in the Hodges yard were an empty lawn chair and a tire swing in a tree out front. A photo shows the not-inconsiderable Mrs. Hodges resting in bed, a bow-tied doctor with glasses pulling back her striped nightgown to reveal a bruise at least twelve inches long, dark and wispy as a sunspot and shaped like a kiss. A sickly and withdrawn woman, Mrs. Hodges found her physical health and state of mind troubled by the stone. Things went poorly for Mrs. Hodges, who, with her husband, had to pay the property owner—the Hodges were renting—$500 for the meteorite. The rock was later given to a state museum. The couple would divorce.

One of the few cars known to be struck by a space rock belonged to a young woman in New York state. Watching television, Michelle Knapp and her boyfriend were at her house in Peekskill when they heard a loud crash outside on the evening of October 9, 1992. Police eventually realized that her Chevy Malibu had been hit not by delinquents but by a meteorite, thus increasing the car's value from the $100 the high-schooler had paid for it to the tens of thousands spent by dealers who wanted the car and the stone. Dealer Allan Langheinrich now owns the vehicle.

In Wethersfield, Connecticut, two houses have been struck by two different meteorites, one on April 8, 1971, one on November 8, 1982. Ballistophobes should stay away from Wethersfield.

In 1984, a mailbox in Georgia received a special airmail delivery in the form of a meteorite. The mailbox fetched about $83,000 in an auction.

As to animals, only one is known for certain to have died from a meteorite hit. A cow in Venezuela was whacked by a chondrite on October 15, 1972. A long-told story of a dog killed in Egypt by the Nakhla Martian meteorite in 1911 turns out to be untrue.

Though a partial chronicle, these entries underscore part of the dangerous appeal of meteorites: the ironic distance between massive catastrophe and a single person, a body in time and space. The distance between a Miocene day in the Ries and a black bruise on a woman's thigh. A *New Yorker* cartoon once illustrated this nicely. Beyond a window, a meteorite had rammed, pointy nose first, into the ground; inside, a child cried, while the mother told the just-arrived father, "Poor Zachary. A meteorite squashed his swing set."

The odds of being killed by a meteorite are, of course, incredibly small. According to the British newspaper *The Telegraph*, an outfit called the Safety and Reliability Directorate of the United Kingdom Atomic Energy Authority calculated that, in the reporter's words, "some poor Brit would be squashed by a heavenly body every 7,000 years or so." The chances that a meteorite will hit you this year are trillions to one. By contrast, in any given year in the United States, you have a 1 in 47,000 chance of being hit by a car.

BUT THE RIES IMPACTOR. Shoemaker-Levy 9 slam-dancing against Jupiter. Canyon Diablo sculpting Meteor Crater in an instant. And, of course, the dino-killer. We are obsessed with these now, the really big ones, the blockbusters. Today we understand that rocks in the sky cross our path, and that someday a close call will become very bad news.

We know asteroids get in our way because of what Karl Reinmuth found on

April 24, 1932, using a telescope called "Das Bruce" at the Max Planck Institute above Heidelberg. Das Bruce can detect asteroids, and that April night Reinmuth found for the first time one whose orbit intersects the Earth's.

He named it 1932 HA, but after deducing that its orbit crossed Earth's and Venus's—and that it swung close by the sun—Reinmuth renamed it Apollo. The term would come to be used for a whole class of Earth-crossing asteroids. Reinmuth's log for April 24, 1932, contains among the subdued numerics of careful entries an exclamation point and underlining. The punctuation—!—of an epiphany. In red.

The mile-wide rock would zing by Earth the next month at 7 million miles. If Apollo's orbit had taken it closer to the Sun sooner than it did, the asteroid's distance at its closest approach to Earth would have been cut in half. Just around the block, astronomically speaking.

The historic glass plate that shows the first detection of an Apollo asteroid is housed with 10,000 other such plates in seven cabinets. The plate archives for the Max Planck Institute are less impressive than they might sound. It's an under-the-stairwell, cobwebby sort of affair, more a closet than anything else, with a concrete floor, paint cans, old typewriter, old computer monitor, rows of mainframe computer magnetic tape, a beat-up bookcase, a mop, some chairs. The glass plates are reverse image. The stars are black, the sky white. A short, faint streak on Plate 6060 betrays the presence of the earth-crossing rock. It is, as one of Reinmuth's colleagues calls it, "The Plate of Detection!" A prosaic streak, really, the way a firefly leaves a trace of light in our eyes after it's stopping glowing.

But it was—is—Apollo, a rock whose name, with its long vowel sounds, rolls across the tongue like the roll of rocks in orbit, skipping over the liquid "l"s like just-missed planets. Apollo, nipped by gravity till it slices in front of some place in space where, sometimes, we show up on our just-missed Earth.

A HANDFUL OF EARTH-APPROACHING and Earth-crossing asteroids were observed in the mid-to-late 1930s, including Reinmuth's Apollo. (A Near Earth Asteroid or Near Earth Object is defined as one that comes within 28 million miles of our planet.) One of them, Hermes, came within a half-million miles of the Earth. The Moon orbits at just half that distance. Because these asteroids all pass so close to the Sun and are often small and dim, astronomers lose them in the glare. Decades can pass before they are seen again.

These NEAs or NEOs are the remains of objects that had been in the asteroid belt; collisions with other asteroids and interplays with the gravity of Jupiter and the Sun alter orbits, sending some rocks toward the Sun and some toward us. Something called the Yarkovsky Effect can also move asteroids or smaller rocks into gravitationally unsteady pockets, or "resonance areas." The Yarkovsky Effect is the uneven warming of an object's surface in space; the warmer side radiates heat and pushes the object toward the cooler side. A more nuanced version of this phenomenon can affect other movements of asteroids and small particles, such as how they spin. It can force unevenly shaped particles out of the solar system and, over time, actually alter the shape of those asteroids that are loosely bound aggregations of rubble.

A few years ago asteroid 1991 BA—the tiniest yet seen, about half the length of a mobile home—passed within 106,000 miles of the Earth, then the closest confirmed approach. Big enough to wipe out a city, the asteroid was discovered hours before its close call. Two years after that, another asteroid was discovered the day *after* it came within a tight 90,000 miles of Earth. The next year a space rock the size of a truck was found to have grazed us at a mere 65,000 miles. The only consolation may be knowing that an object that size typically breaks up on entry. Another truck-sized rock came within 48,500 miles in 2003. It too wasn't seen until it had passed by. A year later the 100-foot-diameter asteroid 2004 FH came within 26,500 miles, swinging over the Atlantic, and it was announced as the closest confirmed approach of an asteroid. By summer 2007, experts had identified about 900 "Potentially Hazardous Asteroids."

Astronomers are also concerned about comets, which are less numerous in the inner solar system, given their widely arcing orbits. Such orbits take them far out to the solar system's fringes before they sweep closer to the sun. Comets are worrisome because of their greater velocities and because they can be hard to detect. In fact, a whole class of comets hiding out in the asteroid belt has only recently been discovered.

In terms of an actual impactor—asteroid or comet—there appears to be about a one-in-5,000 shot of a civilization-killer hitting in any given century.

This works out to two times per million years. The scientific community has even developed threat rankings to help the public understand potential impact hazards. Called the Torino Scale, it sorts impact threats by taking into account the likelihood of an encounter and the potential energy if the object hit. The scale runs from 0—where an object won't hit or would break up in the air—to a 10, which is Kiss the Biosphere Goodbye for Now Time. The scale assumes a 10 event happens every 100,000 years or so, more often than some researchers suggest.

When objects start scoring a 3, I pay attention. At that Torino ranking an impactor can inflict local damage and has a 1 percent or more chance of hitting the planet. Scientists estimate that every two centuries or so, a meteorite hits that is 45 meters, or about 148 feet, across; this would contain the energy equivalent of 10 megatons of TNT. Regional destruction can result from an impactor that is 500 meters, or 1,640 feet, wide; this is estimated to occur once every 60,000 years. The power of such an impact would equal 10,000 megatons of TNT. In 2001, a research team found evidence for an impact-induced regional ecological catastrophe between 800 and 400 B.C. in what is now Estonia. The aftereffects included a century of absence. No people, no agriculture, nada. Every five centuries, a Tunguska-type regional impact is estimated to take place; Tunguska was the 1908 impact over Siberia that colored skies around the world and walloped a forest, livestock, and peasants.

Over the years, I've gathered scraps of paper with numbers and dates written on them. This is what they tell me: The Hiroshima atomic bomb was equivalent to about 13,000 tons of TNT, or 13 kilotons. A kiloton is one-thousandth of a megaton. Now imagine a meteorite that is 2.5 kilometers, or about 1.5 miles, wide, with an explosive force of 1 million megatons of TNT. An object this large hits the Earth every 2 million years or so, causing globally catastrophic consequences.

A rock approaching 13 kilometers, or *8 miles,* in diameter can contain 100 million megatons of TNT, enough to trigger mass extinction across the planet once every 90 million years.

Some 3.5 billion years ago an asteroid 12 to 30 miles wide may have hit the early Earth, killing all microbes except those deep underwater. This event was 10 times larger than the impact that marked the end of the dinosaurs. Other researchers have argued that at least three asteroids hit the planet 3.2 billion years ago with enough force to rearrange the Earth's surface through erupting lava and earthquakes. Cosmic impacts are well implicated in the degradation of the planet's environment about 40 million years ago, in the Late Eocene, when it appears that the planet was pummeled by multiple impacts, possibly from a series of comets. A huge crater dating from this time lurks beneath Chesapeake Bay. Scientists also have found an asteroid called Baptistina, which is surrounded by debris that formed 160 million years ago when it was smacked by another asteroid. Sophisticated orbital models demonstrate that this single collision is the cause of a variety of solar system impacts, including the one 65 million years ago that *T. rex* looked up to see.

If you want to feed your paranoia about future impacts, get on the World Wide Web. There you'll find a radar movie of mile-wide asteroid 1950 DA tumbling in space, sunlight rimming one edge, with gray pixel-hints of the rock shimmering in the black like the shadow of a tumbling iceberg. NASA estimates a 1-in-300 chance of impact on March 16, 2880, though the spin of the asteroid isn't fully understood, nor is the role over time of the Yarkovsky Effect. The odds and the date are likely to change. You'll also find a website hosted by the University of Arizona's Lunar and Planetary Laboratory on which one can calculate a range of impact blows. It's a bit like a video game.

Within my lifetime (I trust) asteroid 99942 Apophis (formerly called 2004 MN4) will pass by Earth at just 22,600 miles on April 13, 2029. That's lower than some satellites orbit. The 1,000-foot-wide rock is not thought to present an impact danger, though when it was first detected it ranked as a then-unheard-of 4 on the Torino Scale. Apophis will be so close that people will be able to see it without magnification as it zips across the constellation Cancer. I'm looking forward to that.

After looking at charts of Apophis's orbit and the mesmerizing film of 1950 DA, you can set up a telescope and keep an eye on the Moon, which records impacts—mostly micrometeorites—all the time. Recently, scientists vindicated a stargazer who in 1953 took a photograph of what he claimed was a meteorite impact on the moon. For decades researchers dismissed this claim. But using photos taken by the orbiting lunar probe Clementine, scientists found a blue crater—blue because the soil had been recently churned. (Weathering from the solar wind and micrometeorites makes soils on airless bodies turn reddish.) The location of the crater matched the location of the photo's flash; energy estimates matched the size of the crater. On November 15, 1953, a 70-foot-wide meteorite did in fact smack into the Moon, with 35 times more power than the Hiroshima bomb, creating a mile-wide crater.

To keep track of all the rowdy rocks up there, scientists systematically survey the sky; one such effort, the Spacewatch Project, uses telescopes on Kitt Peak, Arizona. There are several such surveys around the world. Amateur astronomers have also set up sophisticated equipment at home-based observatories to discover new asteroids and comets. Every few weeks hundreds upon hundreds of asteroids are discovered, but only a fraction of those have been seen more than once. Astronomers need multiple sightings in order to calculate precise orbits.

We're not only looking at asteroids, we're sending probes to visit them. In February 2001 a probe called NEAR Shoemaker (Near Earth Asteroid Rendezvous) landed on the asteroid Eros. Eros is a 20-mile-long, spud-shaped chondrite 120 million miles away from the Sun. The craft had orbited prior to the landing, returning a stream of data that included photos of square craters and big boulders that had moved up from below the surface as they sorted out during the asteroid's own long history of impacts. Eros also yielded a mysterious deficit in sulfur compared with ordinary chondrites analyzed on Earth, but many researchers explain that space weathering zapped much of the sulfur away; a few think Eros may have experienced some melting, which means it and similar asteroids could not be parent bodies for ordinary chondrites. The next few years will be important in answering these and other matters of meteorite science. In 2010 the Japanese probe Hayabusa ("Falcon") may return with samples from a near-earth object called Itokawa that looks undramatically like cat poo covered in litter. The following year, NASA's Dawn should reach Vesta, then move on to Ceres in 2015.

Having demonstrated our ability to find, track, and reach asteroids, we're not that far from being able to move one off an intercept course with Earth (if governments are willing to pay for a demonstration mission). Keeping Earth out of harm's way has fueled movies such as *Armageddon* and *Deep Impact,* in which astronauts fly to meet the meteoroids or comets with explosives and great one-liners. The science in both is ludicrous, though it must be said that *Deep Impact* especially seems more about protecting the world from the ravages of divorce than anything else. Its moral appears to be that falling in love with another person will make the apocalypse less heroic. Be that as it may, while Hollywood and cable television networks churn out their escapist dreck (really, I love it, I admit), scientists and think tanks mull options ranging from firing nuclear missiles at asteroids to hovering a spacecraft over an asteroid's surface. In this latter instance, the small mass of the probe would draw the asteroid off course with very subtle gravitational attraction. It would take a while but it should work. Others have proposed shining light from small mirrors onto one side of a space rock in the hopes the heat might cause gas to escape, thus pushing the asteroid into a different orbit. Thankfully, none of these options require the services of Bruce Willis.

TODAY I HAVE A less ambitious but no less interesting mission to undertake. The day after our long climb up the Daniel, Kathe and I are about to take a sunshine tour of a local apocalypse. I'm excited to see the effects of the Ries impact up close and to pay homage to Eugene Shoemaker's discovery.

"I love to go on excursions," geologist Gisela Pösges tells Kathe and me. "Normally this is my day to clean." Streetside in a sweatshirt and jeans, Gisela is fussing sweetly over her huge dog, a Landseer named Joy who is as reserved as Gisela is outgoing. Gisela's sandy hair catches in her brown glasses. She's unlocking her car. As we arrange packs in the trunk, I look at Kathe in sweet disbelief, uncertain how we—how anything—has ever gotten here. Geology seems to require not only a hammer but wryness, not only a hand lens but some sort of Zen. We make room for Joy.

Interested in geology from childhood, Gisela is an enthusiastic guide, someone who clearly loves the Ries Basin and what its stories say about the violent interface between space and ground. We leave Nördlingen's old town, heading north to Wallerstein, and I note the crater-rim hills to the northwest as we pass fields of corn and wheat. In short order, we arrive at Wallerstein, in particular at a very old brewery. (Beer and rocks again?) Passing pansies, tulips, and daffodils, we enter the property, and Gisela notes with mock seriousness, "Now you are guests of the Duke of Wallerstein." On the brewery grounds is an area of protected woods that grew up around the old bluff, and we're here to look at post-impact limestone, though right now I'm taken with violets and sun and shadow. A wagtail flies off as we reach the treeless, sunny bluff. Gisela, though a visitor here many times, utters a satisfied "ohhh" at the view of the Ries spread before us. There to the southeast is the village of Löpsingen. We're standing on the rolling country of the crystalline ring, which Kathe and I had spied from the Daniel. The crystalline basement, Gisela tells me, isn't as hard as it sounds. The impact and subsequent weathering has made the granite more porous and frangible than granite usually is.

Sun-warmed, Kathe and I kneel beside Gisela, who wants to show us evidence of the toughness of life—stromatolites, fossilized columns of algae. They grew in the lake that over time filled the crater. The post-impact Ries Lake became a haven for these microbes; they layered together in curvy, ropy mats. We see fossil snails in the Ries Lake limestone, their little shells seemingly as usable today as they were a few million years ago. These small water snails were the first postimpact creatures to fossilize, some 14.5 million years ago.

Now Gisela is tutoring us on the impact itself. Megablocks of Jurassic limestone flew out of the crater; so too did a rock called "Bunte Breccia." Bunte Breccia contains a variety of Triassic sandstone and clay, Jurassic limestone, and sandstone, and some of the local basement rock. The Bunte Breccia blew out of the impact and some flowed out like lava. In the ejection and admixture process, suevite landed atop the Bunte Breccia. The gray matrix in suevite—made of granite and gneiss—is home to coesite and stishovite, which is another impact-produced mineral. The black impact glass—the flädle that's inset into the suevite—features vesicles, those little holes where gas escapes. And both the suevite matrix and the black flädle can contain melted microdiamonds. With just a bit of information, these "ordinary" rocks become sudden and ongoing revelations.

Soon we're off to buy Mack-Macks in Wallerstein, burgers to go that we'll eat later, as next we head to see the shatter cones of Wengenhausen. Shatter cones can occur in a variety of rock types and aren't melted by impact but instead record the shock wave of an impact's pressure in the rock itself. The wave hammers through the rock, rearranging it into tiny ridges and channels that show the passage of great force. In the abandoned gravel pit of Wengenhausen, we walk by puddles and mullein to the sheer face of the pit. A farmer disks a nearby field, the tang of manure in the air. Buzzards soar. We harvest shatter cones with picks and hammers, admiring the striations the cracked-open rocks reveal. They're like the ridges of a seashell. The shatter cone I most like is a chunk of rectangular limestone, tannish, that fits nicely in my hand. At one end, I tap off a chunk along a fracture line, revealing a three-fingers'-width series of the ridges created by those supersonic pressures. Here, limestone had been realigned inside itself. It reminds me of snow blown into furrows.

After lunch, which we take onto a hill of the crystalline ring, we head past whitewashed villages and cross the Wörnitz River—"many meanders," says Gisela—to the Aumühle quarry, where suevite is harvested for cement. The place looks bleached in bright afternoon sun. "Joy knows every quarry in the Ries," Gisela jokes. Here we see the fallout suevite—the kind that reached beyond the crater rim, a rain of melted rock. It's layered over a startling, deep reddish-brown swirl of 200-million-year-old Bunte Breccia whose coloration is caused by oxide and hematite. Gisela names this Van Gogh swirl of brown and red "geoart." These swirls derive from the roll-glide motion of Bunte Breccia as it flowed from the center of the impact. It's a colorful canvas, framed by yellow coltsfoot on the quarry floor.

I'm learning as we go, but I'm most anxious for our last stop, which will take us out of the crater, past its rim, to the quarry at Otting where Eugene Shoemaker found his suevite to send back to Edward Chao. Truth is, I feel like a tourist, a geeky Science Channel tourist, the kind who'd skip castles to, of course, drive to a quarry. From the back seat, I ask Gisela about the return of Eugene and Carolyn Shoemaker a few years ago for the filming of a documentary. Gisela drove both of them during their visit, and Eugene used her hammer and her hand lens as props.

"It was a very proud moment for me," she says.

Her story, theirs, the Earth's, mine, all curving together on road B-25—the rim rises steeply, and we're on the course the Shoemakers were on.

I had hoped that Kathe and I might camp at the quarry but our time was too limited and the land private. So instead we're parking on a drive into the quarry, just across from a house made of suevite. Gisela writes in my notebook *Salix weide*—I'm asking about a tree here—as we walk on seepish ground near the old quarrying cliff. The cliff's suevite is grown over with the salix, all these willows, and on our left is a mucky pond and, over all, bees hum as they gather pollen from a second-growth woods of willow, oak, pine, and what I think are beech saplings.

Here, Gisela shows us, are black Jurassic shales—not black impact glass—in some of the Bunte Breccia. This is the precision of a geologist's eye. Black shale, black glass. Not the same. I would not have known.

Here is moss and leaf litter, sun-bleached sky and tall birches. The access road we walk on bends. It's a spring day, cool, in the 60s. A redwing in a birch calls its strident alarm. Joy wades through the muck, and I wander by cattails at the edges of small ponds, glassing a small, yellow bird, but before I can ID it, Gisela tells us: We're here. The grassy flat where the Shoemakers camped. I nod, happy on their ground.

"We now walk on suevite!" Gisela exclaims. "You can see glass bombs here!" We're in the shadow of the small cliff, just under an evergreen. Looking at the cliff, we see layers, suevite some 30 yards thick atop Bunte Breccia atop white Jurassic limestone. We set to sampling the suevite. When the Shoemakers came back for the filming, he took his samples from right here, right here. I chip away at gray-black rock, fancying myself a field technician in a survey of the universe.

But Gisela corrects me. "You are now an astronaut!" she says, smiling, and I grin too. I do feel a bit lighter, having discovered that mood can affect one's mass.

So I practically glide to the top of the suevite bluff, to its grasses and a fence. I've climbed up to be alone, to take in the view of a wheat field on the other side of the fence and, below, the trees, the road, the wide flat where the Shoemaker VW was parked. Birches glow purplish, fuzzed, and spring arrives in a cold wind, dazed and sluggish at the Otting quarry, where grasses and dandelions and mosses grow out of rock made from meteorite and from earth, all this insistence-that-is-acceptance, us and the ants, us and the Shoemakers, and everything beneath particular cumuli. Birds chuff and fly, including a yellowhammer, whose name seems, on this geological foray, perfectly apt. Sunlit bees hum into yellow-tipped catkins, and I laugh all the way back down the cliff, happy with the oddities of aspiration.

Kathe and I smile and exchange the look that couples exchange when it's time to go. We take a few more pictures, and when my photos are developed, the overexposures will make the quarry seem ghosted or lit like the far fringes of Miocene fire.

Now, Gisela, tall and sweet and funny, is very quiet. I envy her having met Eugene Shoemaker, having handed him her hammer and lenses, and when I ask what this place means to her, she says, "This—is my Shoemaker quarry. The astronauts trained here." She pauses. "A great spirit is here, here, let me say, for me."

ONE NIGHT IN NÖRDLINGEN, at nearly ten P.M., back at the cozy room of the Goldene Rose, Kathe and I made love. Haphazardly dressed, we stood afterward by the open window, feeling the night air and waiting to hear the watchman. In this town, watchmen have been calling "All's well"—or sometimes sounding a warning for fire or attack—every hour of every night since the sixteenth century. We leaned against the warmth of the room's radiator while cars below abraded the otherwise clear silence. We could see at the top of the Daniel something like a ball all lit up, a lightning rod I guessed, but it looked like a planet, like a little Jupiter, sweeper of asteroids.

Ten rings sounded for the hour. Then a horn. *So G'sell so.*

"I hear him," Kathe said. I nodded, confused. "It sounded like a horn," I whispered.

So G'sell so.

This time louder, clear, a kind of songlike call, a kind of short yodel from the watchman of the tower. In the crater all was well. And it's true, all was—is—well. I felt content above this village—tidy, green, angular, curvy, ancient, all flowerpots and footpaths, all lark and vines, all *is* and *nevermore,* all suevite and flädle, all astronaut and poet, all going on, going on.

Kathe leaned against me in a universe that seemed suddenly and sweetly improbable.

She whispered, "It sounded like a bell."

Book IX

Life Work: The Biology of Meteorites

· *Chapter IX.1* ·

The Resurrection of Acraman

Just a century. That's all it took for life to work its way back to the Ries following the impact 15 million years ago. Plants began to grow out of the Bunte Breccia 100 years after the devastation, and a cycle began: The crater held rainwater, dried out, gathered deluges, dried again, took in rain, and eventually became a lake. Despite the salt and silt, the lake was a refuge to dragonflies and algae, fish and snails. Like the Great Salt Lake, not far from my home in northern Utah, the Ries had no outlet. Fed only by drizzle, shower, and storm, the crater lake stayed highly saline—imagine the stink!—until the climate itself grew moister. This lowered the salt content, allowing a wider array of life to flourish. Cattails, lilies, mussels, and freshwater snails thrived, and ducks, herons, pelicans, and storks flocked to this oasis.

For many months after visiting the Ries—when Kathe and I saw fossils of those postimpact snails—I think about the tenacity of life. I can imagine pelicans gliding over the fractured ejecta of suevite, sliding down to the bowl of water formed by fire. I can see bats sleeping in water pines, hedgehogs snuffling along the mucky shore, and for about 2.5 million years, the lake endured.

The most profound lesson in this is that meteorites and comets don't always bring just death and destruction. In fact, one might say Impact + Heat + Water = Habitat. Call it the Oz Equation. Canadian researcher Gordon Osinski— "Oz" to everyone in the meteorite community—has spent several summers with field teams in the Arctic, on Devon Island, where Haughton Crater is yielding clues to how impact sites can become refuges for microbial life.

"Impact craters," he says, "still deserve their reputations as scenes of devastation, but as they cool, they become ideal spots for life to re-emerge. And this has led some people to wonder if impact craters on the early Earth provided the environment for life to emerge in the first place." Oz notes that some of the planet's earliest life forms were microbes that seem perfectly suited for postimpact hot springs and pools. He cites studies showing that large craters can maintain hydrothermal vents and springs for about a million years after an impact. "This is getting towards the kind of timescale life might need to emerge," he writes.

Haughton Crater is 39 million years old and formed during an Eocene impact event. Several thousand years after the not-quite-mile-wide meteorite hit, microscopic bugs would have been all over—or in—the crater. "The warmth of the impact heated groundwater, creating hydrothermal systems," says Oz, systems that would have made "perfect homes for intrepid . . . bacteria and algae." Haughton is not alone in such history. Geologists have now found crater rocks reworked by postimpact hot springs and vents at more than five dozen sites. It's even possible that postimpact hydrovent ecosystems once developed on Mars. Oz and fellow researchers such as Pascal Lee and John Spray were the first scientists to map postimpact hydrothermal vents around a crater, though others, such as David Kring and J. D. Farmer, had earlier suggested the connection between impacts and hydrothermal systems.

At Haughton, the rocks tell an interesting story.

"The main heat source," Oz writes, "would have been the thick layer of molten rock that filled the central area of the crater. . . . Groundwater underneath this sizzling mass boiled and rose." Steam encountering that melt rock couldn't penetrate it and therefore spread more or less horizontally to the rim of the crater. This "vapor-dominated regime" was "short-lived." The hot springs might have lasted for thousands of years. As at the Ries, a lake developed.

Fossils from postimpact microbes have not been found at Haughton, though researchers have discovered contemporary bacteria living within rocks, within the lilliputian cracks opened by the stress of impact. Living inside a rock sounds pretty drab, but it's "moist, UV-shielded, and relatively warm," Oz notes. And microbes are tough; they've been found 2.5 miles under the Earth's surface, within hot springs whose temperatures are high enough—300 degrees Fahrenheit—that DNA could almost unravel, and under pressures hundreds of times higher than that on the planet's surface.

Charles Cockell, the lead author on what may become a classic paper, writes that "an increase in [microbial] habitats is likely to be a *common effect of impact*" (emphasis added). Indeed, many of the postimpact Haughton rocks are just transparent enough to allow sunlight to reach photosynthetic microbes. Building on the work of such scientists as Jay Melosh, Cockell and his team write, "Impact cratering must be viewed as a truly biologic process." This is a radical turn of perspective. As recently as 1990, scientists could say that it was possible that "a high rate of cratering hampered early biological evolution on the earth."

Now some researchers suspect that meteorites have done more for life than create local saunas for recent fauna. It just might be that a meteorite—one hitting the area that is now Lake Acraman, Australia—forced the most important step in the evolution of life: the jump to widespread, complex, multicellular creatures.

TO UNDERSTAND THAT JUMP, we have to go way back to the beginning of what science writer Gabrielle Walker calls "Slimeworld."

Arising about 4 billion years ago from chances of gases, sugars, carbons, proteins, and other morsels—perhaps relatively quickly, in a melding of volcanism, carbon, and metals, or more slowly, in a chain of prebiotic chemical events in the ocean—the earliest single-celled microbes (empirical miracles known as prokaryotes, which are cells without a nucleus) toughed it out during, some argue, the late heavy bombardment 3.8 to 4.1 billion years ago, that alleged age of violent impacts when asteroids may have pulverized themselves against a young planet until life's scummy dominion truly began, with mats of single-celled stromatolites. I had seen fossils of them in Germany, and stromatolites once covered the ancient Earth, starting billions of years ago and lasting for hundreds of millions of years upon hundreds of millions more, in layers of living muck clustered upward in hummocks crowding tidal flats and shores, the world's first houses, a yard high, built by the world's first architects, the hundreds of trillions of prokaryotic cyanobacteria upon hundreds of trillions more upon more and more and more. The first green fuse was a wet and teeming tower of blue-green algae. Durable too. Here and there, stromatolite colonies live on even today.

Prokaryotes of all types are everywhere with us. Without the nonnucleated bacteria in our guts, we'd not be around to read, because we couldn't digest food. Without the millions of diphtheroid bacteria swarming on your armpits, the toiletries business would swoon.

About 1.5 billion years ago, more complex single cells arose—the eukaryotes, those cells *with* a nucleus and other specialized features walled in by membranes. Excepting prokaryotes, all life as we know it is based on the nucleated cell. The nucleus contains an organism's DNA, and the development of complex internal cellular structures allowed for biochemical and physiological nuance. Eukaryotic cells have advanced organelles, such as the Golgi apparatus, which secretes waste as a kind of microscopic kidney-and-bowel system. Other organelles include the lysosome, which repairs cell damage, and mitochondria, which convert food to energy.

The difference between the nonnucleated cell and the nucleated cell is the difference between a flame and an engine.

But as Stephen Jay Gould once pointed out, it was a very long time before even these more complex nucleated cells, these eukaryotes, became organized into multicellular animals. That was life's third big leap. Specialization of functions within cells allowed for arrangements of multicellular complexity leading to the array of life around us today. It just took a while.

The first such multicellular animals were part of what is called the Ediacara fauna, which date from 550 to 600 million years ago, during the late Precambrian era. This sudden experiment preceded the first extensive spread of animal species on the planet by about 100 million years. Which means the Ediacara fauna were like a trailer to a movie, but not the movie itself. Or, some say, like a long-lost ancestor who never fathered a child.

ON A NOVEMBER AFTERNOON more than a year after visiting the Ries, I'm in Brachina Gorge, a canyon in the Flinders Ranges of South Australia. Distant thunder threatens a flash flood. I left Kathe in Utah several days ago so I could fly here and start my wee detective hunt for clues to the origins of multicellular life. I'm nearly nose-to-nose with a few organized ripples in stone, a fossil of *Dickinsonia costata.* I've not yet found the fossil, though it should be here, where I'm wedged under a cleft of rock, listening to retired University of Adelaide geologist Vic Gostin tell me where to look. Waiting nearby is Lorraine Edmunds, a naturalist and former park ranger, who has joined Vic and me as a guide. The light is flat, the sky overcast. I can't find the fossil. This dun-colored, blobby shape? "No, no," he says. Vic keeps prompting me and waits. Tired from travel, I grow frustrated. Sand and rocks tickle my back.

Dickinsonia costata was a flat worm with segments spiking out from a bisecting trough, the function of which remains unknown. "Perhaps this was a stiffening rod for its soft body," Gabrielle Walker writes. "Perhaps it is the trace of a gut." Regardless, she adds, "unlike the inhabitants of the Slimeworld, this creature knew the difference between head and tail." Lungless, this worm breathed through its skin.

In our world of emus (we'd just seen one as we drove into the gorge) and kangaroos ("Hell yes," I'd exclaimed when Vic had asked the day before if I wanted to turn around to see my first) and flying birds (galahs, corellas, the melodious Australian magpies) and plants (the muscular river red gum trees) and human beings (new friends in a gorge)—in such a world, a worm, flat or tubular, just doesn't seem, well, all that compelling. Nonetheless, *Dickinsonia* and its fellow members of the Ediacaran tribe, such as sea pens and jellyfish, were that first, most startling development: they were animals.

With the rank scent of Ward's weed still lingering in my nose—the lanky, loose, hairy invasive from the Mediterranean is everywhere in this part of the Flinders—I finally see it. A hand-sized sunburst of many thin rays, a sundial without its gnomon, hash-marked with lines, not numerals. An interpretive sign here in the gorge compares *Dickinsonia* to "a segmented door mat." So welcome home, dude. Set in the reddish Rawnsley quartzite formation, this Ediacaran worm has been right in front of me for a few minutes. I stare at it dumbstruck.

Dusty, I stand up, unsure what comes next. I look about the tight gorge, feeling a curious sense of several almost-translucent dualities. Now and the

Precambrian. Here and canyons back home. Here and a crater in Germany. Here in my skin and floating over the world—which begins at Brachina Gorge, with gray clouds above—and all of it excessively real, exquisitely edged, appearing deceptively, artfully arranged, boulders and the spaces between them, scrubby plants and trees and the spaces between them, the way foreground and background become hyperreal inside old View-Master reels.

Wind rustles the yellow-flowering mulga, an acacia whose wispiness reminds me of the willows of the American West, including, right beside the Blacksmith Fork River, the peach-leaf willow under which Kathe and I put a hammock in the summer. Here and there, first home, latest home. From beige to brown to orange to cinnamon to oxblood, the rocks of the gorge, only a few yards high, close in. I'm being cupped by the world in time.

And I've forgotten the threatening weather, though Vic stands stiffly beside the white four-wheel-drive I rented in Adelaide. He's opened the driver's door. Vic knows I've come a long way to be here. He answered many e-mails before I arrived and even met me at the airport, looking every bit the geologist in his jeans, plaid shirt, bolo tie, and purple trilobite pin stuck in his hat. Affable and eclectic, gray hair swept back as if he's been in wind for years, Vic's a fine tutor.

Lorraine walks me to some limestone, euphoniously called Wilkawillina, where she wants to show me something else. Wilkawillina limestone is 20 million years younger than the rocks that hold the Ediacara fauna. Despite the ten-year drought in this part of South Australia, despite the need to keep water on hand, Lorraine pours some out of her bottle onto this Cambrian limestone, where psychedelic shapes appear beneath the dark wet. We're bending over the cross sections of archaeocyaths, creatures that resemble sponges. They appear mostly circular, but also a little blebby, with alternating bands of gray and black. They have, to my novice's eye, the swirly appearance of agate. Shaped like megaphones, these creatures affixed themselves to the bottoms of shallow seas and formed the world's first reefs. How cool is that, I think.

"The best in the world," Vic affirms of these fossils, adding, "Can we get in the car now?"

The thunder's louder. We're in a gorge. We'd rather not die in a flood. We get in the car.

In a matter of minutes, I've spanned two great ages in Earth's history: the late Precambrian, with *Dickinsonia,* a mundane worm that, along with its Ediacaran friends, represents the first step into multicellular animals, and the early Cambrian, with the archaeocyaths, part of the second but by far much larger—and ultimately successful—wave of multicellular life, what is commonly called the "Cambrian Explosion."

The Ediacara fauna were, many believe, a kind of a cul-de-sac, a box canyon of evolution. Some say all the Ediacara went extinct and, along with them, their unique flat, segmented, and squishy designs, which had allowed for large body area but not much in the way of innards. Others believe a few of these species survived to merge into other ancestral lines, leading to today's complex creatures. Researchers such as Richard Fortey are apt to make the "e" in "Explosion" lower-case, in order to emphasize similarities rather than differences in Cambrian animals, as well as to suggest precursors in the Ediacaran Precambrian.

Later I'll reflect on the irony. My deepest encounter ever with deep time and life—up close and personal with some of the world's first creatures—and we with our eyeballs and spines, our lovers and homes, were in a hurry because of a potential storm. Deep time, quick glance. One day, back in Utah, while looking at a geological map of the Brachina Gorge trail, I'll realize that Vic, Lorraine, and I drove right past the 630 million-year-old Trezona limestones that contained fossil impressions of . . . stromatolites. Older even than the Ediacara! I'll console myself with the prospect of seeing living stromatolites someday, near San Diego or in Yellowstone, and I'll feel a kind of comfort in remembering that bodies sometimes give way to stone and flake, to recorded passage.

In a few minutes, we're out of the gorge and into an expanse of weird storm light. Shadows blanket hilltops only to peel away, then cover a dead-grass plain of mallee and saltbush. Strata of clouds, altostratus above cumulus. We're taking the long way back to our cabin near a circle of mountains called Wilpena Pound—the scenic route, says Vic, though it's all been scenic to me—and I see distant mountains flatten to gray, then dun, then gray-white against the gray sky spilling virga, then actual rain spatter, dust spatter. I'm sitting in the back, mulling distances and affinities, while Vic talks to Lorraine about his book *Environmental Geoscience,* and, unperturbed, drives us through what has become a for-real dust storm.

It's a brief tizzy. The storm ends. The sky breaks into fragments dark and sunny, against which wheel little corellas, a spin dizzy of white flakes with wings. The native hematite of the countryside glows red with slanting light. Our car startles a euro—a kind of kangaroo—which runs alongside a fence before lurching its heavy, cartoon body through a hole, bounding away. We pass Bullock bush heavily grazed by livestock, and I see more exotic birds, corellas, and some galahs, and Vic almost hits one bird in a flock of red-rumped parrots. I spy a small, pale bird of prey, a nankeen kestrel, and all around us bloom tall pinkish spikes of mulla, through which crawl slowly the "sleepy lizards," including this one we've stopped for by the roadside. The size of a cat, the sleepy lizard, or "blue tongue," doesn't move at all. Stumpy-tailed, brown-and-green, it looks at me with its reptilian stare while Lorraine recounts the guilt she felt for moving one such lizard; it kept eating her strawberries. She learned that they are unique reptiles in that they pair-bond and use the same territory again and again. "Now I share the strawberries," she says.

Even if the Ediacara fauna were the dead end some think, the Cambrian Explosion assuredly was not. Five hundred million years ago, 100 million years after the brief rise of the Ediacara tribes, the Cambrian Explosion was the big light switch turned on, the sudden, widespread development and diversification of nucleated, multicellular life. While most of the varied designs of Cambrian creatures have not survived, the Cambrian's wide-ranging accidental "experiments" in arranging complex cells set the stage for evolutionary byways that led to, for example, rhipidistians, the first creatures to move from water to land. Also among the Cambrian fauna was a wormy little guy called Pikaia, the first creature with a kind of spine. Who's your daddy? Pikaia.

"In a geological moment near the beginning of the Cambrian," Stephen Jay Gould once wrote, "nearly all modern phyla made their first appearance. . . . The 500 million subsequent years have produced no new phyla, only twists and turns upon established designs—even if some variations, like human consciousness, manage to impact the world in curious ways."

But why did it happen? Why the expansion of life 500 million years ago? In competing theories that seek to answer these questions, it turns out that the Flinders Ranges of South Australia play a key role.

. . .

THE DAY BEFORE MY FORAY into Brachina Gorge, Vic and I were smearing on sunscreen against the midday glare. We were in the hills past the tiny town of Hawker, amid eucalyptus shrub, spinifex, saltbush, and the ever-present blue-flowering weed called salvation jane. I said that South Australia feels a lot like Arizona only without the cacti. Vic applied a layer of zinc paste to his red nose, whose color had resulted from the loss of melanin after he underwent treatments for cancer a few years ago. Like other such survivors I've known, he has a lightness of being that seems at once accepting and hungry.

In a minute or so, I found myself trying to keep up with Vic while we walked in a dry creek bed and avoided kangaroo droppings. I wanted to head toward meager shade cast by white cypress pines and river red gums, but we stayed in the sun to look at some rocks. We were heading into what was once a Precambrian shallows, a marine delta where the water had been placid, a place from some 680 million years ago, before even the Ediacara fauna had appeared. Vic punctuated geological riffs with the endearing phrase "wow, man," and he told me stories.

It was in this ravine that Australian geologist George Williams had discovered small crinkles in pinkish mudstone. I bent to touch these gentle ridges and depressions. I might have mistaken the little ridges for the scars of passing glaciers, but that's not what these lines of bumps record. They were, Williams deduced, the accumulation of tidally deposited sand—tidal rhythmites. Williams drilled core samples on the small rise just before us and found that the fortnightly tides had occurred at this rate all over the planet.

"Tides are like drums in the orchestra," Vic said. "You can hear music, but you can also find the rhythms underneath." The wind shushed through the brushstroke branchlets of the drooping she-oak, and Vic told me more about the Precambrian world.

The Moon loomed larger in the sky because it was several thousand miles closer to the Earth, which, of course, made for stronger tides. The Earth's orbit and spin were slower and faster, respectively. A year was a month longer than it is now, a day a couple of hours shorter.

Williams found other rocks from that age besides the rhythmites. He investigated rocks whose original magnetic fields were horizontal, and when scientists can determine that original magnetic lines of force are horizontal, it means the rocks originated at the equator. But here's what baffled a few researchers. At the time, the equator seemed to be icy cold. The huge Moon gleamed over glistening equatorial ice and pools of open water where the tides moved.

There was more. Williams found sand wedges—gaps created by freezing and thawing into which dirt blows. Their significance is simple. To get freezing and thawing, you need seasons.

How to explain open water and equatorial ice and seasons? Did all these things coexist in the Precambrian? Some believe they did.

Williams champions a theory called the Big Tilt. He argues that the Precambrian Earth was tilted such that the present-day poles were closer to where the equator is and vice versa. About 4 billion years ago, something whacked the Earth; perhaps another planet hit, which incidentally caused the formation of the Moon. Williams believes this knocked-about Earth can explain the odd conjunction of water, ice, and seasonal changes. Over time the Earth came to its present orientation.

There were other mysteries. Researcher Brian Harland discovered Precambrian carbonates right next to where there had been ice. Carbonates come in a wide variety, but the important point is that they form in *warm* water. Gabrielle Walker compares Harland's findings to "watching a glacier march across Barbados."

So how to explain open water and equatorial ice and seasons and warm-water carbonates? This is a "wow, man" sort of world. The Big Tilt is one possibility. But Cal Tech's Joe Kirschvink dubbed the whole shebang "Snowball Earth." Since Kirschvink, others, most notably Paul Hoffman, have taken up the Snowball Earth theory. It suggests that 800 million years ago, with the fracturing of continents clustered near the equator, "formerly landlocked areas [were] . . . closer to oceanic sources of moisture," according to *Scientific American.* "Increased rainfall scrub[bed] more heat-trapping carbon dioxide out of the air," leading to cooler temperatures, the spread of polar ice to mid-latitudes, and increased reflection of sunlight and heat into space. The effects cascaded into a global ice age, which ended only after carbon dioxide venting from volcanoes led to a thaw. The whole process took millions of years. Another catalyst for the Snowball might have been the Sun, which did not then shine as hotly and brightly as now.

As to evidence of seasons at the equator—the sand wedges, which Williams believes militate against Snowball Earth—others think that far from banishing the seasons, an icy planet may still have had variations in temperatures. Further, Hoffman and others believe the Earth suffered a series of Snowball episodes with the last thawing prior to—you guessed it—the Cambrian Explosion.

Snowball advocates say that microbial life toughed it out in a few places on the icy Earth, such as deep-sea, hot-water vents and patches of open water, and that when the Snowball fully melted, these hardy critters were primed to diversify.

There is, however, the matter of a several-million-year gap between the end of the last global ice age and the appearance of the Cambrian Explosion. In fact, as of 2007, Paul Hoffman had softened his claims: "It would seem that the achievement of multicellularity in (microscopic) animals would be the evolutionary step most closely associated in time with the [last] . . . snowball earth. However, there is as yet no empirical support for this in the fossil record."

So what forced the Cambrian Explosion? The righting of the planet following the Big Tilt? The thawing of Snowball Earth? Something called "true polar wander," which suggests that imbalances in the Earth's mantle forced continental movements that drove ecosystem changes?

Paleontologist Kath Grey thinks it was something else altogether. Grey believes she's found evidence that the Cambrian Explosion has a different cause: a meteorite impact in what is now South Australia. The meteorite—a very big one, some 3 miles wide—hit 580 million years ago, creating a 56-mile-wide crater in old volcanic rock. This was an impact big enough to alter the global climate.

The remains of the Lake Acraman crater were first discovered years ago by Snowball Earth critic George Williams at about the same time Vic Gostin was puzzling over strange chunks of rock in the Flinders Ranges. Now Grey thinks the timing of the Acraman impact—long after the last Snowball melted—sparked the first extensive rise of complex life on Earth. The Cambrian Explosion may have been more aptly named than we knew.

Walking back down the draw where Williams had discovered evidence of ancient tides and where a strange picture of the Precambrian Earth had started to come into focus, Vic and I headed back to our car. We drove next to Wilpena Pound, anticipating a flight over the Acraman impact site, which, if Grey is right, may come to be seen as a kind of fiery Eden.

I WAKE TO FLANKS of billowed clouds, all colored gray, blue-gray, pink, silver, the outback freshly scrubbed with last night's rain. Leaves drip. The "flute-like caroling" of the magpie, as my field guide puts it, echoes and weaves. I'm spending part of the morning alone, away from the small lodge and campground where Vic and I are staying at Wilpena Pound. In the quiet and the song, my mind registers impressions. Ringneck, rufous whistler, red-rumped parrot, white-browed babbler, magpie-lark, and the swamp harrier, which, for 20 minutes, perches across a field in a grove of white cypress pines, as if the stillness that precedes satiation is another form of redemption.

Vic, Lorraine, and I meet up only to wait a few hours for the last of the weather to clear, then, a little after two P.M., we're climbing into the Cessna 180 that Wilpena Pound Resort uses for tourist flights. Vic is delighted to have a chance to fly over Acraman again, and our pilot Melissa Hosking is friendly and professional. Vic and Lorraine settle into the back two seats, while I sit beside Melissa, gripping my notebooks and camera. It's been years since I've been in a small plane; in junior high and high school I belonged to the Civil Air Patrol, took ground-school lessons, and nurtured my Air Force and spaceflight dreams.

We climb to 4,500 feet, and Lorraine tells me that all the crisscrossing lines in the dirt below are 'roo tracks and rabbit warrens. Then Melissa swings us past the southeast corner of Wilpena Pound, which, at about 3,700 feet, is the tallest part of the range. The mountainous Pound is shaped like a bowl that looks dramatically more like a meteorite crater (which it's not) than will the seasonal wetland that is Lake Acraman. The Pound was formed by sediments built up in the Precambrian and Cambrian that folded, then eroded, resulting in its present shape. It takes a full 15 minutes to pass by the massive Pound, and sunlight beats on the plane.

Soon, Lake Torrens is ahead, a strip of white on the horizon. As we pass over orange-red sand dunes, I lean my head around even though we're all wired with headphones.

"It looks like Mars," I say to Vic.

"Bull dust," he calls the fine red sand down there.

Now come the gypsum dunes of Lake Torrens, one of the several seasonal, salty lakes west of the Flinders. To my right, to the northeast, are the Ediacara Hills, first known home to the fossil animals I met at Brachina Gorge. The landscape below remains orange and dun, dotted with black oak, saltbushes, and spinifex. Below us, as we fly west, stretch the sandy deltas of the Torrens Basin, and the ground blurs. Pale sand, cinnamon sand, brown sand. But last night's rains gifted Lake Torrens, where, much to Lorraine's surprise, there's some patchy water. We see a pale gray-blue sheen reflecting the clouds. It looks like the braiding of wet and dry land that fringes the Great Salt Lake back home.

While Vic reels off facts about crust thickness, faults, and erosion rates, the heat and bumpy thermals make me queasy. I can't keep track of his lecture, and the pressure of the headphones pushes my sunglasses uncomfortably into my skull. I can't take notes for more than a few seconds because taking my eyes off the horizon makes me want to vomit, which, most emphatically, I do not want to do.

I force myself to glance at Melissa's charts. We're over the Andamooka Ranges, Burden Hill, others. Pernatty Lagoon slips by, and 50 miles from Acraman, I look down on a striking series of parallel ridges and dunes south of Gairdner, then the salt flats of Lake Gairdner itself.

"It's good to see the water," Lorraine says.

Gairdner's salt flats are so extensive I feel as though I'm visiting a place I've seen before— *Of course.* I'm flying over Thule again. I'm back north, north of the Arctic Circle. Below is the Greenland ice cap rendered in desert salts. Despite my mounting nausea, I am entranced. I'm subatomic, wave and particle, two things at once in two places at once. An outback ice cap.

IT WAS IN 1979 that George Williams saw the circular shape of Lake Acraman in satellite photos and quickly suspected a meteorite impact. A few months later, he collected samples from the 1.6 billion-year-old volcanic rock that makes up the region. The rocks contained shocked quartz. Williams indeed had found a meteorite crater, old and eroded but still Australia's largest. At the time, though, more concerned with mineral exploration for a company, he didn't publish his findings.

Not long after Williams had drawn his conclusions about Acraman, Vic Gostin was in the field, not far from Wilpena Pound, confronting a puzzle. He had found a grainy layer of volcanic rock intermixed with a formation of sedimentary rock that had been laid down by an ancient sea. He thought the volcanic material was similar to the Gawler Range Volcanics hundreds of miles away. It was in the Gawler Range Volcanics that Williams had spied the Acraman crater, though Vic didn't know that yet of course. Might the volcanic rocks be glacial erratics? Or brought in by some other means, such as river deposition? He didn't think so, but was trying to consider all the possibilities. Further fieldwork by Vic and his students, as well as study of samples, showed a continuous layer of volcanic rock through much of that part of the Flinders Ranges. And this volcanic rock had undergone, as Vic wrote in his notes, some form of "gross alteration."

Expecting the volcanic rocks and the marine shales to be contemporaries of each other—perhaps, most simply, a volcano near the ocean had spewed lava into water and sediments—Vic was baffled to learn that the Gawler Range volcanic fragments and the marine shales called Bunyeroo mudstone were vastly different in age. The former were 1.6 billion years old, the latter 580 million years old.

Doubts nagged him. In 1985, Vic had been reading work by the spiritual leader Krishnamurti and told himself *to clear the mind of prejudice . . . let me relax.* He doesn't recall if he was in the field or at home when, wide awake one night, he reviewed everything in his mind once more. *Volcanic material in one layer. Fallen off cliff? No cliff high enough or near enough. River deposit? No, the Bunyeroo mudstone shale is marine—a sea, not a river. Debris flow? The sediments didn't fit that. Glacial? The rocks showed no glacial scratches. There had been a surface explosion, a huge one, in an area of ancient and uniform lithology.*

"Did it fall out of heaven?" he suddenly asked himself. The recent controversial paper about an impact at the Cretaceous-Tertiary boundary, with its claim that the dinosaurs were killed off because of a massive meteorite strike, had very much been on his mind.

The next day Vic and his graduate students put a rock sample under a microscope.

"I wind up the magnification on a thin section. . . . Hey, this is part of the shock, the bang, the hammer hit!" Vic and his students had PDFs—planar deformation features, those sets of lines in glass formed by an impact. "We were delirious."

Doctoral student Peter Haines even found microscopic shatter cones. And the area around Wilpena Pound—where the impact had spewed its nasty mix of remelted old volcanic rock and a mishmash of meteoritic material—was full of big rocks that had been blown out of a crater: impact ejecta. The volcanic rock was early Precambrian, and the impact itself had taken place in the very late Precambrian, when the area was probably iced-up. Vic, Peter, and their coworkers soon learned that Williams had discovered the Lake Acraman impact site.

In 1985, Vic told Williams of his findings. Vic's ejecta blocks, Williams writes, "matched, both in general rock type and degree of shock metamorphism. . . . [the] shattered rocks I had collected from Lake Acraman." The next year Williams and a team led by Vic Gostin published two separate articles—one on the old Acraman crater, the other on the ejecta—side by side in *Science*. Three years later, Vic, Reid Keays, and Malcolm Wallace announced that the Acraman ejecta had high cosmic iridium content—just like the iridium spike at the Cretaceous-Tertiary boundary—and to this day iridium, which is rare on Earth, helps scientists to tag impact events. Acraman was also the first finding of such an ancient annulus of impact fragments.

At the time the reports appeared, Chicxulub, Mexico, the impact site for the Cretaceous-Tertiary bolide, had not yet been found, and paleontologists were still up in arms that scientists from outside their field had invoked a huge meteorite as the ultimate cause of the dinosaurs' demise. The Acraman impact had nothing to do with the dinosaurs (they weren't around then), but the Acraman findings helped swing the scientific community toward accepting impacts as normal parts of the Earth's history, even a collision as bewildering as Acraman.

This huge meteorite hit at a velocity of 15.5 miles per second, producing an explosion far larger than the biggest atomic blast, hurtling rocks for hundreds of miles, and sparking tidal waves. The meteorite had been enormous, 2.5 to 3 miles wide, and carved out a crater just as deep. In his office, Vic has one of the impact's "ejecta bombs," an orangish-reddish-brown rock shaped like a battered and worn American football, but bigger. "The atmosphere moves, man!" as Vic says.

Before I had traveled to meet Vic, I had been in touch with George Williams, who graciously sent me reprints of his Acraman articles. In one of them, he noted that the impact "ejecta of rock fragments, sand and glassy material would have covered . . . nearly three times the area of the United Kingdom." It also cast a global pall of dust.

Vic and Lorraine had shown me ejecta exposures the day before our Acraman flyover. In billboardless country where yellow-flowering twin-leaf and pink clover grow, Vic had driven us around the Flinders, including the ABC Range (26 peaks, 26 letters). At one exposure, where I crushed lemongrass to rid my nose of the musky smell of Ward's weed, the three of us stood in front of a short, steep slope of eroded shale scree. Crows called. White cypress pine stood on the ridge.

There before me was the odd stippled rock, the gray-green shales whose mostly flat tops were faceted by erosion. Within the shales occurred nubbly ejecta—almost like lumpy cookie dough, only finer. Lichen had stained the sandpapery ejecta mostly black, though it also showed stains of maroon or brick red. Where calcium carbonate had impregnated and weathering had taken place, the colors ranged from cream to pale pink. Areas of pink-orange reminded me of the color of bark, of the droughty river red gum we kept seeing. Beautiful colors. I picked up loose rock that was former meteorite, former lava, former ocean. Sun glared, and I sweated up a storm.

It was as if I held some grail. Did the life that led to us begin with this?

I took a sandwich-sized chunk of the grail with me. The blowout, the hammer hit. Like the suevite I'd sampled in Germany, the Acraman ejecta was another relic of devastation. But it might have been something else as well. The origin, the catalyst.

• • •

A WHITE STRIP ON the horizon, a bull's-eye in the Gawler Range Volcanics. Lake Acraman. Ground zero.

We fly toward the edge of Acraman, which is speckled with vegetation and eroded dacite outcrops like scabby blisters. Dacite is a quartzy volcanic rock, here appearing gray from the slow growth of lichens. This was the rock that the impactor struck 580 million years ago.

The Cessna banks above this huge wound in the earth, Vic is excited, and I'm trying to see if my attempts at geological perspicacity can stave off regurgitation, especially now that Melissa is dropping the plane closer to the crater. The temperature climbs as we descend—it's damn hot—and my body feels full of magma.

Now looms Lake Acraman. The entire astrobleme—the worn-out remains of a crater—is about 23 miles wide. It's the eroded remnant of the crater's main zone, including the "central uplift," which is the peak in the middle of any large crater caused by rebounding molten material. Time has pretty much erased the Acraman central peak, and the place, so flat, looks painterly, an Impressionist's modello, with snaking channels, ponds, wet curves along the astrobleme's south edge, depositions of sand that paisley the lake with islands and, there, trees like black flecks of smashed insects. I'm wishing we could land, though there is no airstrip below, and I recall George Williams e-mailing me before my trip to warn in ALL CAPS to NOT drive this area in the HOT Australian spring. That I would DIE.

We dip toward the dacite outcrops on the shore and on the islands where Williams took his samples years ago. I envy him his pick-and-hammer solitude, his under-the-stars notebooks. In the distance the low-lying, eroded hills of the Gawler Ranges partially ring the old crater, and sand dunes nestle closer to the lake bed itself. The area is now a protected geological sanctuary.

In the cockpit, sunlight flares against the dark dashboard, and sweating, unable to write notes, I take photos that will turn out almost as blurred as the "lake" below, that sandy, watery, wattly mixture that seems more swamp than crater. Perhaps the hazy quality of the sky and the land and my body is appropriate, the blurring of this into that, then into now.

After the initial flurry of Acraman impact papers in the 1980s, interest waned. There were other impact craters to find and document. Acraman was becoming yesterday's news.

BUT IN MAY 1998 Vic had lunch with Kath Grey, a paleontologist with the Geological Survey of Western Australia. The Australian Centre for Astrobiology calls Grey the "undisputed world leader" in the study of early fossil planktons and their potential to help date layers of rocks and other fossils, what specialists call biostratigraphy. What Grey wanted to tell Vic was this: The Acraman impact, the fossils were telling her, seemed curiously timed.

For right after the 580 million-year-old Acraman "debris layer," there came a veritable torrent of new species, according to Grey and her coauthors, Malcolm R. Walter and Clive R. Calver, in their 2003 study in *Geology*. Some fifty-seven new species of planktons appeared. And they were different. Before the Acraman impact, most planktons were "simple spheroids," Grey's team wrote. After the Acraman impact spread its globe-trotting dust blanket, something strange happened. The planktons changed. They grew complex spikes, spikes that apparently protected them from hard times on Earth. These "acritarch" planktons were among the first eukaryotes, and they evolved from their more primitive spheroid kin. While there had been a few species with spikes before Acraman, they'd not been numerous. These postimpact spiky planktons were bigger and more complex—and they became dominant.

Grey, Walter, and Calver also point out that this evolutionary burst "did not happen until long after" the end of Snowball Earth. "Post-glacial species [are] identical to pre-glacial ones," Grey writes in an article in *Australasian Science*. "There was no post-glacial colonisation by rapidly diversifying species."

The post-Acraman darkness and consequent cold clamped down hard on bacteria. "Organisms dependent on photosynthesis, such as [simpler spheroid planktons] were devastated, but the event had less effect on a small population of spiny acritarchs, which produce protective shells or cysts when conditions are adverse," Grey says. Eventually conditions eased. That's when Grey's spiky planktons went wild, recovering faster than their harder-hit spheroid kin. This sudden rise in plankton diversity probably widened the bottom of the food chain, in turn helping to spur the evolution of other animals. Grey believes that the last big glaciation prior to the Acraman impact must have served as a crucible for microbial life, but its end was not enough to spark the first great flowering of life on Earth—the "Cambrian diversification," as she puts it. Grey and a few others think this biological cascade was sparked by the Acraman impact in "a baptism of ice and fire."

The work behind this discovery was painstaking. Relying on core samples made across the country in a search for oil, Grey used acid to isolate out tiny bits of organic material. Looking through a microscope, she found that the samples revealed "hundreds of bacterial spheres and filaments, fragments of benthic (bottom-dwelling) mats, and planktonic green algae (acritarchs)," the latter composed of two types—"simple spheres" and "large spiny forms," Grey says.

When she began, Grey "had no idea" of what they would uncover. The project had been an effort to develop dating of Precambrian strata. But she kept discovering heretofore undescribed microscopic fossils. Grey worked with thousands of samples—and did so on weekends and at night, because the research was not directly related to finding fuel and mineral resources, which was the focus of her government position. The study took years, and Grey would go on to publish a massive, 700-page monograph detailing her finds.

The rise of spiked planktons after the Acraman impact was her biggest surprise. "When I first plotted up the results," she says, "I could not believe what I was seeing." Her thesis supervisor, Malcolm Walter, didn't believe it either. So Grey started over, examining all the slides again and relogging data, but the Acraman correlation wouldn't go away. Then she learned from Clive Calver that after Acraman, there had been a global die-off of many other tiny creatures.

Snowball Earther Paul Hoffman wasn't impressed, however. In 2003, Hoffman told *New Scientist* magazine, "It's a great idea," Grey's Acraman theory, "and very testable. If you find one place where there are big spinies before the impact layer, the hypothesis is wrong."

Grey replies that this "idea that a single spiny species before the event would negate the theory is not correct. I would actually expect to see a few spiny species around before the event. Not many, and in low numbers, but they should be there. The [impact] event would effectively wipe out much of the competition, allowing the spiny forms an advantage. After all, the mammals were around but not very significant during the age of the dinosaurs . . . "

Much more remains to be done if Kath Grey's theory is to be proven correct. She concedes this, saying the evidence to back Acraman as an evolutionary catalyst is "largely circumstantial and requires further testing." Moreover, some of her colleagues think the simpler spheroid bacteria themselves might have produced a coating to protect themselves from harsh conditions. But Grey believes she has microscopic evidence that the spheres were "dividing," which implies "that they had soft cellular walls" rather than tough exteriors like those of the spiny planktons.

Some four years after my trip to Australia, with her work proceeding, Grey will tell me that she and others are finding new evidence to support her ideas. Sebastian Willman, a Swedish doctoral student, has studied more South Australian drill holes (either ones not examined by Grey or the subject of only preliminary study), confirming the sequence of events in Australia and identifying new species. Identical species are known to be present in Siberia and near the Ural Mountains, and while the exact age of these successions is unclear, they can be matched to the ones in Australia. The species from these other parts of the world show the same changes in plankton populations that Grey's original data did. At the time of Grey's first study, the actual ejecta layer was known in just three drill holes, but she, Willman, and Andrew Hill will later find the changing plankton in core samples from places in Australia. Researchers will also see hints that the Acraman impact had a major effect on organic chemistry. But it's not yet been possible to match these plankton changes to those of a supposedly similar age in other parts of the world, including China, the Himalayas, and Norway. Grey thinks this could be a function of different forms of preservation, slight differences in ages of the rocks, and varied kinds of sedimentation. Additional study is needed.

In the time following my visit to Australia and my first conversations with Grey, Paul Hoffman will focus on those specimens from China and elsewhere, noting that an "abundant" number of spiny planktons there date closer to the end of a glacial period—preceding the Acraman impact—and that a chemical tracer linked to ancient sponges in Oman also predates Acraman. He will tell me that "these recent discoveries . . . make it more likely that early animal evolution was connected to global glaciation" and probably not to the Acraman impact.

But Grey can't find any other way to explain what she's found. Her observations fit perfectly with the emerging paradigm of craters as refuges—and fountainheads—for life, and even Paul Hoffman will admit, "The story is far from over."

WE'RE LEAVING ACRAMAN BEHIND. I look up to reclaim horizon, seeing a few faraway cottonpuff clouds and the Gawler Ranges to the south and west. Then I steal a few more glances down, seeing the patina of cumulus shadow over beige sand and white salt. Melissa turns back east, and the shore of Acraman recedes.

There my notes end. I fall asleep in the plane until Vic wakes me up for an impressive, late-day view of Wilpena Pound, which, while more immediately stunning than Acraman, does not conjure up the mélange of images in my head: hot, wet water in a silty crater; slides of plankton; a dark, fire-shot column.

We land, and Vic will later tell me of his awe over seeing the structure from the air again. I tell Vic, "I'd make a piss-poor astronaut." We're all happy to be on solid ground.

On our first evening at Wilpena Pound Vic Gostin told me that Kath Grey's research is "the resurrection of Acraman," and, leaving the runway, I recall that the day before our flight, Lorraine and I were examining more ejecta chunked into shale. I sat down, tired and hot, beside a tiny fern that was growing in the shade of the hand-sized rock. I hardly expected to see a fern in the desert.

"What's this one called?" I asked Lorraine.

"A resurrection plant," she told me.

"Oh," I said, reaching down to touch gray-green leaves. "That's perfect."

Resurrection plants live beside and under rock edges and in clefts, to take advantage of whatever water might collect there. That tiny plant, nucleated down to the last drop, had hunkered in among meteorite ejecta, shale, mosses, and meager shade, its thin leaves arrayed against oblivion. Doesn't chance alone seem miracle enough?

"Catastrophe" in its original Greek means "an overturning."

"That's right," Vic had answered at dinner one night, when I asked him if he considered Acraman the most important crater on the planet if Kath Grey is right.

"It's creation out of destruction," he said, nodding vigorously. "Like Shiva."

Chapter IX.2

Old Stones That Can Be Deciphered

Not long ago I learned that some fungus-loving hippies and 'shroom-dropping New Agers believe that life arrived inside meteorites in the very specific form of psychedelic mushroom spores. Talk about wish fulfillment. This must be an eroded form of ancient tales that suggested shooting stars deposited sticky residues on the ground, everything from mushrooms to jelly. Regardless of its origins, this enduring notion must make for some interesting meteor-watching at old communes stretching from the Blue Ridge to the Front Range and on to the Sierras. A little Perseid light show, a hit off some psilocybe, and, woah, the chanterelles are on fire. Or is that my tie-dye bandanna?

Life in meteorites—it's a very old story. The idea that rocks from space have seeded the Earth is older even than the science of meteorites. Anaxagoras raised the possibility 2,500 years ago. He spoke "of germs infinite in number" that were part of the great concoction of all things before rotation spun out each type of matter. "Panspermia"—Seeds Everywhere—posits the unproven, alluring idea that meteorites and comets rained microbes on the Earth, microbes that might have come from other planets, other stars, or even other galaxies. Chemist Svante Arrhenius and, later, cosmologist Fred Hoyle both championed panspermia in the twentieth century. Despite its long pedigree, this idea only pushes the problematic origin of life to somewhere else.

Panspermia has nonetheless gathered serious scientific commentary. As England's Sir William Thomson (better known as Lord Kelvin) put it in the nineteenth century, "The hypothesis that life originated on this Earth through moss-grown fragments from the ruins of another world may seem wild and visionary; all I maintain is that it is not unscientific."

Part of the impetus for Victorian beliefs that meteorites contained otherworldly life or at least evidence of the same derived from the knowledge that plant decay had led to the formation of coal. "Reasoning by analogy," historian of science John G. Burke writes, "if the same type of substances were found in meteorites, they in all probability had a biological origin." And German naturalist Hermann von Helmholtz and others knew that while the exteriors of meteoroids are heated by friction when passing through atmospheres, their interiors remain cold: "All germs . . . that happened to be in the cracks of the stone would be protected from combustion in our atmosphere." Burke notes a fascinating theological impulse behind these notions. Thomson wanted to disprove Darwin's theory of evolution, which seemed to him to be anti-Christian in its emphasis on accident and mutation. If meteorites carried life from other worlds, Thomson thought, this would be evidence of God's hand. Helmholtz himself posited life-bearing space rocks as a form of endurance beyond the eventual death of the Earth. Meteorites were thus caught up in two of the main currents of Victorian thought—arguments over the origin of life and dismay over entropy or heat death.

The boldest nineteenth-century claims concerning life and meteorites came from a German lawyer who had also studied geology. Otto Hahn, like a few others, had looked for organic structures in thin sections of meteorites. Unlike the others, Hahn found them everywhere and in all types of meteorites. He thought Widmanstätten patterns were plant remains, not crystal growth, and he said that meteorites contained everything from fossil ferns to fossil crinoids. Most biologists and geologists considered Hahn's theories bunk; one critic thought Hahn simply didn't know how to use a microscope. Without realizing it, Hahn, his few supporters, and their many detractors had set up the terms of what would become a classic, recurring "this-versus-that" debate. Were patterns in some meteorites tracers of biological origins or merely the result of chemical and geological processes? Further, one of Hahn's allies remarked that some of the alleged fossils were far tinier than species known from Earth. This too would become a repeated theme. How small can life get?

A new question arose in the 1930s: Were claims of meteoritic life based on accidental terrestrial contamination? In the case of bacteriologist Charles B. Lipman, the answer was yes. He had cultured microbes from several meteorite samples, but when faced with withering criticism Lipman backpedaled from his implicit claims of life from space and, in Burke's words, said "he had just presented the factual evidence of having found bacteria in meteorites." As bacteria can find their way into almost anything, this finding was of little use, though it did stir up the press.

In the early 1960s, Bartholomew Nagy reported finding chemical traces of prior biological processes as well as possible fossils inside meteorites. In a paper in *Nature,* George Claus and Nagy reported possible "microfossils" and "fossil algae" structures and even "several . . . organized elements [appearing] to undergo 'cell-divisions.'" Acknowledging that some of their putative microfossils looked similar to Earth species and couching their claims in very careful language, Claus and Nagy nonetheless concluded "that the organized elements may be microfossils indigenous to the meteorite."

Two years after the paper, the New York Academy of Sciences organized a symposium on the Claus and Nagy studies. In one of the talks, researchers presented evidence that some of the organized elements were, alas, ragweed pollen. Different structures were more obdurate, but they had probably resulted from earthly contamination or nonbiological processes. Claus and Nagy disagreed. They still believed their evidence suggested the presence of dormant or fossilized microbes from beyond the Earth.

The Nagy saga was an uncanny precursor to the tale of a Martian rock that made headlines three decades later. In 1996, a team of scientists announced that they had evidence of fossilized microbes in a Martian meteorite found in Antarctica. The meteorite was named ALH 84001. Meteorites found in Antarctica are named by place abbreviation, ALH standing for Allan Hills; by year, 1984 becoming 84; and by sequence of cataloguing, 001 indicating this was the first sample opened in the lab that processes Antarctic meteorites. The two major pieces of evidence foregrounded in a *Science* paper were the presence of something called polycyclic aromatic hydrocarbons (PAHs) and other features known as carbonate globules, which the researchers believed had been excreted by Martian microbes when they were alive eons ago.

PAHs are not in themselves a sign of biological activity. In fact, they're often found as postcombustion products such as soot. Yes, they do smell. Candles, colognes, perfumes all have PAHs. Formed from rings of hydrogen and carbon, PAHs are implicated as being necessary to the development of life. PAHs have been found in meteorites—but that's not surprising because PAHs are all over the universe, it seems. In ALH 84001, their presence was connected to the presumed microfossils as a way of arguing for their biological origins, but other researchers thought the PAHs resulted from that old bugaboo, accidental terrestrial contamination.

As to the carbonate globules, the alleged microfossils do resemble the shape of bacteria, but most scientists believe that these mini-Martians are just too small to have ever been alive and thus could not have excreted the globules. In one of the first critiques of the ALH 84001 claims, meteoriticist Harry McSween noted that the structures are "perhaps one hundred times smaller than the smallest known terrestrial microfossils." In fact, the Martian microfossils—if that's what they are—"would be smaller than most viruses," according to science writer J. Kelly Beatty. Most researchers believe that 200 nanometers is the lower size limit for life—apart from viruses, which are their own special case—but the Martian squiggles are at least half that size and some are much smaller. Too small to contain enough material for reproduction, in all likelihood. Beatty notes, however, that "starving bacteria frequently shrink to as little as one-fifth of their normal size yet remain viable."

ALH 84001 has yielded other controversies. McSween writes that "the likelihood that a 4.5 [billion-year-old] igneous rock . . . would contain preserved microfossils seems remote." Most fossils are preserved in sedimentary rock, which is unmelted. The hydrogen inside ALH 84001 also seems to have come from its Martian environment rather than microbial processes. And researchers have squared off over strings of magnetite crystals in the meteorite. A few scientists say the crystals were extruded by Martian bacteria. Most scientists disagree, asserting that the magnetite resulted from chemical process.

However all these claims are ultimately settled, the fact remains that the original ALH 84001 paper in *Science* "launched," as the magazine itself says, "the modern field of astrobiology nearly single-handedly." Astrobiology—the study of potential life beyond the Earth.

MEANWHILE, panspermia has been getting renewed scientific attention. One group of researchers says the chances are remote that microbes could have arrived in rocks from other stars. But interplanetary panspermia is theoretically possible. Should living bugs from Mars have been ejected inside rocks that eventually landed on Earth, then it's possible that life here was, or rather is, Martian in origin. No one can say for certain. Yet.

Microbes are tough enough to withstand such a journey. *Deinococcus radiodurans* can repair injuries to its DNA with alacrity and can endure radiation comparable to a mind-boggling 17 *trillion* years in outer space. The universe itself is only about 15 billion years old. Space shuttle experiments have shown that bacteria can survive cold, airless, UV-saturated environments. And we know of other instances of microbial survival in space. Inside the probe Surveyor 3, *Streptococcus mitis* bacteria survived on the lunar surface for nearly three years; Apollo 12 astronauts brought them back—the bacteria had been hanging out on foam pads above some electronics. Scientists revived the hardy germs. Researchers have even revived bacteria from the Permian—from 250 million years ago. Meteoriticist Guy Consolmagno hopes to restart experiments that were halted by NASA budget cuts, experiments aimed at testing tiny pores and cracks inside meteorites for their suitability to host microbes under spacelike conditions.

As yet there is no proof of microbial life beyond the boundaries of the Earth, either in the past or at present. No actual or fossil ET microbes—or even DNA scraps—have been found in meteorites. The fact that we are so eager to find life beyond the carapace of the atmosphere says much about our need to belong to the cosmos, our desire for a fecund order, and our fear of loneliness.

This eagerness to find life in space also plays itself out in fears of contamination. Numerous low-budget science-fiction movies from the 1950s and 1960s depicted meteorites as agents of strange toxins, weird rays, and mind-altering germs. Back in the 1960s, the Bartholomew Nagy controversies—real science—helped prompt sterilization procedures for the Apollo astronauts and their Moon-rock samples. It was thought that Nagy's meteorites, which were rare, had come from the Moon.

In time, it was clear they had not. Nagy had actually looked at a form of chondrite called a carbonaceous. These are the meteorites that can look like charcoal for your grill, and they're as fragile as a dry cookie. Carbonaceous chondrites—or c.c.'s, as they're often called—contain olivine and pyroxene but very little metal. The amount of carbon in these rocks varies—anywhere from less than 1 percent to about 5 percent of the total chemical makeup—but this carbon is crucial, for it is an organic compound, an ingredient necessary for life as we know it. The lack of metal, the relative richness of carbon, and the presence of water in c.c.'s distinguishes them from other meteorites. Carbonaceous chondrites also contain other organic materials, including nitrogen, hydrogen, and oxygen. There are only about one hundred such specimens in the world, out of tens of thousands of total meteorites. And because scientists have identified rare-earth compounds in carbonaceous chondrites, they consider these meteorites the most pristine examples of the early solar system. They are the space equivalent to a mountain spring.

So WE KNOW THAT some meteorites and comets, while lacking life itself, contain ingredients necessary for life, a stone soup that's been landing on Earth for billions of years. Take water. Carbonaceous chondrites have a lot of water in them, in some cases up to one-fifth of their composition. In the late 1990s, scientists found an asteroid that might be a spectral match for carbonaceous chondrites—a possible parent body. And the asteroid still has enough water for a couple of big pools here on Earth.

In fact, scientists have found that organics in carbonaceous chondrites are "strongly associated with clay minerals," and clay is produced in association with liquid water. "This association," writes Victoria K. Pearson and her team, "suggests that clay minerals may have had an important trapping and possibly catalytic role in chemical evolution in the early solar system prior to the origin of life on the early Earth." According to what is called the interstellar–parent body hypothesis, the organics we find in carbonaceous chondrites and other meteorites originated in interstellar space and became part of the proto–solar system, and as comets melted, the organics became saturated with water as rockier bodies coalesced from this icy mess of gas, particles, water, and goo. Pearson's team has shown that the crucial mediator between interstellar organics and meteorite-trapped organics is clay, which is known to take in organic material on Earth. Organics were clearly an important part of the presolar nebula mixing bowl, ending up in all sorts of things, including comets, asteroids, and dust. Micrometeorites have been found to contain amino acids.

But just how do organic compounds originate in the vast voids of interstellar space? Researchers have found that amino acids can form using space ice and ultraviolet light. There happens to be a lot of ice in interstellar space, ice that reorganizes its molecular structure after soaking up ultraviolet light. That makes the ice curiously like water on Earth. And that makes it possible for such ice to host organic materials.

A few years ago scientists even found liquid water in two ordinary chondrites—a huge surprise, since they were thought to be utterly dry rocks—one called Monahans (Texas) and the other one called Zag (Morocco). Caged within microscopic crystals of halite (salt), the wee water droplets are ancient. That the Monahans water is caged in halite is significant, for as researcher Michael Zolensky points out, halite forms in very wet environments. So the planetoid on which Monahans originated must have been pretty drippy, or else the meteorite picked up the water by crashing into a comet. The halite formed almost immediately after the first solid bodies of the solar system, which suggests a lot of water was present on forming planets and asteroids, which further suggests that life might have originated extraordinarily early. Researchers think that the Monahans salt crystals came from a parent asteroid with water near or at its surface. Pardon the pun, but let that soak in: *rivers and lakes or an aquifer—on an asteroid.*

One of my favorite images in the annals of the fallen sky is this one: the Monahans, Texas, meteorite filled with salt and water falling near a group of kids playing basketball. The kids were, of course, sweating—exuding salt and water. One boy said the rock "was still warm," scientists report, and recovered samples were bagged by the local police. Because of the quick recovery and for a variety of other reasons, scientists have ruled out terrestrial contamination of the Monahans meteorite; the Monahans water came from space. Millions of years of exposure to cosmic rays has turned the halite from its usual white to an ethereal blue in both Monahans and Zag. Blue skies of salt trapped within meteorites.

Computer models now suggest that wet meteorites or massive protoplanets, not comets, directly brought water to Earth via impacts. Comets may still have played a delivery role, but the water prevalent in those bodies is not the type prevalent on Earth. Yet another alternative for getting water to Earth is a "wet" accretion of the inner solar system's planets. Dust grains in the presolar nebula themselves might have had enough water to supply the liquid to what became the inner solar system's planets.

Meteorites have water. They have organic compounds. They even have amino acids that many think formed in interstellar ices or, some say, in the meteorites themselves (in chemical, not biological, events). Amino acids in a famous carbonaceous chondrite called Orgueil were originally produced inside a comet, suggesting that this and perhaps other carbonaceous meteorites derived from cometary material.

Amino acids are the essential ingredient in proteins, and proteins are crucial to life. Proteins help provide energy and carry out a range of cellular activities. There's a crux here to bear in mind: the difference between life itself and organic material. In somewhat circular fashion, my dictionary defines life as "that property of plants and animals which makes it possible for them to take in food, get energy from it, grow, adapt themselves to their surroundings, and reproduce their kind: it is the quality that distinguishes a living animal or plant from inorganic matter or a dead organism." In contrast to actual living things, organic compounds are a buffet of molecules containing carbon, hydrogen, nitrogen, and oxygen that includes nucleic acids, sugars (which are carbohydrates), proteins, and fats (lipids).

Bill Bryson has explained that collagen, which is a protein, requires the writer "to arrange eight letters in the right order. But to *make* collagen, you need to arrange 1,055 amino acids in precisely the right sequence . . . here's an obvious but crucial point—*you don't* make it. It makes itself, spontaneously." Bryson compares the rise of life from these insane combinations of atoms to the spontaneous generation, in your kitchen, of reproducing cakes. This sort of analogy is, as Bryson himself admits, fodder for the intelligent-design community, who argue that life is so complex that it simply could not have arisen by chance. But wait. Bryson goes on to note that "chemical reactions of the sort associated with life are actually something of a commonplace. . . . Lots of molecules . . . get together to form long chains called polymers. Sugars constantly assemble to form starches. Crystals can do a number of life-like things [demonstrating] that complexity is . . . natural." In other words, "this natural impulse to assemble" seems innate to the physical world, and evolution—not spontaneous generation—shows that molecules hook up as frequently as good-looking singles at a bar.

There's another hitch, however. Proteins prefer dry counties. They don't like to form in water. Might the varied environments of meteorites have been an aid to the evolution of proteins? Meteorites have both wet and dry counties, after all. Perhaps the perfect scenario is that of amino acids forming in interstellar ices then finding their way into meteorites as the solar system formed. That would be a kind of "soft" version of an interstellar panspermia.

Nearly fifty amino acids occur in space—and in carbonaceous meteorites—but not on our planet. But of those amino acids found in both places, how can we know the meteorites have not been contaminated? Part of the difficulty in knowing whether carbonaceous meteorites are contaminated, even partly, is that amino acids on Earth are predominantly "left-handed." This "handedness" is called chirality, so you imagine amino acids constructed in the way our hands are, such that our left hand is a mirror image of our right. Some amino acids in carbonaceous chondrites are left-handed too, though not all, and scientists have been back-and-forthing for years now regarding the origin of left-handedness in meteorite amino acids. Because that's how they are in space or because that's how they became once on Earth? Part of this mystery is that scientists have a difficult time creating an overabundance of left-handed amino acids in labs; they usually come up with a mix of left- and right-handed. Until a probe scoops up a sample from a carbonaceous parent body and returns it to Earth for analysis, the chirality/amino acid question will remain unsettled.

Meanwhile, scientists have reported that samples from various meteorites have been effective platforms for the growth of algae, microbes, even vegetables such as asparagus. This gives credence to the notion that meteorites and other bodies could harbor life and, down the cosmic road, to the dream of humans terraforming other worlds.

For these reasons, meteorite collectors love carbonaceous chondrites. A pinch of clay, a dash of benzene, a touch of water, a swirling in the bowl of the proto–solar system's gravity, and presto, life comes from beyond. Maybe. One of dealer Robert Haag's price lists—this one was from 1999—listed bits of the Australian carbonaceous chondrite called Murchison at $100 a gram, boasting that "Murchison is much sought-after due to its scientific importance. It contains more than 400 organic compounds, many of which are not found on Earth." Murchison also has sugar in it. A bit of stellar evolution under glass, Eden in a case. Haag was then offering 5 kilos of another carbonaceous—one of the most famous, Allende—for a cool $20,000. Allende and Murchison are also popular because they both landed in 1969, the same year we first set foot on the moon.

But if such a specimen isn't exotic enough, consider contacting ICI's Quest International, an English firm that, according to *Sky & Telescope*, "has designed a fragrance supposedly capturing the essence of carbonaceous chondrite." The company said the perfume offered up "metallic, gunpowder notes."

THE MOST EXCITING CARBONACEOUS chondrite fall in history took place on January 18, 2000, not long after I had started thinking I might write a book about meteorites. For a time I imagined traveling up to remote Tagish Lake near the border of British Columbia and the Yukon Territory, but the meteorite was collected so quickly and the timing was so bad—I was teaching—that a trip there was more or less impossible. Over the years, I've read with envy details that made both the fall and the subsequent searches resonate with adventure and scientific intrigue. The ICI perfume designers nailed the carbonaceous chondrite odor, judging from the sulfur smell some Tagish eyewitnesses reported. The meteorite fell from a brilliant, hissing, ground-rumbling fireball with a dust cloud witnessed as far south as Edmonton and the northern United States.

The quick collection of Tagish—right after the fall by a local resident and then again in the early spring by a science team—was very good news, because carbonaceous chondrites are so fragile, friable, and fickle. This one happened to fall right onto the surface of a frozen lake. Once on Earth, these c.c.'s don't last long, getting ground into dust by wind, rain, and the occasional passing hoof. And Tagish Lake is, scientists say, "the physically weakest meteorite known."

Outfitter and pilot Jim Brook was the first to hunt down many fragments from the fall, though he admits to being tricked at times by wolf scat. Brook knew what he was doing. He didn't touch the meteorites. He had worked with Canadian geologists before, retrieving snow for space-dust studies, so he was careful to use a VisClean rag and clean plastic for picking up the fragments, which he then put in clean bags and baggies. He even used sterile tongs to pick up many small samples. Brilliantly, he stored the stones—nearly 900 grams—in a freezer.

Peter Brown, Alan Hildebrand (codiscoverer of the dinosaur-killer impact crater), and other scientists were electrified with the Tagish Lake fall because Brook had worked so quickly and carefully. Further field searches found more meteorites, some encased in ice, though many had become wet and degraded with spring melt. About 10 kilograms were recovered from the meteorite, which is estimated to have been 4 yards in diameter. Scientists from the University of Calgary and the University of Western Ontario even employed a Royal Canadian Mounted Police crime dog in the search for possibly smelly specimens. The dog didn't point on any meteorites, though. Also distinguishing this fall was the fact that military satellite data were utilized to calculate an orbit, the first for a carbonaceous chondrite. The orbit, not surprisingly, places Tagish Lake in the Apollo category of earth-crossers.

Tagish Lake has been compared to some of the most historic falls in the history of science, including L'Aigle. This is due not only to the fast, precise recovery and the orbital information but also to the meteorite's chemical makeup, which is, in a word, weird. Everyone expected Tagish Lake to be full of amino acids. It isn't. Tagish has hundreds of times *fewer* organic materials than Murchison. And only parts of the meteorite, not all of it, appear to have been affected by water. Strangely, Tagish also contains purely interstellar material, suggesting it might have originated from beyond the solar system—a major first, if true.

Analysis has hinted at a cometary origin for the meteorite. Tagish Lake is a carbonaceous chondrite, all right, but may be in a class by itself, somewhere between that type of meteorite and a comet.

For all its seeming rarity, this meteorite may represent the most common stuff that has fallen into our atmosphere. Because Tagish-like materials would degrade very quickly—and most of the meteorite ended up as a massive dust cloud—we may have been skewing our idea of what meteorite falls represent. "If this is correct," one science team wrote, "then the bulk of the water and organics on Earth could well have arrived in Tagish Lake type materials." Later studies have indicated that Tagish Lake "represents an object that physically bridges [comets] and the weakest 'asteroidal' material existing in meteorite collections."

Tagish Lake has yielded one other mystery—hollow globs of organics, little caverns begun in the chilly outskirts of the forming solar system. The structures, compared by scientists to membranes, "may thus have been a common form of prebiotic organic matter" dribbling onto our home world. Not exactly Lord Kelvin's ruins from another world but close enough that he'd be pleased.

I NEVER GOT TO TAGISH LAKE but once, while in Paris, I visited with meteoriticist Claude Perron at the National Museum of Natural History. We talked about another famous carbonaceous chondrite, one that fell a long time ago—Orgueil.

With his glasses and long fingers, his white shirt and salt-and-pepper beard, Perron looked every bit the academic, but he also had the lean build of a runner or bicyclist. He eventually found meteorites "more interesting than cosmic rays," his original area of research, and has been studying space rocks since the late 1980s. Until then, he had never looked through a microscope.

"You're a student all your life," he told me. And Perron is, among other things, a student of chondrites. "You can use them as a memory, a memory of early times of the solar system and before the solar system. The memory is not always clear. . . . The information is there. . . . You have to decipher it. I am interested in the oldest memory," he said.

Orgueil fell on the French village of that name on the mild evening of May 14, 1864, and the National Museum of Natural History in Paris is one of its resting places. With many pounds of this historic fall still remaining and with its deeply primitive chemistry, Orgueil is one of the most requested meteorites in the museum's collection, if not in the world of meteoritics. Studies of other carbonaceous chondrites use Orgueil as a kind of chemical yardstick. This is the meteorite that Bartholomew Nagy believed contained microfossils, and this is the meteorite that established the standards for abundances of solar elements for geologists, chemists, and astronomers. It's a really important fall.

"It's the richest meteorite in presolar grains, presolar diamonds—if diamonds are presolar—[and] in organic material," Perron said. "It may be the most ugly, but is the most precious."

Despite its cosmic origin, Orgueil now travels, like all such lab-bound meteorites, by rather ordinary means—the post office. "I was just typing an e-mail to someone who is requesting a sample of Orgueil," Perron said as I arrived. Staff put a requested sample in a plastic container, rather like a jewelry case, surround it with packing material, and stick it in a box. I like thinking of this carbonaceous chondrite, which once roamed the black of space, being routed by automatic mail-handling machines.

At the museum, Orgueil is now kept in a leakproof container filled with dry nitrogen. For a century and a half, though, samples were kept in a drawer, which contributed to a slow decay researchers want to stop.

The day after Perron and I first talked, I returned to look at samples of Orgueil. "I would have liked to have seen it minutes after the fall," he said without prompting and with a sigh. "It's [like] a piece of coal. You find it on your way, you don't bend to take it," Perron had said once before. That is, unless you know what you're looking for, a carbonaceous chondrite is nothing special to see. But for Perron, and for me, it is.

Orgueil. Cracks in the black crust like a terrain. It's dull like misshapen charcoal, more pitted than L'Aigle. On Orgueil's crumbly, noncrusty surface is a white speckle that looks like mold—a reaction of material with terrestrial humidity. It's carbonates and sulfates growing "like a tree," Perron said. It's transformation. "It falls into dust, organics wash into water, amino acids dissolve," he said. And this is what researchers wish to stop, in order to preserve the sample—though it is this very dissolution that helped transform the Earth into a living world.

I was seeing it right before my eyes.

Then I realized that its dust was coming off onto a paper towel in the box. I cringed. I didn't want this sample to completely fall apart on account of my curiosity. Perron reassured me that this fragment is an exhibit piece, and anyway, dust that spalls off Orgueil gets brushed into vials. Later I'd learn two more oddities about Orgueil. Someone deliberately contaminated a sample in 1864 with seeds of a rush plant; the trickster even *glued* the seeds into the meteorite. The sample was sealed for decades and the act not revealed until 1965, when Edward Anders went hunting for proof of Bartholomew Nagy's microfossils. And unwittingly, curators and researchers—in trying to keep their labs and offices spotless—have introduced organic molecules called terpenes into Orgueil. Terpenes are found in store-bought cleansers. So much drama for so ugly a rock.

I kept looking. *Orgueil*—a black-gray-white-crusted-uncrusted-speckled-fissured meteorite the size of an orange, a broken baseball retrieved from some ancient city, an icky ur-baseball, Gilgamesh's pitch.

I grew a bit lightheaded in the overheated office. The studio lights that Perron had switched on while I took pictures glared and glared. But I kept looking and looking at Orgueil. If black can glow, Orgueil glowed. If a meteorite can open like a spiracle, Orgueil opened.

Beyond the brightly lit office, gloaming in the streets. It was late, so I snapped the lens cover shut, put my notebook away, and said my grateful goodbyes to Claude Perron. Then I walked into the Paris evening, my back drenched in sweat as I stood on the cobblestone terrace of 61 Rue Buffon beside the lit window of some old lab. I hadn't asked to touch Orgueil, though of course I would have loved to feel its grit between my thumb and forefinger, would have loved to have been like Robert Haag and taken meteorite to my tongue.

Beside one of the dingy stone buildings of the museum, I became transfixed by the appearance of everything around me, the green trash bin in a courtyard, the wall and gate whose iron pickets reflected the red glow of traffic lights. At the west end of Rue Buffon, clouds were breaking, showing a bit of late-day blue as children and parents walked by and cars raced past on the one-way street.

Water gurgled into the sewer. Walking, I stared into lit apartments the way I would stare at lit living rooms and leafed-out trees glowing like green brains from streetlamps, all those years ago when I went on long walks in my humid college town in Indiana, where I first mistook novelty for passion, stoicism for wisdom. Behind the Natural History Museum, dogs were walked, strollers pushed, suitcases and shopping carts tugged, and I stepped out of the way of a father and his young daughter who was sputtering, crying. I stood by another sewer grate with its wet tissues and wrappers, more of the city's filth beneath its misty vistas, the water almost roaring beneath my feet. Were there rivers on asteroids? Could I raft one? Could I ride water from here to there?

Drunk on Orgueil, I made my way to meet Kathe, who waited for me in a café. What adventures had her day held? By the time she and I met that night, Claude Perron must have been on his bike, heading back to his life, while the dark briquettes of Orgueil, safely returned to their nitrogen hold, were doing their now-much-slower-burn-on-Earth, their long-disintegrated kin having cooked up a few amino acids a few billion years ago on the giant lava grill that was the early Earth. And I decided that carbonaceous chondrites looked less like charcoal for backyard barbecue and more like hunks of burnt wood left from a campfire back home in Utah, where wind tussles willows and box elders and cottonwoods and the chokecherries candle-stemmed with blossoms that scent the air with a perfume subtler than lilac but no less lovely.

Well, Cliché Man Sings Too Much of Life. Purple Prose Man Meets Deep Time Once More. I walked through Paris, came home, traveled, came home, traveled again and again, thinking of how brimming meteorites fall and scorch, how they drizzle water and salt, how they scatter in the grass and sometimes get found, sometimes get deciphered, and all along, from Kansas to Utah, from France to Germany, from Oregon to Australia, I have wondered if and worried how I might find a meteorite myself one day and all along I have dreamed of the place where I could.

Book X

Old Fire on Blue Ice: An Antarctic Journey

· *Chapter X.1* ·

Above the Clouds, Halfway Down

For there is no escape anywhere.

—RICHARD BYRD, *ALONE*

Today is one of those days where sunlight and wind seem to be the same force, and I pedal hard into it, birds flying about me, streamers of sky, beside a bay of black swans. About a year removed from my experience in Paris, where I was high on Orgueil, and just a few days after flying over the Acraman astrobleme with Vic Gostin, I'm touring on a bike through wetlands near Christchurch, New Zealand. It's an afternoon in early December 2003. New Zealand is my stopover before I do something that seems insane, even impossible: search for meteorites on the high, cold, lonesome of Antarctica. I listen to the gravel spit out from beneath the tires of my rented bike, and my eyes sting with dust, sweat, and sunscreen. When the wind dies, I can suddenly speed up, though sometimes I stop to soak in this spring-to-summer day and to catch my breath. I watch the chop of the bay and the sway of pines. There are New Zealand scaup, paradise shelduck, and shags in ponds, on the bay, in the air. This place and the next—none of it's real, right? I stretch my arms, gulp in the salted sky and I shake my head, happily bewildered, riding the crests of Freud's oceanic, but feeling mighty strange too, unsure of this new verge.

But then the past two and a half years have been nothing but a plunging off edges. I began this book married to a woman I left just months later. Kathe already had left her husband, and she and I were in love. Later that year my mother passed away, and Kathe moved to China for a teaching job but also, in part, to see if our love was honest and strong, which—as it happened—it is. Just before she moved, I tried to get to Greenland for the first time and failed. About a month later and right after my divorce, I visited Kathe in China, my first trip overseas, then spent the remainder of the fall trying to write about meteorites while I looked for a new job and rattled about my apartment. A few months later, Kathe returned when her father fell ill; that winter he died. The following spring Kathe and I traveled to the Ries Basin in Germany and went home to Kansas to pack boxes. By then I'd found a new job, and we soon moved to Utah to start work at Utah State University. We bought and renovated a house and set to repairing damage to our land inflicted by strangers with backhoes. And there was more travel: me to Greenland; the two of us to Paris, Ensisheim, and L'Aigle. My journey to Australia. Now this.

It seems as though I left Kathe back in Utah eons ago, but it's been three weeks, and I keep wondering if I should just sit still here—or anywhere—and simply wait for a meteorite to land at my feet. I straddle my bike, watching stilts and—crazy reminder of home—mallards and Canada geese. Light billows and wafts, things and edges intersect.

And this recounting is more than chronology, more than coincidence, more than a part of the narrative I'm handing over to gentle readers. This is more than entries in a journal some might keep private, as if a writer of history and the world must necessarily be effaced by them. This is a realization and an exhausation. I'm not sure I remember when it was—when I understood that my passions, so long thwarted, and my heartbreak, so often elided, were part and parcel of the hunt, *my* hunt, for the meteorite hunters, and that passion and heartbreak were things I might share with them. When I began this book, I never expected that the characters I'd find—from Hughes to Cragin, from Peary to Nininger—would, in a way, find me out. Their lives interrogated mine, and my answers were found wanting, till I owned up to myself. That I failed to own up to my ex-wife remains a shame I cannot undo. I look up from water and dirt and I know there is fire above that we'll never see.

With the wind behind me now, my map tucked in a pocket, I ride from the marsh toward Centaurus Road.

Maybe I'm more akin to Eliza Kimberly, who stayed on her prairie a good, long time, than, say, to Robert Peary or even several of the folks I'll be living with for the next few weeks. Thing is, I too love inhuman vistas, I too love seeking—and I want to find some shooting stars—but right now I'm pretty much the accidental tourist, a bone-whipped homeboy. Before Kathe took me to the airport for the start of this trip, I wept, holding in my lap a stray kitten that had arrived on our snowy road. We named him Shackleton, and I didn't want to go.

Tomorrow's new verge: If the flight from New Zealand isn't turned around or canceled, I'll be in someplace new again. I'm pedaling back to Christchurch, the salt on my neck warmed by the sun. Back at the motel, I'll meet up with members of the science expedition I've joined. Tonight we're heading to dinner. Tomorrow we're heading to Antarctica.

THIS EVENING I'M EATING with the other members of the Antarctic Search for Meteorites (ANSMET), minus Johnny "Alpine" Schutt and Bill McCormick, the two field safety leaders who, as experienced mountaineers, will train us in the many ways to avoid dying. They'll meet us in Antarctica, where two ANSMET teams, funded by the National Science Foundation and NASA, will gather supplies, train, then deploy to the "deep field" to scan expanses for dark stones from space, old fire on blue ice.

We'll collect every meteorite we can find. They'll be saved exclusively for researchers. We'll be there for weeks, then we'll go home.

I look around the dining table at my compatriots. There is ANSMET's leader Ralph Harvey, a meteoriticist at Case Western Reserve University and a veteran of thirteen summer seasons in the Antarctic, a reader of science fiction and a former jazz musician who is deaf in one ear. He looks as if he could have been a tight end on a college football team. (He wasn't—bad knees.) Ralph has his dark hair cut military-short, but despite the intimidating appearance, he is friendly, smart, and quirky. We're both fans of *The Simpsons*. This alone should make my time tenting with him less stressful. At some comment I don't hear, Ralph booms with laughter.

I think I first spoke with Ralph by phone at my sister's house in Indianapolis right after my mother died in 2001. I already knew that there were two great places for meteorite hunts—the deserts of North Africa, plied by French dealers and well-armed warlords, and the deserts of Antarctica, plied by government-sponsored scientists and cold winds. The latter sounded preferable. Without Ralph's blessing, I'm sure the NSF would have turned down my application to go to "the ice" under the auspices of its Antarctic Artists and Writers Program. So I'll have two roles during my several weeks with ANSMET: I'll be a regular expedition member, helping to find meteorites and to keep camp, and a writer, note-taking to document the experience and to bring the memory home.

Nearby are Tim Swindle and Oliver Botta. Lanky, thoughtful, and articulate, Tim works at the University of Arizona's Lunar and Planetary Laboratory, one of the top such facilities in the world. Tim turned to science years ago from, of all things, journalism. I had started my writing career as a college journalism major, so I already feel a bit of a tie with Tim. Like Ralph, he's leaving children behind for this, his second trip as part of ANSMET. (Ralph always includes some "veterans" each year.) A ruggedly handsome Swiss postdoc, Oliver works for the Leiden Institute of Chemistry, and he's leaving behind his pregnant wife and a three-year-old son. With Ralph's encouragement, Oliver has successfully mocked the accent of actor-turned-governor Arnold Schwarzenegger and spoken Wisconsin, the nasal dialect of Ralph's home state.

Down the way are Andrew Dombard and Gretchen Benedix, friends and planetary scientists at Washington University in St. Louis, where, years ago, I completed my master of fine arts in writing. Andrew, who has the taut build of a runner, has delighted me with a vividly corporeal sense of humor that would be at home on *South Park.* Gretchen, though she must be younger than I am, has lovely silver hair, as well as the fetching habit of repeating lines from the nonstop stream of jokes we're generating. Like Tim, Gretchen's also a veteran of ANSMET.

Ralph's second-in-command of ANSMET, another planetary scientist at Case, Nancy Chabot, sports a deep tan from having just been in Australia. With her dark hair, bright white teeth, and an effusiveness even when tired,

Nancy gives me pause, as if the reserve in her dealings with me is also a form of judgment. Then there's Barb. With multiple piercings in one ear, one side of her head buzzed, the other side sporting shoulder-length hair, Barbara Cohen is a spunky, affable meteoriticist at the University of New Mexico who would have been comfortable at the venerable punk club CBGB. An English minor as an undergrad, she has engraved on her portable music player one of the great lines in poetry, William Carlos Williams's, "Say it! No ideas but in things."

At the table's far end is Monika Kress, a physicist by training who is cobbling together teaching and research assignments until she lands a permanent position. With her short blond hair and infectious smile, Monika is charming and ebullient. She also can't stand the idea of eating anything that lives in water, but as we won't be sailing to Antarctica, this is not an issue. Finally, the quietest of the group, at least from my perspective, is Gordon "Oz" Osinski, he of Haughton Crater fame, the up-and-coming impacts researcher who, with his field experience in the Arctic, should be a happy camper at the planet's other end. Oz, who wears wraparound sunglasses, has the kind of calm reserve that can bring out the paranoid klutz in me. He's kind but scary-good.

No paranoid klutz was Sir Robert Falcon Scott. The second man to the South Pole was, Ralph tells us, "the tormented one." Roald Amundsen, the first man to the South Pole, was "the driven one." Tim, Oliver, and I weigh in, and after I note with surprise that Amundsen's conquest did not satisfy him—after all, though Peary had apparently grabbed the North Pole from him, Amundsen had taken the South Pole first—Oliver turns to me, very serious, and says, "You can't expect that after achieving the goal, he's happy for the rest of his life." Taken aback, I nod quietly.

Soon Ralph is talking about his Greenland skiing trip when he looked for but failed to find meteorites on that ice cap. Then he jokes. He really ought to look for meteorites in warmer climes, he really ought to start BASMET. The Bahamas Search for Meteorites. We laugh, but it's slightly nervous. After all, we're itching to get as far from warmer climes as we can—to the emptiest, highest, driest, coldest, iciest, windiest continent on Earth—and three times this week we've failed to do so.

. . .

AFTER EVERYONE ELSE ARRIVED in Christchurch—I'd gotten there several days before—we gathered in the motel room that Ralph and I were sharing and where Nancy gave us a briefing, complete with maps and photos. One picture was an aerial shot of where I'd be staying for my first two weeks in camp—the LaPaz Icefield, named for Lincoln LaPaz, rival of Harvey Nininger and mentor of William Cassidy. Cassidy is the founder of ANSMET and was its leader from the 1970s until Ralph took over in the early 1990s. Cassidy was one of the first researchers to understand the significance of a 1969–1970 Japanese discovery of *nine* meteorites of *five* different types in *one* place in Antarctica. The discovery indicated that some process was concentrating meteorites from different falls at different times, a process no one had seen before, anywhere. Cassidy was keen to find out more.

Now we know the basic mechanism: Ice flows out from high points on the polar plateau—think of giant glaciers—but the ice sometimes encounters mountains, above surface and below, forcing the ice to pile higher and higher. Eventually, though, the topmost layers are slowly stripped by powerful katabatic winds and by the odd summer day when it's warm enough to sublimate or melt the ice. Eventually rocks that were encased in the ice, including meteorites, are revealed at these "stranding surfaces." Meteorites that have been in or on the ice for more than 3 million years have been found at some stranding surfaces, though many such areas are "only a few thousand years old," Ralph writes. Some ice fields also simply preserve the "direct infall" of meteorites that happened to have landed right there. Further, the dry cold slows down the weathering of Antarctic meteorites. The blue ice fields where some or all of these tendencies occur are actually pretty rare, but having used aerial photos and helicopter surveys, ANSMET knows where to find them.

Antarctica has become the most productive meteorite-hunting grounds on the planet, a place where, since the 1970s, U.S. and Japanese scientists have collected more than 25,000 meteorites, or some 85 percent of the world's total number of specimens. The continent is so fruitful that other nations have sent search teams too. As Ralph puts it on his website, "The ANSMET specimens have been the only reliable source of new, non-microscopic extraterrestrial material since the Apollo project and will continue to be until future planetary sample-return missions develop and succeed." Unlike the specimens collected privately in North Africa, all the ANSMET meteorites are mapped and named, and samples of every single one are available—free—to researchers.

Antarctic meteorites have helped advance planetary sciences first and foremost simply by being described. Like North American ornithology of the eighteenth and nineteenth centuries, meteoritics is still in a cataloguing phase, trying to fill in chemical and petrologic gaps. In fact, what with the thousands of ordinary chondrites made available by ANSMET recoveries, scientists can feel less queasy about destroying such samples in order to study their constituent materials. Antarctic meteorites have been found to contain elements from Wolf-Rayet stars, which are the professional wrestlers of the stellar world, big and hot-tempered. Wolf-Rayets are about 20 times larger than our own sun, but they seem bent on a Weight Watchers diet of shedding as much mass as they can. ANSMET chondrites also have demonstrated chemical links among previously separate chondrite groups, suggesting, as Ralph says, "a relatively smooth variation in nebular conditions . . . instead of distinct nebular zones." A type of carbonaceous chondrite, the CK category, was "partially defined by ANSMET discoveries," according to Ralph; the CKs got hot enough to hint that they were well incorporated into a big asteroid. ANSMET has recovered enough space rocks to show that the asteroid belt is "a set of complex, miniature planets, exhibiting features consistent with gradational levels of planetary processing, involving both traditional and decidedly exotic geological activity." From somewhere out there, an entire class of meteorites, the ureilites, seems to have come into existence through more than one process, and the majority of ureilites come from Antarctica. So the simpler model of an asteroid belt with lots of homogeneous rocks and a few geologically complicated ones is out the window. ANSMET has helped to reveal the asteroid belt as a complicated place. Perhaps best known is that researchers have recovered pieces of the Moon and Mars off the ice of Antarctica and recognized them as such.

Sitting on the itchy carpet of the motel room, I stared and stared at the picture of LaPaz. A sweep of light turquoise—ice everywhere and flat—with a few patches of snow and a few specks. The latter were tents and snowmobiles, the microscopic evidence of the small, temporary ANSMET camp that had scouted the place in a prior season. John Schutt had deemed LaPaz a good place for systematic searching, so most of us would be going there. It looked like what Ralph Harvey had called the entire East Antarctic ice sheet: "the Earth's very own glassy windshield."

THE DAY AFTER OUR MOTEL BRIEFING, we were driven to the Christchurch airport, to the Clothing Distribution Center of the NSF's U. S. Antarctic Program for "check-in." I was nervous.

At home, preparing to go to Antarctica had tapped into a mind-spinning trait of endless planning, printing off Ralph's copious ANSMET web pages, and reading them with a glass of wine or two or three, reviewing to-do lists. There was much sorting and worrisome concern over, oh, the number of batteries to take. And while Ralph and Elaine Hood, my ever-patient contact at the NSF's contractor, Raytheon Polar Services, had both been helpful, I felt caught in a vortex of my own obsessiveness. There were times when I felt nearly paralyzed before the open maws of my suitcases.

The new orange tabby named Shackleton had left no space in my luggage unexplored. Rolling my long johns in tight bundles, placing books of poetry in among jeans, and setting aside sweaters I'd decided against, then decided I'd take, then decided against, I felt the weariness of moving again. This would be a trip, not a move, I told myself. I'd come back to a life better than any I ever had. Still. This felt like more like loss than it should have. When I'd left my wife, I paused a long time to look through my study windows at hackberry trees where northern flickers had nested. A year later, when Kathe and I left for Utah, I stared at the apartment's white walls and thought of all the silences we had broken there. As I prepared to leave our new home in the mountains for this Southern Hemisphere sojourn, I thought, *This is adventure?* I gave no consideration to asking to delay my participation for a year or two, when my life would be more settled.

The night before I left, Kathe and I ate carryout pasta dinners by candlelight, the cat snuggling between us, and we quietly talked of love and separations. After all that had happened, what was three and a half or four months? Besides, how many people get to go to Australia and New Zealand then live in Antarctica? I felt more than a little abashed, and Kathe knew that, even as she was telling me what I was telling myself: You're about to experience something extraordinary.

So the day I stood in the lobby of the Clothing Distribution Center, I felt both grateful astonishment and a background swarm of *uh-ohs*. Holding a clipboard, Marlene arrived in her green blazer and matching green glasses. Behind her, parkas and insulated pants festooned a display board. Marlene was there to make sure we correctly picked our ECW—Extreme Cold Weather gear—and had our two orange bags of the same packed and ready to go for departure the next day. We'd also have to wear ECW on the plane from Christchurch to the ice. Along with other scientists and support personnel scheduled to leave on the same flight, we ANSMETers divvied up by sex, and the men headed into a room that was a cross between a warehouse and a junior high gym. Cold floor, high ceilings, and a partition on which were mounted big fans and a mirror (as if what we were going to look like really mattered). Soon a bunch of nearly naked men were trying on gloves, yazoo hats, insulated overalls, long johns, boots, jackets, balaclavas, and parkas. I fretted over insoles and socks, head gear and weight limits as Marlene marched to and fro, barking instructions, oblivious to the many manly pale thighs. Eventually, I made my selections, stowed my personal luggage in a storage room, and packed my orange duffels.

A few hours later, Andrew and I stood outside at the motel. He joked that surely he should have an entire chapter in my book devoted just to him, and I laughed in agreement. Then he got serious.

"It's really setting in, isn't it?" he asked.

I said yes, yes it was, it was really setting in.

Months ago I'd been on a trail above the intense blue of Bear Lake, about an hour's drive from our house up relentlessly scenic Logan Canyon. I was sitting under a tree reading a book called *The Springs of Adventure,* by English mountaineer Wilfrid Noyce, wondering if he was right, wondering if "the very poignancy of renunciation renders more exquisite the pleasure of achievement" and if "the thought of achievement in untrodden fields is a delight that makes renunciation worth while." I hoped so. A few paces up the trail, wind soughed through a 600-year-old limber pine and rustled Indian paintbrush and curl-leaf mountain-mahogany. So if I was possessed of any single-mindedness—Noyce's "divine madness" of obsession—was it less to go to Antarctica and more to record the going? For in writing, the having-done outlasts the doer.

That evening we all thought we'd be in Antarctica the next day. Alone in the late night, I leaned against the window of my motel room and listened to the ticking of my watch and stared at a gray corrugated wall and a strip of gravel right outside, all suddenly too real. I stared at a cabbage tree, its one flowering stalk among branches that had died back. It looked like a cousin to the Joshua tree. Unseen traffic deepened strangely all these clarities—each wave, each wave, each wave more than words could bear. I took myself outside, strolling to a tiny park with grass and roses and cabbage trees at the corner of Mandeville and Riccarton. Above me, an old friend upside down. Orion. I pressed my hands onto the bark of a cabbage tree and tapped it twice, only to scare out a bird. *I'm sorry.* Backing away, I said, "Go home, you can go home."

THE NEXT MORNING RALPH TOLD ME about his dream. He was being quizzed, though he was the teacher of the class, about "beandorf." He was stymied, having no way to clarify "beandorf." Beandorf? What could it be? Beandorf was insoluble. Ralph and the inexplicable beandorf.

Bustled back to the Clothing Distribution Center, we all loaded up in our selected ECW—there's so much of it, about 25 or 30 pounds, one feels loaded more than dressed—and we got on a cargo plane and took off. That afternoon we disembarked, but not in Antarctica. We had turned around, trumped by bad weather. A boomerang—or maybe a beandorf? Back in Christchurch—"Cheech" or "ChCh" in ice lingo—we turned to errands, televised sports, naps, and books. Attempt to Get to Antarctica 0, Antarctica Itself 1.

The following day our 7:15 A.M. flight was pushed back three hours. Then our flight was canceled before we even boarded the plane. McMurdo Station—the main U.S. science base and our point of arrival—was still being battered by a storm. Attempt to Get to Antarctica 0, Antarctica Itself 2.

So today, earlier this morning, we tried once more, but the flight was canceled yet again, which is why I set out to bike among the birds and marshes in afternoon frustration. Attempt to Get to Antarctica 0, Antarctica Itself 3.

During one of our waits at the Christchurch airport, I sat in a café and stared at the menu: Shackleton Soup of the Day, Scott Pasta, Amundsen Quiche, and Mawson Stuffed Potato. The cooks were kidding, right? Ernest Shackleton was the last explorer of the Heroic Age and is best known for his leadership during his ill-fated three-year Antarctic expedition, which began in 1914. After his ship, the *Endurance,* was crushed in sea ice, Shackleton and his men fled in lifeboats to Elephant Island, from which he led a smaller crew on a makeshift boat journey to landfall on South Georgia. There, he and two men crossed snowy mountains before reaching a settlement, where they summoned help for the rest of the *Endurance* expedition. Shackleton Soup of the Day must be rancid seal meat seasoned with equal dashes of ambition and recompense. Scott? He reached the South Pole on January 17, 1912. He and his men found that they'd been beaten by the Norwegian Roald Amundsen's team, which had gotten there on December 14, 1911. Scott's party all perished. Amundsen's did not. Scott Pasta, then, would be best served with bitters. Amundsen's quiche would include the meat of his dogs killed along the way to the Pole. As to the Mawson Stuffed Potato, it sounded least appetizing of all. Douglas Mawson, a geologist who would go on to study the Flinders Ranges in Australia, is best known for surviving a trek that killed the two other men with him. Starving, snow-blind, and poisoned by vitamin A—from eating too many dog livers—Mawson traveled 100 miles solo. At one point, he sewed the rotten bottoms of his feet back on. (Was I worthy of a continent where men sewed the bottoms of their feet on?) He also fell into a crevasse, where he considered the attractiveness of cutting his rope. Instead, he crawled back up. Eventually Mawson staggered back to camp—just as his rescue ship was leaving. He and his men wintered through some of the worst weather the planet can dish up, and over the years, Mawson endured something else: rumors that murder and cannibalism might have aided his survival.

Antarctica hadn't even been seen or touched by humans until the 1820s. Since then just 100,000 people have experienced that place of alien cold and ice. Of them, about 600 people have died there. I was not in either category. I passed on the Mawson Stuffed Potato. Instead, wearing long johns under my shorts, I bought Antarctica-themed merchandise at a gift store. Not having been there, I had evidence of going.

I thought of the past spring and summer, as Kathe and I felt more at home in our house and on our four acres. On warm afternoons, I'd wade across the Blacksmith Fork River, just a few paces from our back door, and get in a hammock under some willows to read—among other titles, Thomas Henry's *The White Continent*, unable to believe that his narratives of exploration, science, and despair had taken place on the same planet where I was, the planet where I was in a hammock listening to dippers and warblers, with a view toward Millville Canyon's juniper slopes. I learned things. Antarctica covers 5.5 million square miles—it's bigger than the United States—and 98 percent of the landmass is covered in ice thousands of feet thick. At the South Pole, the ice is almost 2 miles deep. The ice sheet has locked up 70 percent of the planet's fresh water, and meteorites are sprinkled on the surface like a giant's sortilege.

So I would go, I thought in my hammock beneath those willows, I would explore Antarctica and find meteorites. I would map obsession and chart its topography against a different landscape, that of acceptance, trying out my Richard Byrd against my Lao Tzu. Right. Me. A Webelos dropout. Sometimes I'd snap the book closed, set my jaw, and look at the river passing by.

TODAY, after this morning's failure to get to the ice, after the van ride back to the motel, after the repacking and our afternoon excursions, including my bike ride in the marsh, we've broken bread and blown off steam at dinner. Be flexible, Nancy keeps saying, expect change. This evening we're leaving tips, stretching our legs, and straggling out onto the dusky city streets. We amble through Hagley Park once again, walking back to the motel. Every time we've walked through the park, I have secretly yearned for another serious birder among the geologists, but to no avail. But we spot something else: Mercury, just below Venus, both planets hanging above a steeple and briefly visible in a clearing of the clouds. Will this be the last night I see in weeks?

• • •

REALLY, THE ONES WE should honor most aren't Scott and Amundsen, Shackleton and Mawson, but the nameless engineers, the makers of sledges and motors, radios and compasses, forgers of better metals and smoother ball bearings, builders of this New York Air National Guard LC-130 Hercules. We're on it, it's flying, and the weather hasn't turned to shit.

We've been aloft for how long? I'm 2–2 at chess—Ralph reset my electronic game to idiot level—and I'm reading *The Odyssey* in snatches. *And the roads of the world grow dark.* In our lunch bags are Mars bars, which is seen as propitious, for we'd love to find a Martian rock on the ice.

ANSMETers, workers for Raytheon Polar Services, other scientists, and DVs ("distinguished visitors") are dispersed along the sides of the Herc, either sitting in a sling seat or standing beside strapped-down duffel bags in the middle of the plane. Having shed our parkas, we're all squishy enough in our insulated bib overalls and clodhopper clumpy in our white vapor-lock boots—bunny boots, they're called. I sit under kidney-shaped, red plastic "Shield Assy" trays—I have no idea what they are—and two trash bags hang up front under a strapped-down cooler. Next to that a curtain hides the urinal and next to the urinal sit the DVs, who, Ralph had noted gleefully on another flight, will be the first to be decapitated should the propellers come unhinged.

The plane is hot or cold, the plane is all wiring, metal walls, metal floor, metal ladder, conduits, insulation, canvas, straps, netting, hooks for packs, vents, lights, latches, safety signs, bolts, and bags. Jonahed in the roaring Herc, we lean into ears to talk or we resort to mime and primate facial expressions.

I bide my time not only with Homer but with the *United States Antarctic Program Field Manual,* which tells me how to make show shelters, read the weather, and fashion giant letters in the snow to attract rescue, a skill I won't need at McMurdo, where we'll be staying in a dorm. Nor do I think I'll need to resort to big letters at my first field camp, at LaPaz, where the main ANSMET group—eight of us—will be deployed. But after two weeks or so, I'll join the four-person recce, or reconnaissance, team, John Schutt's scout party moving along a stretch of the 1,800-mile-long Transantarctic Mountains. The larger team systematically searches and collects a single area. Recce seeks out promising sites for later thorough collection. And recce worries me—just me, John, Monika, and Gretchen. What if the radio and satellite phone fail? What if the other three die in a crevasse? Will I have to spell out letters in the snow that say, "I have no clue, please help"? I set the manual aside.

So when Barb Cohen leans over, I'm just absentmindedly eating a peach pastry. I take out my earplugs to listen.

"I think I see ice," she yells.

I undo my seat belt and walk to a porthole window and see channels of ocean below. Not ocean. But *channels* of ocean, ocean as dark-blue cracks *between* white, with just enough surface relief to see that the white is, in fact, ice. Barb and I grin. We're flying over a sea of ice. Sure. A sea of ice. Cropped by the window, the ice/land/sea looks like a Joseph Cornell entry to a Norwegian science fair. It makes me tingly-jittery-happy.

After Barb ventures to the flight deck, I do too, asking the same question—where are we?—and the navigator points, "Out there, land ho!" showing me the chart, a straight line from Cheech to Mactown—McMurdo. We're about to cross over Victoria Land, with its steamy young volcano, Mount Melbourne, which I'll not be sure I actually saw. "Very scenic," says the navigator.

Very scenic, yup, if that includes connotations of "Cripes, that's actually Antarctica, with sea cliff, fast ice, mountains, whitewhitewhite." Scenic in the way that the event horizon of a reverse-image black hole would be scenic. Monika comes over, leaning against me to steady herself to take photos. The icebergs are bleached mesas. Antarctica is an albino Utah.

When later I say to Barb, "Thirty minutes," her eyes widen, and she smiles like a kid. "This is crazy," I say. She singsongs, "We're GOing to LAND a PLANE on ICE." Earlier, Barb and Nancy had bounced beside me when I admitted that I had to pee. I just couldn't go behind the curtain, near the propeller-unhinging zone and its potential distinguished victims.

Skitters of disbelief. The loadmaster—a National Guard overlord of cargo and passengers—collects our lunch bags, and we're all putting our ECW back on. My parka feels warm and sharp-edged, stuffed as it is with books, chess game, pills, batteries, Clif bars. I put on my fortieth-birthday Chicago Bears cap, sitting down to face nineteen green Exposure Suit Bags and five orange bags for the DVs. My bookmark in *The Odyssey* says, "Antarctica."

Flaps lower. The Herc banks slightly, and the white light of the sun shifts across our moving room. The heat's on. I've lost a liner glove. It's bumpy air, and the fellow to my left looks at me. At 5:28 P.M. on Thursday, December 4, 2003, he gives me a nod, then with his right hand flat in the air, goes, "Shhhooo." I can *feel* skis on ice. Barb and I shake hands.

"You can be the first to mark your territory," she says.

Year and hours have come to this, a few minutes of taxiing and the scrunch of snow underfoot, having stepped the three steps down from the LC-130. We squint behind dark sunglasses or yellow ski goggles. It's overcast but the brightest gray, and I'm standing on snow, looking at my first for-sure live volcano, Mount Erebus, all 12,450 feet of it and cloud-covered at the summit, and the snow expanse leading to those mountains and on which gray-and-red airplanes and yellow generators simply are. All around us, miles of flat seasonal sea ice and banners of green, red, and orange marking danger and safety, like prayer flags becoming a map and the place itself. The Ross Sea Ice Shelf—a plank of ice bigger than France—where in our red parkas we now stand.

THERE'S LITTLE TIME FOR parka-puffy awe because we're ushered into a high-clearance van, hot-rod red and pimped up with mondo tires and an overactive heater. We're being taken toward town but have to stop for a pickup truck at an "intersection." We're all craning, *ahhing*.

The driver says, "You're driving across several hundred feet of ocean."

Ralph points and says, "That's McMurdo."

Founded about fifty years ago, Mactown's a clutter of metal buildings on steroids, the illegitimate offspring of a mining-town-and-a-community-college one-night stand. It reminds me of Thule, all industrial, but later I'll see wooden sidewalks with spray-painted visages of actor Gary Coleman, more than one set of dreads, and a sign on a dorm-room door that reads, "Do Not Disturb We Gettin' the Freak On." McMurdo is labs, a coffeehouse and wine bar, a dorm called Hotel California, The Chapel of the Snows, a gym, a hospital, a seaport once the ice is melted out of Winter Quarters Bay, a post office, and a library.

It's hard drinking for some and safety warnings for all. It's bowling, lectures, art contests, and live music. McMurdo is about 1,000 people each summer, many of whom are scientists, the town's raison d'être. They do penguin surveys, ozone studies, ice sampling, and more. Each year an entire book is published just to summarize all the science projects NSF funds down here. Because Antarctica has been, by treaty, set aside for scientific research, several nations have summer and year-round facilities. For the United States, McMurdo is the primary community, but there's also Amundsen-Scott South Pole Station and another station on the Palmer Peninsula. To help the scientists' projects, Raytheon Polar hires hundreds of workers—"modern nomads," one person will say—from twenty-somethings seeking adventure to fifty-somethings honing the Age of Aquarius on an ice edge, from lawyers on leave so they can live here by scrubbing toilets to backwoods Alaskans driving trucks to get health insurance.

As we near town, Ralph tells us a story. Many years ago support personnel dispensed with an excess of unneeded sausages by burying them in a hillside. The hill was later blasted open for a pipeline, hurtling a shrapnel cloud of frozen sausages.

So here we are, riding from a sea-ice runway to a town atop basalt past a hill that once exploded with flying death-weenies.

"WELCOME TO ANTARCTICA!" ANDREW Dombard exclaims.

"Let's have a meeting!" I exclaim back.

We're standing in The Chalet, the A-frame building that serves as headquarters, and the NSF representative gives us a welcoming lecture in a dark wood room. "It's hard for you to think about leaving right now, but . . ." We set to filling out paperwork related to planned and preferred departure times and flights. Antarctica—Brooding Land of Many Forms.

Afterward, there's time for dinner in Building 155, McMurdo's communal heart. Here is the spacious galley adorned with massive landscape paintings of the ice (which include hidden renderings of candy); here are dorm rooms, the barber, the office of the town paper—the *Antarctic Sun*—and "skua" bins, named after the scavenger bird, where one can leave unwanted items and scrounge for new ones (a cap? a paperback?). Here is Highway 1, the busiest hallway on the continent, since along its length you'll find a laundry, a gifts/liquor/movie shop, e-mail terminals, and the housing office.

By ten P.M., I'm stuffed from my salad, Swiss chard, pork, carrots, squash, chickpeas, and couscous. The food is terrific, and I'm sitting like a bloated potentate on the bed in my windowless room. Someone's marked up the metal wardrobe, writing, "Fuck this place." Someone else has written, "Love this place." Next door "Amazing Grace" is playing so loudly on some movie that I'm not in Antarctica, I'm in college, my geeky freshman year.

Pulling back the red bedspread, I cringe to remember that I had to have Andrew help me set the alarm on my watch. Beside me is a lamp, an orange, an apple, my notebook, my wallet. I've put my small portable stereo by the phone, which happens to be on the small fridge. I've stowed books in a nightstand. My little home. Then I look at a small photo album with its pictures of Kathe and me, pictures of the house, of the river and of Harvey Nininger, meant to inspire. My little journey. And my body still vibrates from the Herc, so, without having to move, I am for the first time in years rocked to sleep.

THE NEXT FEW DAYS are a flurry of rushing to the galley to make mealtimes, sitting on my dorm-room floor punching numbers to get past jammed outside phone lines, shopping, doing laundry, taking walks, and—primarily—meetings and supply pulls, in preparation for a "shakedown" camping trip this week and for the real thing, for several weeks on the polar plateau, living in tents and driving snowmobiles as we hunt for meteorites.

Our first meeting is with Johnny Schutt, that veteran of many ANSMET seasons, of Arctic expeditions (he works with Oz at the Haughton Crater site on Devon Island), and of mountain climbing around the world. He's a legend down here. After he hands us our Mactown schedule, we notice—typo or not—a "Team Beating." Compact, bearded, with owl glasses and a ponytail, John wears an "Extreme Drilling" baseball cap and is more talkative than I expected, even playful. Though his hands shake from the crud—a cold—John exudes a quiet, unsettling confidence.

After reviewing the schedule, we spend most of the day in Building 160, the Berg Field Center, or BFC, which is a bit like a giant sporting goods store. It's here the tall "Scott tents," sleeping bags, fuel, stoves, and other necessities are checked out by expeditions before they're flown to sites on the ice. ANSMET has some of its own gear in a chicken-wire cage, where, attached to the door, a stuffed penguin hangs and a sign reads "ANSMET Crevasse Specialist"—a drawing of a penguin on skis upside down in an ice hole. Cuter than Mawson.

We do kitchen boxes first. They're named after pets John and his wife have had, including "Bedlam" and "Calico." The wooden kitchen boxes are painted lilac-purple, and since Ralph and I will tent at LaPaz for the first two weeks, he and I select cooking utensils and silverware while other tentmates do the same. Well, I watch him do this while he explains that first off, soap takes up too much energy and time. We'll just wipe dirty dishes clean with paper towels. Leave the strainer behind. We'll need all the water we can get, he tells me. We clatter forks as we pack.

Next we bustle over our "sleep kits"—sleeping bag, pillow if you want, fleece blanket if you want, thermal pads. I'm beginning to grasp something here in a muscular sort of way: Where we're going, it's going to be cold. This seems odd because the temp at McMurdo is warm enough that I can wear a sweater and khakis, no long johns.

When we're finished we head to the mild outside, walking across snow and volcanic gravel, to load the kitchen boxes and sleep kits into strong "triwall" cardboard boxes the size of Dumpsters. These will be flown with us, our tents, and our snowmobiles to the field camp sites. Here we're joined by Bill, tall, silver-haired, and attentive, who will be the field safety leader at LaPaz, while Johnny fills that role with recce. Standing by the Tri-Walls and pallets on a rise across from the BFC, Bill points out the landscape. The Dailey Islands, which look higher than expected because of an inversion layer, and the Royal Society Range across McMurdo Sound. Bill's tutorial gives me that I'm-a-speck-on-the-planet feeling again, but it's not altogether unwelcome.

Next we meet Peggy Malloy, the chipper manager of the expeditions' grocery store, with its slanting walls and gutter on the inside to catch leaks. Shelves brim with boxes of noodles, oatmeal, potato chips and much more. Drew leads me reverently to a shelf full of Clif bars, where I nod like a sage, and Peggy explains we're allowed two candy bars a day.

"Are we there for two hundred days?" Oliver asks.

"You'll want to minimize work," John says, meaning we should find stuff that's easy to cook. After a day of driving snowmobiles and looking for meteorites, you'll be tired.

We divvy up and consult. Ralph naturally takes the lead on compiling our list—the man has a hankering for chicken patties, while I veto any shrimp. "Allergic," I say, and Ralph nods. Above Peggy's stereo, which unspools folk music, there's a painting of a woman in red peasant garb harvesting fresh fruit and veggies. Tomatoes and bananas. Nothing that we can take along.

When John tells me and Drew the tale of a stove-heated Vienna sausage can exploding in a tent, I wonder, "What's with exploding sausages down here?," while Andrew pretends to pick a piece out of his hair, concluding, "You always have a snack."

Tonight, back in my dorm room and wiped out, I think about how Ralph quietly absolved me of camp chores. I'll demur because I can't imagine not doing them. I need to pull my own, I think. Then after a shower, I find myself staring at a picture someone cut out from a magazine and stuck by the sinks. Purple coneflower, a bee. A prairie picture, a close-up. I have to tear myself away.

THE MORNING SNOWMOBILE BRIEFING: easy enough to drive these things, Ski-doos by name. But watch out, 'cause Johnny sometimes turns on the kill switch to befuddle the driver, who can't figure out why the engine won't start. The food pull: fetching items from our lists, using bar code scanners, stowing the food in boxes. The afternoon: sweeping up, tossing a football in the Berg. The evening: Kathe's letter: "I miss you sorely. And I know you miss me. But promise me that you won't spend all of your time wishing away what will be one of the most extraordinary experiences of your life. . . . What are a few short weeks when we have all of this?"

THESE FEW SHORT WEEKS will include such questions as "Do I cross this crevasse?" and "Is my companion freezing?" John, Bill, Nancy, and Ralph are talking to the rest of us in an open-kitchen area of the BFC where we sit in worn-out sofas and chairs. For the first time in days, no one is joking. Monitor fingers and toes. Be able to feel them. Take advantage of breaks to adjust clothes, to pee, to eat snacks. Look for white skin. Look at manual dexterity. Do arm twirls. Swing legs. Run. Eat. Stay hydrated. Wear sunblock. Wear your sunglasses or goggles.

They tell us that LaPaz is at nearly 5,000 feet and 250 miles from the Pole, while recce will be working at elevations ranging from 6,500 feet to more than 8,000 feet. Temps will average 0 Fahrenheit, with -30 to -40 windchills possible.

Fortunately, none of our leaders have seen any major injuries, only some frostnip, some frostbite, a fair number of mild hypothermia cases, a broken leg, a case of snow blindness.

We're told to walk carefully because much of the time we'll be on uneven ice, which, on "warm" days, develops a thin layer of water. "One slip . . . can make your field season agony," Ralph says. Other falls are more insidious. John tells us that we need to stay "positive" to avoid the "third-quarter blues," when almost everyone runs low, a fourth of the time yet to go.

I want to hear more, but we're almost immediately on to a frustrating blur of knot-tying and carabiner lessons, which reinforce one of the reasons why I dropped out of Webelos, and when Ralph and Monika try to help me with a knot, I snap, "I learn better without people talking to me constantly."

Calmer at dinner, but still abashed about how poorly I did with all the ropes, I meet a Mactown worker, a young writer. We talk about favorite books, graduate programs, travel. But when she says, "You're going on my dream trip. It will be so intense. It will be an adventure," I can only nod and smile weakly.

"YOU WERE BORN BREECH, weren't you Ralph?" asks Gretchen Benedix as she watches Ralph Harvey squiggle through the tunnel entrance to a snow shelter. Ralph laughs as he does this, wriggling in his yellow jacket, struggling to egress.

We've stopped for a break after leaving McMurdo thirty minutes ago on our snowmobiles. While others examine the structures, I put on fleece and take a few notes, the first chance I've had this busy morning. We began by loading supplies and tents onto sledges so we could drive out to an area called the Windless Bight. The snowmobiles are a snap, and because we're following flags, there's no danger of getting lost or plowing headfirst into a crevasse. No knots for rescues! This is the start of our shakedown overnighter, and we're all in high spirits. Showing ourselves to be good students, we check up on one another, look at faces and fingers. The sun glares down, bright even through my silver-tinted Darth Vader ski mask. Motioned to get going, we sit astride our black cushions, restarting our yellow, red, and black snowmobiles, all of us looking, I think, pretty damn cool.

In the early afternoon we arrive at the campsite on the spine of the peninsula, in sight of Erebus and ensconced within the Windless Bight, aptly named by Edward Wilson's Winter Journey Party, part of Scott's South Pole expedition. It is, in fact, without wind.

"Is everybody clear we're on a glacier?" Bill asks.

Now come tests. Ralph and I have to set up the tent. A Scott tent is a tall, double-walled pyramid of cloth that you set down on the ice with the double-tubed entrance facing up. Then you tip the point into the wind, which should help raise it, but hold on to the guy ropes if it's really windy. Then you stake it, shovel blocks of snow around the base for stability and insulation, then throw in a tarp as the floor. Scott tents don't have floors, so if the tent blows away, you don't go with it. Inside, there's about nine square feet of home. I'm messing with the ropes when Ralph says, "That's, uh, sort of a figure nine. It's ugly but it'll work." Praise enough I guess.

"John will do things faster," Ralph adds, "and that's okay."

This makes me sulk while Ralph is on the satellite phone to family in Ohio. After finishing his call, he turns to me to say, "It's going to be much colder at LaPaz." *He's trying to get me ready.*

Still, I'm feeling a kind of whole-self gratitude—I mean, I'm camping in Antarctica!—so when we're walking over to the snowmobiles later I say, "Hey." Ralph looks over, puzzled, and I pause, then offer, "Thank you."

"You're welcome," he laughs, pleased, and I'm pleased too.

The rest of the day we spend at a crevasse big enough to swallow cars, whose far side is higher than the near and jags with pointy snow hoodoos. Nancy confides that ANSMET has never had to do a crevasse rescue in the field, which is good news. Bedecked in parkas, with ropes attached to our waists, we spread across the sloping glacier. Brian, a McMurdo safety instructor helping us on the shakedown, says to me, "It'll get cool in a second." A beat later, Johnny "falls" into the huge crevasse. I'd expected it, and everyone is yelling.

While others set up a pulley, I'm sent to the edge to see how the "victim" is doing. My ropy "skills" surely gave me this job, and, truthfully, I'm grateful. It's more exhilarating to belly-creep up there with my ice pick and test the snow lip and peer down into a sheer canyon of ice, a deep crack in the blue world. From the edge, I relay John's "injuries"—busted leg, twisted ankle. I run my ice ax under the pull rope to prevent it from digging too deeply into the snow, thus keeping the victim from having to be dragged into the snow once lifted out. The pulley ready, Bill tells me to kick excess snow from the ledge, and so I lie back, feet uphill, and kick, but really gingerly.

"Hey Chris," Bill shouts. "Boom! Boom! Boom!"

I'm kicking hard and fast till the edge is clean, and the team begins to pull, and as John emerges, I pull on the rope too.

"No!" John yells. "Pull *up!*"

I've just dragged my future tentmate headfirst into the snow. *Not good. Not good.*

At our next knot-tying session, Nancy is kind, telling me she had to be told the story of Mr. Ropy in order to memorize a simple figure-eight-on-a-bight. She tells me, "Look at Mr. Ropy! He's our friend. Now we'll choke him and poke his eyes out!" The ropes follow the story, and the knot materializes like a miracle. I hadn't expected her good cheer with me. I appreciate it more than I can say.

And in their ways, others are struggling. "Yay," Barb scoffs. "I'll spend six weeks doing this." Barb, who is short, finds the ECW so bulky she has trouble moving. Secretly, my spirits lift when I hear one of us lost a sledge on our drive out and another drove over his bags.

My spirits also lift at day's end as we face The Wall of Death, a deep, steep windscoop that we get to drive in as a reward for enduring the rescue training. I watch as people zip their snowmobiles down the middle then up, up, up the sides. "Fuck yay!" I say, waiting for a clear interval, and I'm suddenly an environmentalist's nightmare, a screaming, hollering, slobbering machine-riding goober, cutting giant loops on the sides of The Wall of Death, held in fast by g-forces, taking steep angles, caroming circles of sheer speed till I'm the last one on it and I accelerate up, up, up the steep side till my right track might be at the very edge of the scoop's top, then I hurtle across the middle—velocity as joy—and I climb the other side, shallower and not as high but perfect for hitting the lip and, oh yes, hell yes, going airborne, landing in a rocking thump.

Andrew says, "You were scaring me back there." Behind my ski mask I smile a big smile.

THE NEXT MORNING, I write in my journal, "Slept okay. Chilly but not too bad . . . . Hopefully vistas, people, and routine will make the time go by fast not slow. Two 3-week chunks. That's manageable. Don't worry too much about John. Let him do his job. You do yours."

Throughout the day, we review knots, dead-man anchors, and pulleys. The squeaking of the latter sounds like sparrows arriving, and while they call, I tally small vindications. Last night I used a pee bottle for the first time. I was able to tie and untie the double-tubed fabric entrance to the tent. I tell Bill he hasn't locked one of his carabiners. When I confide my frustrations with the ropes to Monika, she slaps my arm.

"Chris, I have a Ph.D. in physics. . . . Did you see me?"

On the late-day drive back to Mactown, Ralph attaches a sledge between his snowmobile in the lead and mine behind to keep the sledge from veering and tipping. I have to match my speed with his to keep the rope from falling slack. And I do, I do.

"Nice job," says Ralph.

WHAT IS THAT LINE FROM ROBERT HASS'S POEM?—"half the world's work, loading and unloading." At Mactown, loading gear, boxes, folding tents, hauling them, hauling pallets to be stacked on, weighing boxes on scales, marking weight, Monika keeping records, our frozen food pull (it's minus 14 in the freezer—a freezer in Antarctica?), Tri-Walls, sleep kits, coiling ropes, Oliver strapping down cargo Tri-Walls, taking cargo-laden sledges by snowmobile to the ice runway, sometimes gunning fast on the ice roads, and Tim and I briefly testy over whose bagels belong to whom.

SEVERAL OF US WALK on the snow-patched scoria road that curves under the radar dome of Arrival Heights, past stacks of pallets, a shelter called "the Tower of Power," elevated pipelines with five spigots, and huts of gold, orange, and red. We hear distant beeps and rumbles of machines. The sky is gray, the rock that much blacker against the snow.

We reach Scott's Discovery Hut. Assembled in 1902, the Australian-design shack is made of jarrah, with a wraparound porch, wide eaves, and a shallow pyramidal roof. Variously a living quarters, a theater, and a rough lab for preparing animal and bird skins, the hut was used by several expeditions. Men who searched for the corpses of Scott and his dead Pole party lived here.

Oliver, who retrieved the key, reads the rules, which include not touching anything for obvious historical reasons and because "Anthrax DNA may be present." We go in, breathing an era that smells of ammonia and hay in a place that is dark and dirty. Andrew disappears around a corner then announces, "Hey, Chris, there are some Clif bars here." I don't crack a grin this time.

The evidence of lives: containers of Robinson's Patent Barley, whose packaging retains in places its original cobalt blue, Hunter's Famed Oatmeal, Spratt's Special Dog Biscuits and Special Cabin Biscuits, Beach's Golden Plum "Warranted Pure"—all labeled in the fonts of bygone commerce, foodstuffs nearly a century old—along with rinse basins, rocks, empty bottles, stacks and piles of boxes, nails, a gas stove, animal carcasses, and dingy cloths draped over beams.

The only time I smile is beside the plum tins. Kathe and I love this little poem of Basho's: "Unknown spring/plum blossom/behind the mirror." Plum, pum, this is what Kathe and I call each other. And to think of what these men had left behind—

THERE'S ALSO BEEN TV watching—*The Simpsons* and *Sex and the City*—at the bar, plus my last taste of wine for weeks. Everyone's told me it's too much trouble to thaw wine each night back in camp. Like most of the ANSMET men, I got my head shaved (less fuss in the field), and we had a party to send off recce before the rest of us leave for LaPaz. We've had "bag drag," at which we had our duffels weighed and tagged for the flight tomorrow, a flight to the South Pole. That's where we'll spend the night before staging to LaPaz.

At a meeting, Ralph had said, "This is a team project. Don't take ownership of any specimen. They're not yours. They're not ours. They're the world's."

Collecting the meteorites will be as clean an operation as possible, as our skin won't actually touch the stones. Still, we can count on "contamination events," such as snotty noses dripping on a meteorite and the occasional running over a meteorite with a snowmobile, which is called "roadkill."

Showing us a picture of a typical, weathered meteorite with a fusion crust, Ralph said we'll develop the ability to pick out meteorites on the icescapes. At LaPaz, no terrestrial rock has surfaced, so there it'll be easy. If it's dark, it's either a meteorite or a shadow. When looking at earth rocks versus meteorites on recce, we'll get a sense of what Johnny called the "particular patina that develops with ordinary chondrites."

"Like a black mirror but on a meteorite," Ralph added.

We'll find meteorites by forming a lens shape with our line of snowmobiles, traversing, driving across blue ice till one or more of us finds a specimen—just by looking, just seeing it—then we'll wave people over. Once we're out there, we'll be shown the complicated collection procedure itself. For all the high tech—satellite phones, camp radios, snowmobiles, planes on skis—when we're out there we'll be using our Paleolithic skills. We'll be collecting seeds on some far savannah. We'll be drawing on our deepest selves.

In my room tonight, I make calls, thinking they'll be the last for a while, because it's not clear how much time I'll have on the sat phones in camp. I wish Kathe a happy birthday and make my dad laugh with tales of knots. I'm edgy and wistful as I write postcards and stow excess luggage in the ANSMET cage. I play music as distraction. *The South Pole. LaPaz. Recce.*

In "Dominion Road," the Mutton Birds song I've been listening to ceaselessly, the protagonist is slowly recovering from a girlfriend's abandonment. I've identified with both him and the girlfriend, but it's the jilted lover who is, of course, the most bewildered. He finds that crossing a street is akin to an alpine trek: "He tests his footing like he was up 10,000 feet." Maybe, I'll think later, it's the juxtaposition of the ordinary (crossing a street) and the extraordinary (crossing a glacier's face) that appealed to me, for I knew well the dangers of the former and was about to know versions of the latter.

"Above the clouds," the song continues, "halfway down Dominion Road."

· *Chapter X.2* ·

I Crap Through Disco Night at the South Pole

This morning we're waiting in a van on an ice runway near McMurdo for our flight to the South Pole. Tim sleeps. Others wander into a little nearby galley. Andrew, who confessed earlier to some difficulty with "lung butter" or what we otherwise would call phlegm, writes postcards. I write some too. We plan to mail them from the Pole. In the van, *Science Friday* plays on the radio, and when I tire of correspondence, I stare at two Herc props set up on girders down the way or else watch languorous clouds on the Royal Society Range. Most of us did not sleep well, and one ANSMETer's spouse "had a meltdown" on the phone.

"Come fly with me," Sinatra's singing on the radio now.

For four hours, we're not. We slowly move from the van to the galley, where a few other scientists and the flight crew wait. It's serve-yourself—toast with p.b., cereal, plums. Plums! We loll, we're bored, we read, we don't.

"You must be important," a crewman announces, gathering his jacket and things, "because they granted us a waiver." The cockpit voice recorder isn't working, so if we die no one will know what the crew said, which seems no great loss.

Into a fumy, cavernous LC-130 we nestle—there are only eleven passengers—and Barb gleefully hops in her seat.

This humming web of logistics stretched over a continent like straps across a palletized Tri-Wall, all to get humans to places where we're not evolved to live so we can find things on the ground, pick them up, and learn. I'm not quite feeling afraid and not quite sure how I feel: tired, yes, bewildered—that's a word that keeps coming up. Bewildered, yes, and this far? The Pole, then LaPaz, a place visited by what—*four* people? I'll literally walk in places no other human has walked, a very peculiar thrill. And if the physicality of it will be less fraught than that of the rotten-feet heroes on skis and sledges, the dimensions of this experience are as big as the solar system itself. With my big white bunny boots, I'm a faux astronaut again and a homeslice—as in Drew's catchphrase question, "What's up, homeslice?"—riding ice all the way to the asteroid belt and back (more Mrs. Chippy than Sir Ernest), bagging rocks and words and maybe some flexibility in the chill. Mrs. Chippy? The cat aboard Shackleton's ship, the *Endurance,* and actually the tabby was pretty content, pretty sweet. A good role model, that cat.

The Herc hums around us like a determined metal womb with just one word, *omm.* After a time, Oliver holds up one finger. One hour.

And after an uneventful flight—for Hercs are a second home now—we land, clapping, eyes big, bags tumbling. It's 2:47 in the "afternoon," and, suited up, we are disgorged for our extravehicular activity, silent, onto the flat, bright white that is the planet's end.

Magnetic lines of force arc up and out of the South Pole and land in my body, I swear I can feel them, and a van deposits us at the new South Pole Station. The station's on thick stilts, a misplaced Key West motel with its unfinished wooden sides and a sign by metal stairs that proclaims "Destination Zulu." On its columns, nearly 50 feet high, the new station is the tallest human thing on the continent between the Pole and McMurdo, some 850 miles away. This slabby rectangle will replace a clutter of metal buildings tucked under an unheated geodesic dome built more than thirty years ago, a dome that squats nearby like a cyborg's eyeball.

We enter through insulated meat-locker doors, climb stairs past red-and-gray walls, then straggle to a dining hall with windows. There I take pictures of the flags and footprints that crescent the South Pole's ceremonial marker, a mirror ball, not unlike the ones folks back home put out in their gardens.

Tonight we'll stay in the gym, because no regular rooms are available, and after watching a video on the Pole, we ask about phone calls, about e-mail. Hauling our duffel bags and backpacks through an unheated stairwell shaped like a beer can (it goes by that name), we move through more unheated corridors and arrive at the gym, with its Ping-Pong tables, punching bag, basketball hoop, and scuffed-up climbing wall. We throw down our sleep kits.

Soon Oliver, Tim, Barb, Gretchen, Drew, and I meet at an icy T-junction of corridors leading into and out of the old dome, from whose roof icicles dangle and snowflakes fall; the ribbed metal ceiling's frosted up like an old fridge. Trundling in our ECW we walk up the snow slope out of the dome, part in sun, part in shade, snow piled high on the right side. Later, this scene will make me think of the astronauts in *2001: A Space Odyssey* on the ramp in Clavius, there to investigate the mysterious black monolith, the monolith that presages some transformation. Behind us now, the dome, with its cluster of antennae, looks like a spaceship half-buried in the ice.

We're heading out to make orbits around the Pole. But just walking winds me something fierce. We've gone from sea level to 9,301 feet, which, because cold temperatures efficiently lower the air pressure at the poles, seems to our lungs like an elevation of 10,000 feet or more. And though it's just -10, the windchill hovers at a hefty -40. Not even close to the winter temps, which can drop to -100 Fahrenheit, but I still put my gloved hand over my mouth and nostrils, the only parts not covered by clothing. Air whistles in my goggles. *This is nuts.*

Yards away some stairs lead to a viewing ramp, but from here the steps seem to lead to nothing, an Escher climb into air, and we stay near to the half-circle of poles from which whip the flags of Antarctic Treaty nations. We're trying to absorb the history summarized in painted letters on signs that note the arrivals of Amundsen and Scott. To this contemplation comes a man wearing a pullover and a Santa hat. Gloveless, no parka, no heavy pants.

Drew sizes him up. "May I ask you a question?" Andrew says. *"Aren't you frikkin' cold?"*

He demurs—this is warm for the Pole—and he wanders off, point made. We suck.

A sundog glows above us—Oliver's pointing at it—and the light of one true sun and its phantom companions falls on us and on a series of U.S. Geological Survey metal medallions set in the snow. I examine them, curious about these designations. Each one marks the geographic pole on a certain date. Because the pole's ice creeps outward, every year a new marker is erected, while the old medallions—like meteorites trapped in ice—creep outward toward the coast. They'll get there in about 100 millennia.

I walk a few paces over to a fence made of five bamboo poles with teal flags (Reliable Racing, Queenstown, NY). The flags flail in the wind. A yellow tape stretched around the makeshift fence announces, "Entry by Permit Only Confined Space." In the middle are three crossed poles with two black flags and a sign stating it will be the position of the geographical South Pole on 1 Jan. 2004. Nearby is the mirror ball—with it, people can take pictures of themselves—and, beyond, the metal dome and the new station. My single shadow partly merges with that of the mirror ball's wooden base, so we're two shades joined, two legs about to start walking, though just where I couldn't say.

I'm cold and feeling queasy, but I insist we reach around the world, glovetip to glovetip, arms outstretched around the yellow tape, and we whoop it up, we globe-spanners, only to yell, "Do it again! Do it again!" No one starts singing, "We are the world . . . " but we could have and we wouldn't have cared because luck this capacious can carry any off-key tune.

At the brass marker for the 2003 geographic pole, Drew circles the planet's axis five times, I, once, as Oliver unfurls flags from the European Space Agency and his hometown so we can take photos of him. Then we talk of Scott, of how he must have seen, must have kept seeing, at some distance before arrival, Amundsen's tent, which the Norwegian had left with supplies for the Englishmen he thought might follow. To see your fate, to walk up slowly to it, after such struggle? Tomorrow, we realize, is the 92nd anniversary of Amundsen's conquest of the South Pole.

OVER A PIZZA DINNER, Ralph issues warnings. The last two weeks, when you're down, go outside and think, *This may be the last time you'll breathe air this clean, see sky this blue, drink water this clean. And you may never come back here again.*

"Antarctica?" he had said to me once. "I love it. The thing that pains me most is knowing I'll have to give this up someday."

He'd felt a different kind of pain two field seasons ago, when he left Ohio not long after the birth of his second son. Ralph will tell me that leaving then was a mistake, that his emotional outlook the entire time was "melancholy. I had too much time to think about what I'm missing at home." Last year he didn't even come to the ice, though he managed to send a poem to ANSMETers that went like this: "ANSMET: RAW! RAW! RAWS! THE METEORITES COME FROM MAWS! WE'LL PICK THEM UP AND WRAP THEM UP ALL DAY AND NEVER PAWS! WE'LL SNAG 'EM! WE'LL BAG 'EM! WE'LL DO A LITTLE DANCE! AND IF IT IS A MARTIAN, WE'RE GONNA WET OUR PANCE!"

This year he's back for just two weeks. When I switch out with Oz, and Gretchen swaps places with Nancy—Oz and Nancy are on the scout/reconnaissance team right now—those two will finish out their season at LaPaz. Gretch and I will then be with recce till the end. At the same time, Ralph's arranged to come out of the field and be replaced at LaPaz by Rene Martinez, another researcher. Ralph will go home before any of us.

After dinner, it's time to explore—the store. Fighting my bellyache, I wade with my teammates into the three-room shop crammed floor-to-ceiling with clothing bins. I buy a ball cap and a pullover, both announcing, simply enough, "South Pole Station." Then it's time to work. In bright, bright sunlight, we spend two hours lining up cargo from the LC-130 so that we can more efficiently load the smaller Twin Otter planes that will ferry us and our supplies on tomorrow's staging flights to the LaPaz Icefield.

It's after we settle in the gym that the diarrhea hits.

Which means I'll miss disco night at the South Pole.

It doesn't take me long to see Molly Hutsinpiller, the South Pole's doctor for this summer season. (The sign on the clinic, beside large canisters of pressurized who-knows-what, reads "Club Med.") She suspects that a pill I took in McMurdo to help me to acclimate to the Pole's thin air has unsettled my system, and since it's not exactly life-threatening and we're both seemingly starved for new people to talk to, we talk. As I sit on the edge of the diagnosis bed, in what could be a doctor's office in any small town, Molly and I speak of home and travel and curiosity. I say curiosity is a form of love, but she's not so sure, offering as evidence a lecture by one of the astronomers here at the Pole. Someone asked the astronomer a simple question: Why try to detect neutrinos? And he couldn't really say. That didn't seem like love to her. Before I'm able to come to the scientist's defense, Ralph arrives to see how I'm doing, then shows me, on the walk back, where a restroom is.

Soon, snug in my bag, I try to sleep but my gut grows seismic. The windowless room is very dark after the lights are out, but I can see my fair beacon—the glowing red exit sign to which I frequently waddle in long johns and untied boots, a desperate penguin on an unsavory journey. Out of the gym, I waddle past a patch of sunlight from beyond an open bay of the dome and tunnel, past metal doors, clomping on a metal-grid floor just above the snow, past pipes and braces and conduits like the insides of a submarine or the bowels of a robot. I waddle past the heavy doors leading to the power plant to the room I need and once inside I lean my head on the wall, dizzy.

The sharp cold in the corridors at least keeps me from going in my pants, and the one time I feel like vomiting, I just break into a cold sweat. Thus I spend what will probably be my only night ever at the South Pole. Which makes me think of what killed the men who strived to come here. Which makes me think that I wish I could rake some micrometeorites off the bottom of the South Pole well—they've found a trove of them for research—and swallow 'em whole, because they couldn't make me feel any more rotten than this. What's the worst that could happen? I redeposit some cosmic spherules? All of which makes me think, hoarder of irony though I am, that, in truth, I could use less of it right now. Maybe the doctor's wrong. Maybe irony's my condition. Irony, the official disease of English majors.

On my last bathroom run—when? the hours are confused—I wend my way to the "Comm" hut, where I'm shown the phone. Satellite coverage limits calls at the Pole to odd hours, but using a standard calling card like one you could buy at a grocery store, I put the call through.

"You sound exhausted," Kathe says.

I choke back the heat in my throat, behind my eyes.

We don't talk for long because I need to go back to the gym to try to sleep a bit longer. But at least we talked. I wish her a happy birthday, forgetting I'd done so already at Mactown.

When morning comes, I speak to no one till Ralph gently asks how I am. Staring at my uneaten cereal, I whisper, "Crappy."

LOADING AND UNLOADING, yes, half the world's work. Bill instructs Tim, Barb, and me on what next to pull from the Tri-Walls, which are too big to fit onto the Twin Otters, so we place things in sleds and refine our cargo lines for easy loading. I'm not racing to the john anymore and pitch in as best I can. And between flights we each have some time to be alone.

On the ninety-second anniversary of Roald Amundsen's arrival at the Pole, I stand by myself with my back to the new station. It's nine A.M., and I am blissfully solitary. And I look. Between the austerity of sastrugi, those rows of wind-formed snow as far as the eye can see (think the innermost circle of Dante's hell rendered large as an icy Llano Estacado), and the industrial nature of human activity here (boxes and boxes of cargo outside, pallets of snowmobiles, and rumbling snow tractors, giving the place the feel of the world's most remote loading dock and warehouse, which, in fact, it is), the South Pole is less a pinnacle of exploration and more a weirdly moving testament to the human need to carry itself, its entirety, to every farthest point.

No more tattooed naked men run around the various Pole markers as one had done this morning. The -47 windchill bites through my ECW, and I think of Scott and his men, stagger at their pain, twirl my arms to stay warm, and I'm *happy* to be here and glad I won't stay.

I put structures behind me, out of view to watch only snow skitter and go. I turn in all directions, white in all directions. I consider that serious cold punching me again and again in the back, the South Pole's fuck-you fist landing and landing. Molly said last night, "Sastrugi are like rivers, always changing, always the same." Above them is the purest air in the world. I walk around the Pole again, seeing myself reflected at the bottom of the world. Again I twirl my arms like windmills to keep my hands from freezing. I climb the steps of the viewing ramp, then I have to force myself to go back down, to go inside, the experience of a lifetime truncated by cold, exhaustion, and tasks.

At the galley, I have some oatmeal and juice. I go to the gym for a nap and more packing. I'm too tired to consider what I'm about to do. At times, I just stare at the scuffed and empty gym.

When Barb and I meet up to check out "Comm," we learn a Twin Otter is returning. For us. To fly to LaPaz. I brush my teeth and zip up my bags. I think, *This is it. This is really going to happen. I'm going to go now.* Barb and I walk outside.

After the plane lands, we watch one of the pilots drive a snowmobile up a ramp and into the hold, but the snowmobile's butt-end is still sticking out, so Tim, Bill, and I have to muscle-fuck the machine to straighten it up. Then it's time. Barb and I climb in, settling into tiny seats behind the snowmobile, a white plastic sled, two tents, and many boxes. *The bracelet?* I'm wondering about a bracelet I ordered for Kathe back in Christchurch. The bracelet is silver, to be engraved with a Maori proverb: *Turn your face to the sun and the shadow will fall behind you.*

The plane pounces into the sky.

Blurred horizon, cloudy, almost like a smog, sastrugi like cat-claw marks, like furrows, scalloped edges of lace afar.

"It's like flying over Ganymede," I say.

"Only brighter," Barb says.

Cross-hatching on the sastrugi now. An entire ocean of wind-whipped, suddenly frozen milk. And that's all, that and sky as far as the eye can see, nothing but small furrows of ice, splays, unending, nothing but the weather-worked plaster of a continent. In that whitework, I can almost see fossil grasses, stalks, seed heads, and delicate slender leaves, some fan-shaped like ginkgoes, some a frondy ice forest. I draw the shadows and the sastrugi in my notebook and decide: This is not snow-angel snow. Now the sastrugi look like bird tracks impressed in ice cream.

How high are we? Who am I to be here now? Barb had said everyone's saying six weeks is too long. I do and don't need to hear this, but I'm too tired to care and fall asleep.

When I wake, it's to rougher sastrugi below and a gentle rise and fall of the snow sheet. Nap again and wake to smear cream on my nose and lips because the sun is shining hot on my face, a counterpoint to the mean draft of cold air behind. Over smoother patches are long lines that look like crevasses. Clouds now and cloud shadows. I'm staring as hard at this place as I can, and it's not hard enough.

"Any ice yet?" Barb asks, meaning blue ice, the blue ice where we'll camp and find meteorites. The ice goes blue because all the air has been tamped out of it by the weight of snows. Snow to firn to ice to suffocated ice. Airless, it doesn't filter the blue portion of the visible spectrum. A drowned turquoise. This is the ice she means.

I shrug. I see crevasses like lines made by a drunken plowman.

Ice yet? Maybe. Maybe there's blue ice to the left, ice that's more like ocean rollers than the white sastrugi I've grown used to seeing. Blue ice now? Blue ice finally? Blue ice with patches of snow roughed in at places, the firn at the edges of the bluer surface we are bound for?

· *Chapter X.3* ·

Bedlam

The LaPaz Icefield—an enormous plain of snow, blue ice, and the firn patched in between, horizon to horizon, nothing but white and light blue rimming the world and reaching right here. Barb and I, stepping out to hearty welcomes from those who have preceded us, catch glimpses of the weirdness, like that ice pick yards away stuck in uneven blue ice. No—

Already? Yes, Drew found a meteorite just after his arrival! *Too easy,* I think as Ralph shows the stone to me and the two pilots—mainly them, because he wants such support crew to see the point of this work out here. And to dissuade, if dissuading is needed. He tells a cautionary tale about one pilot who was caught with a stolen meteorite. The message is clear. Don't mess with the rocks. Ralph had said in a briefing that long ago there had been "souvenirs" from expeditions, but not anymore. Still, I'd been told by many sources—none on ANSMET—that pilots, some scientists, and maybe others have illegally "poached" meteorites over the decades.

Though private dealers and collectors have yet to find the means to mount a major hunt down here ("Hoovering," someone called it), a few scientists with private foundation support have looked for meteorites, and there are tourist flights to "observe" meteorites in situ. ANSMET and NSF aren't happy about either, and I can't believe that someone who has paid thousands of dollars to "observe" a chondrite sitting in the Patriot Hills isn't going to pick it up for a keepsake. Not that anyone would tell me, of course.

So Ralph's making sure the Twin Otter crew won't be trouble. As he explains what meteorites are, how they get surfaced in blue ice, and why they belong to science, I stand in the wind off by myself. The plane's several yards away, and the first two Scott tents stand beyond. White clouds come down to the horizon, though some pale sky shows higher up. I hardly know where to look, at the tiny brown stone so plainly obvious on the ice or this planet of ice I'll be calling home. The stone: not much bigger than a fingertip. The ice: bigger than anything I've known.

I need to get to know both, so I scootch on my belly toward the rock. I'm eyeball-to-fusion crust with a meteorite in the wild for the first time, here in the middle of Antarctica. This burnt pecan just nested in a scoop of ice is startlingly actual and oddly disappointing. I really wish Drew hadn't found it so quickly. Doesn't he know how narrative tension is supposed to work? Scientists! Unhappy with myself for feeling this way, I stand and stare.

Brown stone, pale ice, men in parkas, a blue ice pick that marks discovery.

WE'RE CAMPED ON HARD SNOW—snow that's pocked or flat or broken like flakes—near dimpled blue ice. We're surrounded by an Antarctic prairie that in one direction gives way to low rolls and swells that remind me of the Smoky Hills of Kansas. The wind-shaped snow extends to the horizon—sastrugi, those hard-packed furrows that metaphor can make into hundreds of shapes. Inches or feet high, the sastrugi become fishhooks and mesas, plowshares and ridges, catchy fins and talons. Some sastrugi have the aspect of tears. Where snow gives out, blue ice begins, not smooth but scalloped like a mold of mammatus clouds. The ice itself is the color of blue eyes with cataracts. Compacted snow inside the ice and crevasses filled with snow on the surface cross this blue like a Cubist rendering of crazed kelp.

The low hills not far from camp are the only relief on the horizon, a low hummocky ridge topped with nubbles of ice at regular intervals. Pinnacles, they're called. They are the round ice peaks, Barb says, of the below-surface topography that slows the flow of ice here, sort of like the exposed tip of an iceberg.

But again it's time for work. We unload the plane, using a snowmobile to pull a sled piled with boxes of food, emergency supplies, and isopods. The latter are white metal bins in which we'll store the bagged and tagged meteorites for transport out of the field, when the searches are over. The isopods hold my attention. They were used during Apollo to carry lunar rocks here on Earth. Ralph will tell me later that he doesn't know if ANSMET's isopods were spares or if they may have held, for a time, samples of the Moon. I like to think they did.

A stiff wind slows us down, and I have no idea how cold it is. Cold enough that when I stop lifting my sweat chills fast. We haul over more tents, food, sleep kits, and the survival bags of snacks and extra gear that we'll carry on our snowmobiles.

We pitch the last two Scott tents as we did at shakedown, and I feel only moderately more comfortable with the process—until I dig blocks for the outside base of the tents. I'm pleased to see that I excel at digging. My efficiency buoys my spirit. *At this,* I think, *I do not suck.* We throw in red plastic tarps. We hand over sleep kits and kitchen boxes through the tube entrances. A few paces away, we set up "cargo lines"—lines of boxes with food and spare equipment. Much as we'd like them right by our tents to minimize our exposure to the cold, anything set down on the polar plateau attracts drifts, which we don't want building up by the tents. The cargo lines are the camp's outdoor garage.

I'm too busy to feel entirely the magnitude of this endeavor, but it comes in flashes, as when my brain nearly sings, *You're lifting a box of frozen food from a sled and setting it down in the snow. You are doing this 250 miles from the South Pole.*

This afternoon, after the last Twin Otter grumbles off and we've each had a chance to settle in our tents, several of us gather in the one I'm sharing with Ralph. We eat potato chips, sip cocoa, listen to acoustic covers of AC/DC tunes, and marvel at Andrew's munificent belches. We sit with our backs to the yellow walls of the tent, where the sewn-in pockets on my side are stuffed with books, goggles, and gloves. Oliver and Drew crowd closest to the draping inner fabric tube of the entrance; the outer tube flaps in the wind. (Upon "birthing"—our term for entering or leaving the tent—someone quickly ties up the tubes to keep the wind out.) Between the sleeping bags, Ralph and I have placed our kitchen boxes, and beside those, a plank on which our two lit stoves sit. The stoves and their little flames—crucial. They're kept on when we're awake in the tent and do an admirable job of keeping us comfortable. Indeed, at the top of the tent, which has a vent for fumes, the air is actually hot. So it's all rather cozy, and we talk of the weirdness of being here and Drew's blisteringly fast discovery.

After everyone disperses for the evening—"evening" and "night" are subjective terms, since the sun won't set—I follow Ralph's directions. I saw up frozen steaks, put in water we've melted from chipped ice, add frozen corn and Alfredo sauce, heat it up, and presto, dinner. Cleanup is simply wiping out the pot and my all-purpose coffee-and-food mug, then stuffing the paper towels and trash in a baggie. I bide my postdinner time organizing my books, binocs, ski mask, CDs, gloves, and snacks, trying to ignore the tent's old stains of blood and soot.

Then, when Ralph kills his stove and turns over in his bag to sleep, time seems to stop. I stick a bottle of warm water at the bottom of my sleeping bag and turn my own stove off, wincing at the fumes, and at nearly 10:30 P.M., my nose goes cold, my breath steams. Ensconced in all my ECW except the parka and wind shell, I busy myself adjusting my hat and the position of my head and the flap of the sleeping bag so I can breathe but not feel trapped. I cover my eyes with a mask.

Outside, the sun's as bright as the winds are loud. I put earplugs in, I take them out. I rip the mask off. I stare up at my yellow walls rattled by the wind. Years after trying to sleep on the polar plateau, I will read what the explorer Carsten Borchgrevink said of these winds, that they sound like "centuries of heaped up solitude."

MORNING. My summer job begins at LaPaz.

Ralph's up, his stove is going, and I'm fumbling with starting mine—applying burn paste to the metal, lighting the goo to warm the stove, then putting a match to the gas vent—while the man who calls himself Mongo speaks to civilization.

"Gulf 058 at LaPaz." Ralph is calling the South Pole for our daily morning radio check-in. "Good morning, South Pole."

The South Pole asks if our camp, Gulf 058, is happy.

"Well," Ralph replies, "happy is a relative term."

There will be no meteorite hunts today. We're expecting flights to bring in more supplies, there are several fuel drums some distance from camp that must be retrieved, it's windy, and two of Ralph's radios are not working. Even the satellite phone is acting up. Mongo isn't pleased.

Leaning in a small V-shaped camp chair atop my sleeping bag, I feel a bit guilty for enjoying my instant coffee while Ralph grumps about and I'm surprised that I actually slept about eight hours last night. Perhaps I'll be more comfortable than I had expected.

I pull out some ice from a white bucket marked "Human Waste" (new and clean, as yet unused for its labeled purpose) to melt more water in a pot on the stove, then begin the ungainly process of birthing from the tent to go pee. I crawl out and stand up and amble past the cargo lines and snowmobiles and hitch about till I'm done. So many zippers, so little time.

After an oatmeal breakfast, I get onboard a snowmobile that Oliver's driving and hold on to the back as we head out with Barb, Gretchen, and Bill to the fuel drums, which were dropped by parachute before we arrived. As I have absolutely no sense of direction here, I'm glad Bill knows exactly where we're going. I'm watching the scenery—snow, ice, horizon, sky, snow, ice, horizon, sky—when suddenly Oliver hits a rough sastruga that quickly tips us to one side. *Shit, shit, we're falling.*

As we lean, we leap, and the machine slow-mos itself in a hard thump back on its treads. We're both on our feet but dumbfounded. Oliver and I just stare at each other.

"You okay?" I shout.

Oliver's okay. I am too. The snowmobile still works.

Adrenal with surprise, I remember all those words about false steps, about accidents, about field seasons that end in agony. On my first full day at LaPaz, I could've slipped as I landed on my feet, wrenching a knee or ankle, breaking a bone. We get back on the machine.

My heart beats hard all the way to the fuel drums, and I grip the back of the seat tight enough my hands hurt. The rest of the morning we dig out snow from about the orange drums, pallets, and parachutes. We pull out cords and pick up trash and loosen straps and tip-slide the barrels onto our sleds. Throughout, I curse my icy, fogged-up mask.

Back in camp, more chores. Drew's been helping Ralph with the fritzy radios till they finally work. Others of us unload planes and gas up the snowmobiles with a pump on one of the big drums. I slap duct tape banners on the guy wires of our poo tent, since some of us—well, me—keep tripping against them. After helping Tim put on the snowmobile covers—this involves much rolling about, grabbing at the lines and snaps under the machines—we stand together by the tent he's sharing with Oliver. We look past the orange poo tent toward the long line of white horizon.

Tim, who had said earlier that he was addicted to Antarctica, asks himself out loud, "So *why* did I want to come back?"

He doesn't answer but goes on to tell me that it's colder and windier here than when he worked with a team in the Queen Elizabeth Range back in 1997–1998.

We still haven't asked about the temperature but it doesn't matter. It's cold enough that, despite feeling restrained in my ECW when working, I'm grateful for its woven exoskeleton and my long-john force field.

And we agree. Neither one of us could do what the early explorers did.

"A form of insanity," Tim declares.

He turns to look back out at the big white nothing that is LaPaz, where the clouds and broken blue sky mirror the sastrugi and pale-blue ice. We stare in silence for a goodly while.

Tonight, back in our tent, Ralph tells me that the recce team has already collected 94 meteorites, including one possible "wow"—it'll be a Martian—and while he's happy for John Schutt's scouting party, he's pissed that we've not gotten out to collect anything yet, not even the small meteorites near where Drew found the first.

So Ralph lifts his spirits by playing songs on his computer, recordings from *The Simpsons,* including the advertising anthem for "Canyonero."

Can you name the truck with four-wheel drive,
smells like a steak and seats thirty-five . . .
Canyonero, Canyonero!

Well, it goes real slow with the hammer down,
It's the country-fried truck endorsed by a clown!

Mongo and the English Prof are reduced to fits of teary laughter. What would Scott and Amundsen make of this surreal, comic scene? Not much, I suspect.

When Ralph breaks into a fine imitation of Smithers, the sycophant ("Sir, my legs are on fire"), I do my best Mr. Burns, the evil corporate mogul ("Well, Smithers, can't you pull yourself about by the arms?").

The sotto voices, the one-liners, the chuckles—all help to domesticate even briefly the inexplicable weirdness of being here.

"PRIMITIVE MAN ERECT. PRIMITIVE man in dirty socks."

Thus self-announced, Ralph is ready to head outside to use the facilities and to alert the other tents about our upcoming nine A.M. briefing. Drew and Bill, Tim and Oliver, Gretchen and Barb are all tending to mornings in their tents.

The radio's not working—again—and the windchill hovers at -8 Fahrenheit. Still, the winds are under 20 knots, so we might be collecting meteorites today. Anything windier, Ralph says, and snow skims so hard it reduces surface visibility.

"Would you like to try some chai tea or is that too effete for a rugged naturalist writer like you?" Ralph asks upon his return.

I decline, distracted because I can't find my Swiss army knife. When I do, I slap it dramatically on the box named "Bedlam."

"Note to self," Ralph remarks without looking up. "Writer slamming sharp-edged objects down on surfaces."

While Ralph was outside, I successfully used the satellite phone, reaching my sister, who, along with her office staff in Indianapolis, was delighted to take a call from Antarctica. But I still haven't reached Kathe, so I left a message on our answering machine.

This failure nags me as we all meet at our briefing outside the tents, some of us hopping from leg to leg to stay warm. We're decked out in ECW and haven't collected a single meteorite, and already we're cold. I fog up just listening to Ralph's explanation of our meteorite-collecting plans. We'll head over to the handful of small meteorites near camp in order to get our collection procedures down pat.

Then Ralph gives us a break in which to tend to clothes. Inside, I put on a tight fleece pullover (it will become my smelly savior) and I cut holes in the foam pads of my ski goggles to vent condensation. I leave my glasses off and remind myself that I can change socks at lunch and warm my naked feet over the stove. Shimmying and half standing into my parka, then birthing on all fours from our tent, I'm anxious to see how this first harvest of meteorites will go.

We bustle about our snowmobiles, stowing extra gloves or thermos bottles in the black cargo boxes mounted on the back. We adjust hoods, tug at gloves. People are zooming off to follow Ralph—everyone must be anxious, because they're driving fast—but when I try to start my engine, nothing happens. That's when Barb comes over and wordlessly turns the kill switch off. The engine kicks to life, and I shake my head.

Our snowmobiles bristle with bamboo flagpoles strapped to the sides, their banners whipping in the wind as we drive to the first of four such markers near camp, each flagpole set up by a meteorite so we can locate them right away, and we're there in a jiffy. It begins!

A kind of choreography: A person who spots a meteorite is to wave both arms till others converge; we're to leave an ice pick to mark where we've left the traverse. Ralph will pull up to take a satellite reading of the location coordinates. (The Global Positioning System antenna will make it easy to pick out Ralph's vehicle.) Anyone who happens to have one of the collecting kits is to plunk the backpack on the ground, while another team member kneels, often on gloves, to pull out a mechanical counter. One of the collectors will pull out a metal tag and announce the field number imprinted on it, while the other person turns the counter dials to match those numbers, then uses the measurements marked on the counter to quickly size the stone; he or she must announce its length, width, and estimated amount of fusion crust, all of which Ralph records in a notebook. Then the counter is to be held near the meteorite for yet another team member—it will usually be Barb Cohen—to take digital photos. That done, someone fishes a plastic bag out of the backpack, scoops the rock into it or picks it up with tongs, folds the bag over, inserts the tag, folds the bag again and again, then seals it with white freezer tape. Meanwhile, someone must chip ice deep enough for a flagpole, on which the field number is written in black ink. If the GPS needs to be taken again, the spot can be relocated. Several steps but each simple enough.

The collection kits are kept, I notice happily, in backpacks identical to the one Kathe took with her to China two years ago. Coincidence, I know—I don't believe in fate, and even if I did, I doubt fate would care about backpacks—and Ralph is unimpressed when I quote the poet Reg Saner on "the hidden affinities in mere coincidence."

We conduct our trial run with those four small brown-black stones near camp. I stare at them intently, as if trying to communicate with them telepathically, but they're sending no signals, and I'm not sure I'm actually here, actually doing this, anyway. Four brown rocks from space? They're scooped up, bagged, tagged, and dropped into the backpacks.

For the next three hours, we ride snowmobiles in our ragged lens formation, searching. When I hollered at the briefing, "We're collecting meteorites! Whoa!" no one responded, and the silence stung. Are we too cold? Hell, isn't this what we came for? In any case, we find no more meteorites this morning.

Back out after a lunch of noodles and pepperoni—and a chance to warm cold red hands and feet—I write in my notebook, "Traverse 2, after lunch." I'm watching part of the team collect another meteorite, a new find. "At diff. points, peeing, eating gorp, had white freezer tape on glove for hole . . . Arm twirls. Leg kicks. Running. Running in place. Pools of light in sunlit blue ice. Pale green low sky opposite sun. Mostly cloudy, some blue. Andrew experiences 'conservation of cold feet,' first left cold, then the right."

Suddenly one of the Twin Otters roars overhead, flying low, overtaking us from behind, a welcome strafing, a hoot-'n'-holler moment. I forget the cold and the distance and the exhaustion of changes, I forget my longing to find a meteorite today, and I say aloud to no one, "I'm in fucking Antarctica man!" Ralph points his mitt toward camp; we drive back home, ice flying beneath us, silver-masked ass-kickers for science.

While we unload the plane—I hand over a letter to mail—and refuel the snowmobiles, I pause to consider the solar panels, the wind turbine, the line of poles on which we've strung wire for the radio antenna. All this equipage, more real than it should be. I crawl inside the tent, feeling crammed with images. Surely someone else's head has been shoved into mine.

Ralph has less transcendent things on his mind, however.

"I can't express to you how many cable failures we've had," Ralph is telling John Schutt over the satellite phone while we sit in the tent in the late afternoon. "I can't believe we haven't done a full traverse yet," he adds.

But we've nabbed some cosmic plunder—eighteen meteorites, from the size of a chunk of honeydew melon to the size of a peanut.

And I didn't spot a single one. I try to shake off doubts about whether I'll ever spot a meteorite by remembering how, at lunch, I stood outside and watched low light breaking in under a cloud deck, lighting the sastrugi, sharpening shadows against the clear strop that is the polar air.

"Look at the light behind you," I said to Ralph.

Stopping from some chore, Ralph smiled, saying, "That's why I keep coming back. I'm a sunshine freak."

Once today, while we were collecting, Ralph said aloud to no one, "I can't believe I only have eleven days of this."

Eleven days? Last night when he told me that the reconnaissance team often pulls out sooner than scheduled—information he warned me not to share—I silently cheered. So we're both paying attention to time, though our clocks appear to be running inverse to each other.

This evening Gretchen and Barb made chocolate silk pie for Oliver's birthday, and we stood outside his tent, all bundled up, to eat it.

"This is the weirdest birthday I'll ever have," he said.

Tonight I'm trying to finish the last helping of Ralph's good chicken rice; he doesn't like to have leftovers, and I hate the idea of throwing out food.

"How's it look?" he asks me, turning his laptop to show me one of Barb's photographs of a browny-black shape on snow. Above the rock gloved fingers hold the counter box with its digits dialed to the appropriate field number.

Good, I say, good. I'm leaning in my camp chair with my boots still on, jotting down notes and listening on my headphones to James McMurtry's "Too Long in the Wasteland."

"Water?" I ask, and Ralph gives me an empty bottle to fill.

I lift the pot off the stove and pour.

"Thanks," he says, adding, "I'm tired of having 'Human Waste' facing me."

I turn the bucket away.

The wind's picked up, and I can hear the slight whine of the turbine, which I welcome for its high-pitched and querulous notes, not unlike those of a screech owl.

How did I feel on the traverse today? I kept my eye on Ralph and tried to keep well in formation. I kept an eye on sastrugi. Once, dropping an ice ax to mark my place, I gunned my snowmobile over to watch one of the meteorites being collected. Venting some sudden and previously unnoticed frustration, I deliberately hit a sastrugi hard and fast with my left skid and the whole machine flew up in the air three feet or so, banked right, then landed hard, giving my left arm a shot of pain, a shock, but I liked it, it felt good and no one saw.

The meteorites—tucked in a scoop of ice, say, some snow built up, just so, little glitters of snow crystals on the fusion crusts, white sparkles on black. I'd turn from the red-parka clan to see sun burning in cupped ripples of blue ice, white ice, gray ice, like sun on windy water, a filmy shine, and the pinnacles off in the distance, our surest landmark other than ourselves in all this accidental beauty.

When I went out to chip ice, meeting up with Drew several yards from camp, he directed me to the best part of the blue ice—our frozen reservoir. He was getting ready to return to his tent but I convinced him to take a picture of me pretending to lick my "Human Waste" bucket. After the laughs, he turned to leave, and I saw his receding figure in the middle ground of this severity. Between us stretched the ice and patches of firn, the snow on its way to becoming ice. Above hung the cumulus deck with a few ponds of blue. Beyond Drew, not far really, but seeming impossibly small, was our camp breaking the horizon with yellow Scott tents, silvery poles strung with wire, red and black fuel drums, the yellow snowmobiles. I snapped his picture, a lone red figure walking away, and the reflective silver patch sewn on the back of his parka—we all have them—looked like a rectangle of ice, as if we carried a piece of this place on our backs.

ACROSS SASTRUGI ARRAYED LIKE miniature canyons, we drive single file, then space into our lens shape to search blue ice for meteorites. We have yet to find a really productive stranding surface. The sky is clear, bright, and plenty breezy, which sends windchills into the -20s, but Ralph understands now. This is how it'll be at LaPaz, cold and windy, so to hell with tent time, we're working and, yes, it's cold, and I'll spend the rest of the time outside here constantly messing with my parka hood, yazoo cap, and balaclava in an effort to breathe easier and keep my face from getting too cold and my goggles free of fog. I'll do this even as we drive.

Astride the snowmobiles, we turn our heads back and forth, looking for dark spots and watching our formation. We signal one another to back away, come closer, slow down, speed up. Ralph stays on the outer end on one side, Bill on the other, spying landmarks—such as they are—and from time to time our leaders set down an orange barrel to serve as a signpost. More than once, Ralph halts our search to reshape the traverse line. We're not exactly the Blue Angels here.

At one stop Ralph tells us there could be crevasses near the pinnacles, where we're heading now to get a better view of the area. Crevasses half the length of the snowmobile, he says, that's okay to cross. Anything more, don't. When we get off our machines, use an ice ax to probe snow and watch your step.

We drive uphill, and the pinnacles swell into shapes like breasts or blank faces, lunar hills or ocean rollers, and the crevasses helter-skelter everywhere, dips and lines like many frozen freshets, most of them narrow, most of them crossable. Into the undulant pinnacles, I look left and right, gripping my handles and tugging at my hood strings to keep out the stinging air. I can see snowmobiles climb, then descend into the troughs between the pinnacles, the crests, a rising up, a sloping down, appearance, disappearance, so beautiful that I gasp. We're off-roading on Europa. When we stop, the pinnacles stop their oceanic rise and fall. Now they are static sine waves of silex, and I face the sun. Camp is two miles distant. We're about 200 feet higher than the ice prairie below. And we're agape. Snow snakes, wisps, tendrils, and smokes across the ice.

"This is gorgeous," Tim says.

"The Little Switzerland of LaPaz," Bill says.

I pull my camera out of an inner parka pocket—and find that the cold has sucked the batteries dead.

Ralph surveys on high but not for long, because we're moving again into the long traverse he's dying for.

Soon we're searching in places we keep straight with fictive subdivision names such as Kozy Katabatic Kove and Terrapin Landing Royale, Pinnacle Vista Estates and Rue du Monkey Bar—that last to honor the restaurant in Christchurch that kept us waiting for hours before eating, so everyone got to see me become "Monkey Bar Mad," a slow burn of hostile politeness directed at an unctuous, cagey waiter. It's weird to think of a dinner out while driving a snowmobile in—

I see a meteorite—no, two. After nearly four days at LaPaz and a handful of weather-delimited traverses, I see two black chunks cut in the middle by snow. No mistaking it!

Revving the snowmobile, standing, I squint and hope and yell and speed to the spot, my words lost in the din of the engine and the clattery-smacks of snowmobile skis on uneven ice. The stone—the size of a tiny squash put out as countertop decoration in the fall—sits behind a bit of snow that made it appear to be two rocks. When I get off my snowmobile, I squat and peer at the meteorite—an ordinary chondrite, it seems—and I am disbelieving and I am thrilled. Then I stand up, waving briskly, people arriving. I forget about the persistent pain of throttle thumb in my right hand and, absurdly gloved, smack the cargo box, saying, "Hot damn!"

"Is this your first?" Andrew asks.

I respond with an enthusiastic, "Oh yay!"

I do a meteorite dance as Ralph pulls up fast.

My first find and my first bagging. Gretchen has a kit and pulls a field tag—16706, she announces to Ralph—and I set those numbers on the counter, then hold it above the meteorite.

"Closer," Ralph advises while Barb photographs it, then me.

"It's a vanity shot!" I say. I measure the stone with the counter and announce the dimensions: 8 by 6 by 2 centimeters.

"Fusion crust!" Ralph asks briskly, wanting me to hurry.

No one has been on this ice before.

"Uh . . . "

Ralph and Gretchen say simultaneously, "You can move it./Use the scissors." I lift the stone with the tips of the scissors.

"Eighty percent," I say. A bit of gray interior shows, not much.

Gretchen hands me a baggie, and I try scooping the meteorite up, then try again, heavy end first. It goes in! I want to sit here quietly with my new teacher but I can't.

Gretchen holds out the tape for me to press the bag against, and we start to wrap the tape around the clear plastic.

"Sticky side up," she says. "Don't cover the number."

Wrapped like a tiny mummy, field number 16706 is plunked inside the backpack.

"What's your event number?" Ralph asks. He writes down the number of my Artists and Writers grant to confuse the staff at the Johnson Space Center, where the meteorites go for initial classification and curation.

At my snowmobile, I lose my happy face. *Shit.* Had I ventured into Barb's search territory, veering into her line of sight, something we were to avoid? I start to walk over to apologize when . . . I spot another. Two in a row, after being afraid I'd never see one! The second one is tiny, smaller than a chickpea. I wave and holler, and when Ralph roars up on his snowmobile he says approvingly, "That sets the record." The two quickest finds for the season so far.

I don't mind being teased about how small the stone is as I begin to collect a second rock from space. Small meteorites have earned the acronym POS—Piece of Shit. Very small meteorites have earned the acronym NIPOS—Nearly Invisible Piece of Shit.

But I love it, how it is wedged into the ice, how some snow surrounds it, another tiny cosmic find. Two meteorites from somewhere in the asteroid belt and, tweaked by chance, they'd rounded closer and closer till they couldn't fall anywhere else but where they had to. Into the air, miles per second, ionization flaring behind like certitude. Into the ice, inches and millennia, then surfaced here, right here.

When I sit atop my snowmobile's cold black seat, I wait a moment before starting the engine.

LATER IN THE DAY I find a third meteorite and, for a time, I'm responsible for three out of the four finds that afternoon, which, though I know this is a team venture, gives me a much-needed sense of accomplishment and belonging.

On a late search, however, I miss a pallasite that Bill notices in my lane, which chagrins me greatly. A pallasite discovery would have been a great find for my first-ever successful hunting day. A loop back to Eliza Kimberly and F. W. Cragin, back to the Kansas prairies I left a year and a half ago.

"Real fractal kind of looking," Ralph says, as he looks at it closely. "With metal sticking out all over the place."

With Tim, I help collect this piece of astral shrapnel. Such lovely light is trapped inside it. The size of my fist, it's one of only a half-dozen or so pallasites ever found on the polar plateau. Why did I miss it in *my lane*? We wrap it up, drop it in the backpack.

To add to this frustration, I keep getting waved over on traverses, as I'm not keeping my spacing between my fellow searchers. Their hand signals insist and contradict. Come closer, go farther. Cold, I curse behind my mask.

On our ragged ride to camp—we don't have to keep a lens shape now—I go airborne off some sastrugi, trying to loosen up. It's the centennial of the Wright brothers' first flight after all. The landing jostles loose a jerry can of fuel, so I have to turn around to retrieve it.

Back in camp, as we refuel, Oliver chastises me for treating the equipment so roughly. I'm nearly speechless, but a few minutes later I'm collecting trash from all the tents, so I apologize to him through canvas walls.

"That's nothing," Ralph scoffs when we're together again in our tent. "Are you sure you weren't raised Lutheran? You've got this guilt thing down well."

Despite a cold night last night, the missed pallasite, and my guilt trip, it's been a great day, my best yet in Antarctica. And before heading out this morning, I finally reached Kathe by satellite phone. My stepmother's been ill, a source of worry for me, but her medicine's being changed, Kathe tells me, and she herself and the new kitty are fine. Shackleton was resting on the sunlit kitchen floor. Kathe's voice was warm, the call too short. Best of all, today I found what I believed I had come for. Meteorites. Was it already a kind of culmination?

When I'm not stirring tonight's pasta, I'm staring at a map labeled "Pecora Escarpment," which shows virtually no landmarks, just a wash of blue expanse. Ralph chuckles when I ask where our camp is. Our camp, Ralph says, isn't on the map.

"There be dragons," he intones.

I nod, say nothing, mulling how I watched others cross ripples of ice on the slope to the top of the pinnacles and how I was captivated by a band of sunlight shining on blue ice and a yards-long, light-brown stain beneath the ice—faint subsurface streamers of volcanic dust. I think of how we probed crevasses with ice picks to make tunnels of blue light.

This evening, beside the steady whoosh of our stoves, Ralph busy reviewing field notes and photographs, I recall a conversation we had when we talked about the sense of connection and dislocation one feels when traveling across the ice prairies of LaPaz. Ralph had stunned me.

"You see what's in your heart," he said, "not just what's around you."

On some traverses today, the sunlight just right, I could see my eyes reflected on the inside of my goggles. I stared at myself staring at the sterile sublime.

TWO IN THE MORNING on Christmas Day, awake again, exhausted in fact, and I debate what to do. Should I open the small gift from Kathe now? I've brought it all the way with me from Utah for these holidays away from home. Will I cry and lose a chance to try to get to sleep? Should I open it later, waiting for a time when Ralph is out? I don't want him to see me choked up.

Since that heady first day or two of successful meteorite hunting, much has been weighing on me, unreasonably, but weighing nonetheless: difficult nights, groggy mornings, more cold searches for meteorites, sometimes beneath an overcast that flattens the light, which makes seeing contrast harder, which means you can hit unseen sastrugi hard, which means you hurt. I contort in the bag, trying to avoid cold spots, trying to rest muscles sore from moving in all this clothing. My knees ache from kneeling on my gloves as I help collect meteorites. Day and night, cold stings my nose like a big, blunt wasp, and I yearn for cushion from the hardness beneath.

Those are the things I try not to think of each evening. Before bedtime, I might clean myself with a towelette—though in some places I've learned they cause, alas, a rash—and I might move a fleece cap from my dash bag, which I carry on the snowmobile, to have near at hand in the night. Each night, I make sure I have my water bottle and pee bottle on either side of the sleeping bag—I really don't want to mix them up—and I've taken to contemplating what I scratch off my head, wondering if dandruff can accumulate like névé. I fuss over the duffel bag at the head of my sleeping bag and push the dash bag into a corner by my feet. I can be a fidgety sleeper even at home.

Before I close my eyes, often after I've turned the stove off, the stinky tent chilling quickly, I read *The Odyssey* or look at the pictures of Kathe and the house that I've clothespinned to the tent pockets. I jot notes, put on headphones for love songs, chants, Bach's cello suites, I take Benadryls and aspirins, I futz with my mask and earplugs, with my hats, a fleece throw, a sweater, anything I can drape over my head to avoid pulling myself all the way into my bag, which I hate. I actually love camping but this, this is a different beast.

A few minutes after two on Christmas morning, and I'm hungry after last night's meal of chicken patties, cheese, ketchup, and buttery English muffins, a meal whose grease prompted much laughter. But the easy humor of the day doesn't help when, as now, I'm awake too early with aches and growls, with my own dragons, uncertain of what comes next and how to like it.

Last night we all exchanged little presents, but I don't think Ralph really appreciated the blank specimen tags I gave him—tags I'd found in Harvey Ninninger's desk drawers at the American Meteorite Laboratory in Denver. We all drank a bit and laughed, though. I really had no idea going into this that latent juvenile senses of humor would be so important and welcome.

One cloudy afternoon, the temp hanging in the single digits, Barb and Drew scraped a bit of crust off a meteorite for a researcher back in the States who is interested in knowing if microbes from the Antarctic ice—believe it or not, there are some—might colonize meteorites collected on the stranding surfaces. Marc Fries will find in the years ahead that we didn't gather enough samples for a clear result and will plan to conduct the experiment again.

"Drew," Barb said, after seeing the purple latex gloves he had to don, "we'll need to get your core temperature with that." Thus began the day's scatological humor.

At the next meteorite, an ordinary chondrite the size of a baking potato, we made wind breaks with our snowmobile covers so Drew could scrape more meteoritic material into a vial.

"So, Chris," Ralph said, "why don't you help Drew?"

I too put on the cold purple gloves, requisite gear for minimizing contamination—then held my hands up like a doctor in an operating room. I sighed in satisfaction, pretending to admire the stretchy latex. "Ah, just like my sheets at home." The team groaned.

So, yes, jokes help stave off cold and grumpiness, cold and nerves. Enchantment and little victories help too.

The last meteorite of one day, field number 16186, was a carbonaceous chondrite that I'd spotted. It looked like and was the size of a small, round, burnt cream puff, with a fusion crust Ralph described in his notes as "frothy." The fusion crust was beautiful, and I was delighted to have discovered a meteorite whose kin are implicated in bringing to Earth the stuff to help make life. The meteorite will be classified as a CK carbonaceous chondrite, a relatively new category that recognizes shock veins indicative of impact. I celebrated by hitting a sastrugi so hard, catching some air, that I broke off one of the panniers on my snowmobile. I broke the weld.

"It's a point of pride, man," Ralph exclaimed. "*You broke the weld.* Wait till the guys at the BFC get this!" He meant the workers at the Berg Field Center responsible for the snowmobile baskets. I thought, yay, why not, it's a point of pride, smiling at the praise.

Well past 2:30 on Christmas morning, I hold the small box with Kathe's present to me, caressing the ribbon while I warm an energy bar between my legs. My mind races through recent memories, new routines, lingering problems.

I wrote in my journal this past week that "I hate that I keep waking up to Ralph and a tent instead of Kathe and home. I can't believe how long I've been gone. . . . Morning is the worst—dreams are over, workday not yet under way."

Mornings are the clattery pumping of our stoves—I'm never sure just how cold it is in the tent after hours without the stoves, maybe in the teens, maybe 20, cold enough I don't care for numbers. Mornings are perhaps the hardest time to imagine the Antarctica of ages past, a warm land full of *Glossopteris* trees. Mornings are slow waits for water to boil for coffee and brief solitude when Ralph goes out to the poo tent (I pee in my pee bottle then). Mornings are the warming of towelettes, deodorant, sunscreen, and Clif bars in stages by the stove. Mornings are slow minutes drinking coffee out of my mug, then fixing oatmeal in the same, then the laborious donning of ECW—layer upon layer of underwear, sweaters, insulated overalls, parka, balaclava, hat, yazoo cap.

Still, once the coffee kicks in, the mornings can improve. Sometimes Ralph and I talk about teaching or families or some science fiction he's reading. One morning we listened to Brazilian jazz on the shortwave. The day he gave us off from searching—the morning wind-chill was about -12—I talked to Kathe for nearly fifteen minutes. "Antarctic plum!" she said, as she put down her brush from painting the utility room.

"This wind is very disappointing," I told Ralph that morning, though when a breeze snuck in and sent soot or lint flying about the tent, I pleased myself to think it a bug, something alive in the deadness. It took me days to figure out that the source of lint in our water was the socks hanging at the top of the tent to dry.

Each morning I take notes about the night, the past day, what the new day brings, and, mostly, the same yearnings to speed up time. I've read the *Tao Te Ching* off and on at LaPaz, trying to immerse myself in the moment and detach myself from the desires that arise from my exhaustions, the physical ones of just being in this extreme environment and the emotional ones of riding the tail end of several life changes all the way to this dirty tent.

I've even played mind games involving how many fresh pairs of underwear I have till the end of the field season, sometime around January 26. A couple of days ago, the first thing I wrote was: "I should not have said it. My hope to be on a flight to Chch on Jan. 24. Ralph said, 'That's a wish best kept to yourself. Not that everyone else isn't feeling the same way.'"

It's five goddamn hours before I should be up. I carefully unwrap the energy bar, the paper crinkling; I don't want to wake Ralph. Thawed enough to bite into—barely—and I chew on it the way I sometimes chew on being here. Awkwardly.

Much of our time is taken up with chores. Usually, the Antarctic veterans let us novices know when we are not quite getting the jobs done right. One afternoon I put on dry socks, which improved my mood, until Ralph interrupted my reverie with some comment about the stoves, I don't recall what. Perhaps I'd spilled some fuel earlier or let a stove run out of gas. He didn't say anything until lunch. I responded testily, "I can't read your mind, Ralph."

"Maybe you should try harder," he replied.

"Maybe your brain waves aren't strong enough," I countered.

Smiling, Ralph eased the tension: "Good, good comeback."

Ralph also had characterized my attitude about refueling stoves as one marked with "trauma." This was overstatement, but it's true it's taken several days to master the process, reducing my time to refuel the stove from thirty minutes (in snowy wind on the first afternoon at LaPaz—Ralph timed me) to a matter of seconds. Refueling involves moving an unlit though very hot stove through the tent's cumbersome entrance, then pouring gas into the stove while one's body is half-inside, half-outside. There are caps, funnels, fuel cans, the planks to set the stove back down on, burn paste, a pump. Only once have I caught a bit of spilled fuel on fire, which Ralph patted out without censure. The worst, he said, was when someone lit their crotch on fire, then their sleeping bag, then their sleep pad. The first time I refueled and relit my stove quickly, without using burn paste, Ralph and I exchanged satisfied looks.

After that testy exchange about stoves and brain waves, I went out to chip ice. At such times I feel a yearning for home, along with the tingly strangeness—not unwelcome—of learning how to live on the polar plateau. Sometimes I remember the ends of things and the beginnings. I might think of the tiny half-bath off the bedroom in the house I had shared for years with my ex-wife, where once I stood as she told me that a book I'd bought to help us work through one of our failures was "stupid." I might think of the trickle of water from a spring near White Pine Lake, where Kathe and I have gone to camp.

Until now, the longest time I'd ever spent in a tent on the trail had been my last backpacking trip with my ex, three nights in Wyoming's Wind River Range. On that trip, she had stripped by a mountain pond to swim. She had stretched out on a sunlit rock. But I did not, could not, go to her.

When I bustled back into the tent with a bucket of ice, Ralph said, "Your next Antarctic lesson will be in how to chip out perfectly square blocks of ice so they nestle up against one another in the pot." Always, another Antarctic lesson.

Somehow I fell back asleep and now's it nearly seven as I wake again, deciding to open Kathe's present before Ralph is up. I quietly tear off the paper and open the box to see an oval, polished stone. It's a Hindu lingam stone, considered to have components of an ancient meteorite and a sacred combination of male and female energies. Shaped like a stretched-out egg, the color of mud and the color of wet sand. Kathe's note makes me cry, I hunch on my side, feeling that I'll always be here where the wind is a roar and the wind is a moan and the wind is something in between and where the wind is not the place I live.

DESPITE THOUGHTS OF HOME and quests for home and nights of meager sleep, despite the achy rankle of chores and the quiet following the button-click that ends a satellite phone call beyond the ice, despite these things and my own chagrin, the days give tokens, surges of adrenal benediction I know will last even as I start to falter.

Here I am standing by the fuel drums, cutting ties to pallets while Bill and I talk of the poet Richard Hugo. I hadn't expected to consider "Degrees of Gray in Philipsburg" on the polar plateau of Antarctica but I am happily startled to do so.

Here I am beside the tent, watching a wall of snow rising in the distance and coming closer, sun shining through the scrim.

Staring down, here I am, as katabatic snows scamper past my boots like fast-forward clouds, like heaven on speed.

And here, cutting my engine on a traverse and cocking my head at a new sound. A tinkling. Like chimes. Ah, the wind blowing bits of ice my treads just cut.

DAYS ARE NARRATIVES: MORNING, afternoon, evening, night.

But searches are blurs of cold happenstance, and memory resolves itself in shards.

My overbite exhalations to get warm air on my neck.

Red parka arm, black pants, rope handles of the cargo box, cold feet, Windstopper gloves wrapped around the heated snowmobile grips.

Pulling flags that marked the boundaries of a search formation, harvesting the flags that marked finds once we're done.

My head crooked to let in just enough air to keep my goggles from fogging.

Thus we search, frozen cods on long traverses, how they repeat themselves. Running around the snowmobiles, twirling arms, drinking hot instant soup from a thermos, playing French cricket or acting out scenes from Ralph's Bad Samurai Theater—a riff off *Saturday Night Live*—all to stay warm, though the latter also elicits much-needed laughter. Drew, who often pulls a flagpole from one of the snowmobiles, wields it like a light saber from *Star Wars*. I do jumping jacks, which heats me up but almost always steams my goggles or glacier glasses.

Sometimes, standing while driving, I'll shift my weight to one leg and stretch the other one out behind me, shaking it the way a dog shivers off water. On Saturday, I was the coldest I'd been yet. Winds at about twenty miles per hour, air temp 5, the windchill about -15. We were out forever.

We search the downwind ice edges where the firn builds up and where the wind can shove meteorites. We scan. We watch Ralph wave us forward when we're ready to start again, after a collection is finished. We strap on ice stabilizers, like crampons, when the ice is especially slick. We complain about throttle thumb. We're chilly eyeballs dressed up for the hunt.

We search, and sometimes it's good. At the start of one traverse I find a grapefruit-sized ordinary chondrite in an area we'd passed through before but that's now scoured clean of snow. I do another meteorite dance, all pelvis thrusts and arm jabs, Ralph wheeling up right by me with arm extended for us to slap hands, saying, "Good job," wrapping an arm on my back, saying, "This is two thumbs up. I feel good about this one. It means we're keeping our eyes open." One day we're scooting about Pebble Beach in a biting wind, a literal pain in the neck, and we hit pay dirt. Lots of meteorites, including big ones—I find one the size of a football, and it's the first of many large ones. Oliver and I collect them, and that day was our best by far—thirty-one total.

We search, and sometimes it's cranky. I'm hunched over another bland brown stone and, glove off, finger licked to crinkle open the bag, I let my finger briefly touch the meteorite. I have touched a wild meteorite!

"I think my finger touched it," I say.

"Did it or didn't it?" Ralph asks.

"I'm not sure."

"There's no maybe for a contamination event."

I had touched it—and my first touch of a wild meteorite was "a contamination event," a cold, cosmic zero, a mistake and less than the pressure of a chickadee on my palm. And not as tuneful.

ON CHRISTMAS DAY, I help inventory meteorites outside, squatting by the isopods, while inside the tent Ralph checks our field-tag numbers against those on his computer. Then I reach Kathe on the phone; she's staying with her aunt in Green Bay. We each wish each other Merry Christmas and Merry Christmas Eve.

"Do you know what next week is?" I ask.

"No."

"It's January . . . "

"And then you come home."

"And then I come home."

Perhaps I'm failing the test of Taoism and Kathe's admonitions to me. And yet how much more present can I be? I feel the continent in all my sinews, on my skin, all the time.

SO FAR, we've found 150 meteorites.

At the nine A.M. briefing on December 26 there is sun and blue above us but on the horizon blowing snow, gray and wispy, as tall as clouds. Ralph shows us a map in the snow using the end of a broken bamboo pole.

"This area is a bit wavy, so it will be hard to keep things straight," Ralph says.

"Wow," Tim responds laconically, prompting much laughter.

Ralph pokes Tim's bunny boot with the pointer, then Tim kicks Ralph in the butt, and we're all quiet.

"Ow. I have tendinitis in that cheek."

No one says a thing.

To end the briefing, Bill and I engage in a Mime-Off, in honor of his prior job as a professional mime. Bill and I are to imitate Drew, so I pretend, uh, various bodily functions. I couldn't see Bill, but he won that round, according to the crowd.

At the next Mime-Off, during a break in a traverse of slippery blue ice, Bill imitates me, and I imitate him. He pretends to be scribbling notes furiously. I pretend to scan for orange barrels with mock binoculars, then take a nip of invisible whiskey. I win this time. All good fun, but we'll never have a chance to settle the draw.

BEFORE GOING INTO THE tent for lunch—it's the following day, December 27—I squared away my cargo box and paused. "Remember this," I said, looking at bright gray sky, level horizon, the only relief being the low headland of pinnacles. "This is the heart of Antarctica."

It's after lunch that the switch-out flights are canceled. Surface visibility is poor right now, so recce has to wait. Gretchen and I are ready to fly to John Schutt's team, swapping out with Oz and Nancy. Ralph is unhappy to be leaving LaPaz so soon. His replacement will arrive soon too.

I'm disappointed. I'd been looking forward to a shower at the South Pole and getting to recce—to feel that it was the home stretch.

"*Tentative* plans," I had said to Ralph.

"Wishful thinking," he replied curtly.

This really pissed me off, but, selfishly, I wasn't thinking about Ralph's own desires for home and his desires to stay.

To work off the grumpiness, to give Ralph some space and me something to do, I join the folks heading to the fuel site again. I follow Drew, whose sled comes unattached twice—his hitch is missing a locking pin—so we switch the sled to my machine. When I pull up hard, tight, and fast to four black drums—fuel for the Twin Otters—Bill and Drew are already there.

"Whoa, yay man!" Bill blurts, as I position the sled right up beside the drums, perfect, slick, like I'm some sort of professional stunt driver.

"Are you a repo man or something?" Bill jests.

After Bill tells me recce has 35-knot winds I say, "Here or there, whatever."

"You're sounding very Zen."

"I'm either Zen or pissed off." Immediately I regret saying this.

"Pissed off? At what?"

Tim and Oliver are pulling up too.

"That's too strong a word," I say.

"Are you getting sick of us?" Bill asks.

"Yay, I've just had it!" I say in mocking fashion. "No, I'm just homesick. I was homesick before I left home."

"How can you be homesick before you've left home?"

Now I'm pretty sure I don't want to say more. "Well, we had only been in the house for a few months."

"Oh."

"And it's just all the chaos of the past three years."

We set about muscle-fucking the drums. We lean into them, tip them onto the sled, and strap them down.

"Like this, Drew?" I ask, tightening a harness.

"That's it."

Then I pull my sled, driving behind Oliver, who has his own sled, and at least I feel competent right now, turning over my shoulder to see how the sled is doing, modulating speed, and steering to avoid any hard whacks. Over sastrugi toward our tents, I think two-week chunks, two-week chunks, two-week chunks. Two weeks at LaPaz are nearly done. Four weeks left to go. I can do this, I think, I am doing this. The bad feelings of the day are left in the snowy wake of the trail back to camp.

I AM ERIK KRAMER. I am Bobby Douglass. I am . . . *Doug Flutie.*

We're playing five-step football on the snow of LaPaz. After the ball is snapped, each player is allowed only five steps. You can take them all at once or stop and start. My love of the Chicago Bears and Oliver's admiration of Green Bay's Brett Favre had led to good-natured joking, but now we're on the same team. Oliver, taller than me, would be a natural at quarterback but he vigorously deferred. Having played soccer—real football to him—he didn't feel confident throwing the ball.

I pull away from the line of scrimmage, lurching back to heave a throw into the crowded end zone, a Hail Mary pass. The ball is actually my glove stuffed with a frozen box of juice. It will later burst. No matter. A perfect loft, a perfect fade. Botta scores! I whoop it up. This is the highlight of my football "career," which I thought had ended the day I walked off the Fulton Junior High practice field in Indianapolis as my despised coach swore at me. What if Coach Bertalon could see me now, me throwing a touchdown pass in a place that makes the frozen tundra of the Packers' Lambeau Field look like a tropical beach?

Our teams tied. Fitting, given the mostly good feeling of our camp.

And tomorrow I leave it behind.

· *Chapter X.4* ·

Massif: An Unmaking

I am sitting in a tent the color of piss.

I am sitting in a tent the shape of a pyramid, no, of a rocket nose cone.

I might as well be drowning. Encased in rock. On another planet.

I am at the Otway Massif—my third camp with the scout team. I'm in the middle of Antarctica. I'm with three other human beings, and no one else is alive anywhere. I have been here forever and will never leave.

The Otway Massif: an island of mountains riven with glaciers, two of which, nameless, flow down to the far end of the pale-blue ice on whose near-firn we camp. We are a mile and a half from the massif. Our two tents face one end of the Otway, where dark mountains and ridge line curve like a wind-snapped cape or, I'll later think, like an arm about to embrace or strike.

The day we arrived at the Otway, shredded cumulus crossed a blue sky, throwing black on the brown dolerite, animating the otherwise lifeless edifice. I might have said the clouds were like puffs of milkweed. Sequences of sandstone and lahars stain the massif, giant tan patches that I wistfully thought could be some heretofore unknown, prodigious Antarctic lichen. Rising some 3,000 feet from the ice to more than 10,000 feet, with its eroded sides and wrinkled peaks and occasional columns—one stands alone like a finger pointing to escape—the Otway is the most sublime place I have ever seen.

But I am in the tent now, again.

My tentmate Johnny Schutt asks me, "Flu?"

I've just told him I'm not doing well.
"No," I say. "Depression."

I HAVE BEEN WITH recce since leaving LaPaz on December 28, when Gretchen Benedix and I arrived, via a quick visit to the South Pole, at the Davis Nunataks, which crop up out of the ice near the end of the Dominion Range. When we got out of the plane there, our first recce camp, I was heartened by the sight of mountains. Behind us was the tail end of the Dominions and Mount Ward, and in front of camp, across a wide ice field, was a nunatak that rose like a big, sloping mesa, looking a bit like Dundas Mountain in Thule. I was grateful for that sense of connection. And I was heartened by my being heartened. It meant, I knew, that I was now more at home in mountains than on prairie.

But the recce camp itself—two Scott tents, one poo tent, four snowmobiles—felt paltry compared with the long line of four Scott tents at LaPaz, that ample town of eight folks. Not only is recce half the size but the Davis camp was higher, at about 8,000 feet. And windier. Johnny told us it had been zero recently, with 40-mile-per-hour winds, putting the windchill down to a respectable -29.

In the tent with John, which was much neater and cleaner than my quarters with Ralph, we chatted amiably while I put up photos from home and set about to organizing my books.

"Good Lord." John chuckled when he saw all of them.

"I don't know what I was thinking," I say. "Tent days, I guess."

With a manner somewhere between wryness and stoicism, John was quick to help me that afternoon. I needed to rinse my gloves from the juice that had soaked them during the football game, and John got right up and retrieved basins from the cargo line. When I was finished, he said I got most of the water out of them. There. I knew it. He'd been keeping an eye on my skills.

SOON, JOHNNY SCHUTT AND Monika Kress led Gretchen and me to the blue ice field directly between our camp and the big sloping mesa of the Davis Nunataks. We drove a short distance to a moraine strewn with dolerite to practice "defocusing" for rounder, darker stones. A volcanic rock, dolerite typically sparkles and looks more vesicular than meteorites, unless it is weathered, so one learns to avoid what glitters. Monika had found a meteorite there and urged me and Gretch to pick it out from all the earthly clutter, a splay of dolerite rocks the size of fingernails and wing nuts.

"That's it," I said, pointing to a rounded rock stippled with fusion crust.

"Yay," said Monika, grinning.

We'd all occasionally mistake dolerite for meteorites, usually when the light was flat and we were looking from the snowmobiles. Even scooting close to a rock sometimes wasn't enough. We'd wave everyone over, only to be told "dolerite," at which we'd laugh or swear or sigh.

Gretchen has been in charge of the collection kit, with my assisting her with the counter, fusion-crust estimates, and taping. Monika does the photography, while John does field notes and GPS readings. Typically, either he or I chip ice for the flagpole on which Monika writes the field number.

On my first full day at the Davis Nunataks, a three-hour traverse cut short by high winds, I found one ordinary chondrite, after staring hard at coals and dolerites. Just before I found it I was thinking, *God, what if I don't find another meteorite for the rest of the season?* As I held the stone up with tongs, I told Gretchen this, and she laughed as she opened up a baggie.

"Statistically, you *have* to find some," she said.

Ah, I thought, neuroses trump statistics anytime.

That day I saw the ethereal blue of ice beyond the end of the moraine we searched, the same blue I saw in Greenland, the same blue that poured out of holes we poked in the snowy crevasses of the pinnacles at LaPaz. We stopped frequently, walked to rocks, lifted up our goggles to look; we drove up over a swell of blue, then down, and found wide sprinklings of stones, mostly small, angular dolerites. By the moraine: boulders on and in the ice, V-shaped deposits of rocks just sitting on the ice, like tesserae in an unfinished mosaic. It was beautiful.

When the winds picked up, I could see stones quiver like aspen leaves in the 35-knot gusts, just 10 knots lower than the worst John's experienced in Antarctica. A -30 windchill, he said.

Back in the tent, after my short, first traverse with recce, John and I napped.

He was slumped over copies of *Wired* and *The New Yorker*. If I thought I'd experienced wind at LaPaz, well, this was wind. The wind roared and whistled, the tent pulsed and rippled. The world was a long, bright grumble, and fabric snapped like gunshots. My body vibrated, it hummed and it wavered.

One evening, while Johnny transcribed field notes and downloaded GPS coordinates, he said of the wind, "*Que será, será*," and at same instant, we both sang, "Whatever will be will be," then laughed. It was a good moment. Doris Day, Lao Tzu. But it didn't take me long to miss the throatier sense of humor at LaPaz, a wackiness that had been keeping my spirits up.

On New Year's Eve at the Davis Nunataks, a blowzy snow-film raced off the nearest nameless mountain. A few high clouds, no lenticulars. The stark sloping shape of the brown mesa. Clouds across the range leading to Mount Nimrod.

Sitting with Gretch and Monika in their tent, I told them, "I'm stoked over New Year's Day." Gretchen pulled out her calendar showing a count of days and of meteorites.

"Come on," Monika admonished. "We're not in prison."

Back in the tent with John, I cooked our holiday dish of chicken in pesto sauce and, while waiting, I wrote, "I can foretell nothing but right now the light is beautiful. Here, there is nothing alive that I can sense. Here is a grand indifference. Who knew that adventure involves sitting most of the time? Antarctica's colors: white, blue, brown, black, and gray."

After dinner, we played Scrabble, and proud as I was of my "maven" and "cringe," they crumbled before John's "quist." We joshed, and time felt more affable. At bedtime I snuggled into my bag knowing that I was one hour from January.

Our second recce site, the Scott Icefalls. We were camped some distance from that feature, though it still looked like three oceans of angry teeth flash-frozen after a collision. We arrived there on January 3 well away from the crevasses, with mountains on the horizon, including the Otway Massif, which would be my next camp and, though I didn't know it then, my last.

Twin Otters flew us in stages, and Monika, who of course had gained more experience with camp moves prior to Gretchen and me arriving, conveyed instructions.

"No, Chris, don't worry about those," she said of the guy ropes of a tent we were breaking down. I had been trying to loop them up. I was cautioned to not dig snow blocks too close to the tent and to keep my sleep pad and camp chair in my sleep kit. Tired and anxious not to make a mistake, I quietly bristled. In any case, we arrived and set up without incident.

One windless, sunny evening—warm enough to sit outside on my snowmobile and listen to Bach's cello suites on the headphones—I watched as Gretchen took photographs. When she finished, we chatted, and she compared the place to Idaho in winter. Only a geologist would say that. No ravens in firs here, no chickadees among the maples, no deer track. I said nothing along those lines and didn't notice how internally short-tempered I was becoming. I let the calm air soothe me back to something more social and I hid without knowing it.

Nights at the Scott Icefalls camp, I heard my first serious icequakes. Groans and cracks and thwacks that woke me in the sleeping bag, my cheeks and nose and arms very cold.

In the mornings, increasingly obnoxious music on my headphones, fast and loud. AC/DC. Jimmy Eat World. Warren Zevon's "Detox Mansion." One morning, Ralph called from Ohio; he'd been home for some time. He gave me a football update—the Bears were not doing well—and asked, "How ya doin' buddy?" I was suddenly wistful for Mongo, for Primitive Man.

I'm sure I said something like "great" or "terrific" or "ex-cellent."

On our traverses, faster work with just the four of us and an occasional compliment. "Mighty fine," John said of a chipping job I'd done for one of the flags, and "you have the hot hand" as on one search I spotted meteorite after meteorite after meteorite.

Beauty too. That gray day, the Otway Massif hulking in the distance and the backside of the Dominion Range snaking away, I waited for John to finish a

GPS for another chondrite, another blackish bit of ancient space, utterly familiar now, so instead I looked at the mountains and watched the Supporter Range just as sunlight suddenly flared on its flank and I gasped.

I tried to hoard such moments against my growing impatience and raggedness. Sipping hot cider when we'd take a collecting break, I tried to register everything, impossible, but everything—the tighter pinnacles we sometimes collected in, the cupped blue ice, a few crystals of snow sliding by, a meteorite I held by tongs beside John's snowmobile, a meteorite, a meteorite, a meteorite, and the black letters of my parka name tag: Christopher Cokinos.

I tried to hone my already stronger muscles by doing sit-ups outside in the evening. I tried to hone something, anything, against the darkness I felt inside. I kept looking at a calendar in my planner, counting days gone, days to go. I had even asked for some St. John's Wort tea to be delivered by the pilots, but McMurdo didn't have any. I imagined Bears games to pass the time, especially while we worked, long play-by-play commentary that always ended in a victory. I listened to John talk to Mac Ops—the flight control center at McMurdo—as he planned the Otway move. Because he never briefed us on contingencies, I passed along whatever I learned, mole-like, to Monika and Gretchen.

Finally, I hit upon a strategy that seemed winning for a time. Each evening I ended my notes with two categories, "Savor" and "Studly." The first was meant to be poetic and connect me to the land, the second was meant to be self-boastful and assuage my insecurities. As in "Savor: A huge curtain of cloud and snow, a squall, over two glaciers of Otway, wind apparently bringing that weather to our blue ice field, [but] which warmed up calm and clear in p.m. The squall thinned and stopped over the mountains." As in "Studly: Just being here. For the rest of my life, I will have been here."

And we found meteorites. I counted days ahead, and we found meteorites. I fueled my stove, and we found meteorites. I called Kathe and fought back tears while outside on pleasant, sunny, windless evenings, and we found more meteorites. I took notes, and we found meteorites.

The silliness I yearned for rarely erupted, but after breaking down much of our Scott Icefalls camp to get ready for staging flights to the Otway, we waited in a tent, sacked out, for the plane to arrive. Thankfully, John had said nothing about how I wrapped up the bamboo flagpoles using bungee cords instead of rope, a concession to my knot-tying ineptitude.

Out of a stupor, John erupted with Mongo-like weirdness.

"I don't care what they say! Nietzsche was wrong!"

"Nietzsche was wrong about what?" Monika chirped.

"What," asked Gretchen in mock solemnity, "about Kierkegaard?"

There was a brief pause.

"Trapped in their spaceship," I narrated, "the astronauts argued about Nietzsche till they ran out of oxygen."

We burst into giggles.

AT THE OTWAY MASSIF, I can't stop counting.

Each day of the trip I track the numbers of days till weekends, till Wednesdays, till day-after-tomorrows, till new destinations, till finally coming home. I keep score of how many changes of underwear are left, how many calls to Kathe, how many nights might be left in the tent, how many search days. I say to myself, on a Monday, "The day after tomorrow is hump day, and then you can say the day after tomorrow is the end of the week." I have done this obsessively, ceaselessly.

This morning we set out under a sky glittering with crystals and refraction. Tangential arcs, halos, parhelia, shimmering opacities gifting me on this, January 13, my forty-first birthday. We collect meteorites beneath lovely chances of light, but I am stolid with the deadness of the polar plateau, the quietness and irritations of camp life, the nature of our traverses, and most of all, the creeping consequences of fitful sleep.

It's true that to about 84 to 86 degrees south, some lichens and mosses eke out an icy existence. I'll not know this till months from now. Had I known of Antarctic endoliths, I would have considered cracking open every rock I could, just to believe that something other than the ANSMET recce team was alive. When the LaPaz camp found, of all things, a headless skua, I felt a surge of envy—a bird! Had it encountered the moving propellers of a Twin Otter? I longed to see its body, this evidence of life.

The first time I used the word "depression" in my notes was an entry that described how, "apart from the green world, I am forlorn."

Though John and I have talked more and more, ranging over weighty subjects, our tent time is far quieter than my time with Ralph. The adolescent bent at LaPaz suited my stunted sense of the comic. As if to compensate for the quietness—and perhaps to underscore our comfort—John sometimes reads from Scott's journal of his doomed South Pole expedition. This serves only to accentuate my mute rancor and my guilt at watching the clock's slow ticks. I sometimes, therefore, dread "tent time," when John and I are stuck inside because of high winds or just to take a midday break. I feel less and less inclined to spark conversation, and John—who of course is also worn out—naps or pours over information to plan the next traverse or reads his magazines. John is also single-handedly producing maps of all the ANSMET search sites and meteorite finds for future researchers. While he works in the tent, I do too, taking notes, but I'm finding it hard to concentrate on words, my own or Homer's or anyone else's.

It doesn't help that we work in harsher conditions than at LaPaz, with windchills sometimes between -20 and -30. Once the wind seared a portion of my forehead. Not exactly Mawson stuck in a crevasse, I know. On such days, Johnny seems well suited to the role of hard-core explorer, with his ponytail and patched-up pants, his unflustered efficiency and self-reliance.

But this self-reliance also manifests itself in, at times, a lack of instructions, especially during camp moves. After arriving at the Otway Massif, John questioned why I had a half-size flagpole with me, one that had been kept under a tripod outside our tent. A GPS antenna was mounted on the tripod.

"Oh, no," John said, "that was supposed to stay."

I felt a surge of anger. "Well, I didn't know that. The stones around it are still there marking the spot."

Before he got on the plane to replant the flagpole and to get the last snowmobile from our Scott Icefalls camp, he called out, "Chris, after you get the tent up, you should get the radio hooked up."

The tent was no problem, but as we three looked at the radio boxes, wires, and antenna, Gretchen asked me, "So . . . do you know how to do this?"

"Me? Are you kidding? He said you did."

Gretchen blurted, "Me?"

We were all frustrated, and the wind had picked up considerably. So we agreed. We'd just set it up the way it had looked. Monika and Gretchen unwound poles and wires. I augered holes in the ice for the several antenna-wire poles and the solar cells. Gretchen climbed into the tent, and we fed cables back and forth, plugging things in; then I moved the solar panel farther from the tent's shadow.

When Johnny tried the radio it worked. "Good angle," he said to me after looking at our setup. I scooted up to Gretchen and slapped her on the back.

"The radio works!" I proclaimed.

"Great!" she said.

These small victories have not had a lasting effect, however.

The camp moves at least break routine, which has become a polite word for "bored." Even the traverses themselves—which, though I sometimes yearn for them so I can escape the tent—have at the Otway revealed a sometimes achingly dull strewnfield with many, many fragments of just one, once-huge ordinary chondrite. We've been working blue ice along the 10-mile stretch of this massif. The Otway is at the northwest end of the Grosvenor Mountains, plunked between two giant glaciers, Mill and Mill Stream, though those are not the ones we can see from camp. It is an impressive setting, but we are all hankering for something else, for any other kind of meteorite.

"Another crudite," John said once, upon arriving at another Otway chondrite fragment. At one briefing, he actually told us, "We'll have some tedious searching followed by some tedious collecting." This frankness surprised me, and because he wisely sensed our flagging interest, John took us off the ice and into some hills bouldered with weathered red dolerite. Gretchen pulled out an inflatable TV-character toy and took his picture: "Mr. Krabs on Mars!" Suffering from a massive sugar crash from having eaten a startling amount of thawed raspberries that morning, I wandered off alone and, while noting the alien beauty of the place, also dug into the thought that these were not my mountains. A few minutes later, John started back to his snowmobile—a sign to regather—and he asked us all, "Didn't you always want to walk on Mars?"

On the traverse that day, a productive search, we found thirty-seven meteorites, each one just another sample of the busted-up Otway chondrite, each one its own piece of science magic, though I wasn't finding much awe in that anymore. Here we were collecting meteorites *in Antarctica*—that had been my ultimate goal, right?—and I couldn't even muster the energy to ask my co-workers on the ice any questions about what this chondrite could tell us, about meteorites they'd studied, about, well, about anything.

Frost's lines ran in my head. "Earth's the right place for love:/ I don't know where it's likely to go better."

EVEN AN EARLY TRAVERSE at the Otway had revealed frayed nerves. We had mistaken John's waving with one arm—he was far ahead and way down an ice slope—as a sign to gather up with him, to collect a meteorite. At that point, we'd not found any. We three drove over, quickly realizing that the slope was so steep that we couldn't easily brake the momentum of our snowmobiles. We were heading right toward boulders that might damage the snowmobiles or worse. This was suddenly very dicey driving, and I was scared.

I swung my snowmobile around so it was parallel to the base of the ice and I kept braking, till finally the snowmobile clattered to a stop. My heart pounded. The machine had been right on the verge of being out of control, and though I'd managed to halt its slide, I dismounted gingerly, afraid of falling or tipping the snowmobile or both. Treated to intimidating views of the violently creased Scott Icefalls and the snow mountains of the Dominion Range, I walked carefully over to Johnny, who stood in front of a long, dark ridge. And who was pissed.

"I DID NOT DO THIS," he said, doing the two-arm wave to signal a meteorite.

That morning we all grew more irritated because we hadn't found what Antarctic meteorite pioneer Bill Cassidy had characterized as a big strewnfield. He had come to the Otway in the mid-1980s and collected three meteorites, only one of which was part of the strewnfield.

"Where are all the meteorites?" John had asked, each of us silently wondering if Cassidy had mistaken volcanic rock for meteorites (unlikely) or if the original team had garbled the correct directions and descriptions (possible).

Then I found one. That first Otway chondrite specimen and all the rest we begin to find in droves all look like cold, desiccated meat loaf or old Mexican chocolate. Pieces of an ordinary chondrite main mass, a weathered, beaten meteorite, 4.5 billion years old and sitting for unknown ages on and in and on the ice.

Monika eventually will lead a research team that will study the Otway stranding surface, finding it to be perhaps, in her words, "one of the oldest intact strewn fields ever discovered in Antarctica," one that preserves the "aerodynamic sorting" of meteorite fragments when they break up, with the larger stones at the far end and the smaller stones clustering at the near. The fragments will range in weight from a half-ounce to more than 60 pounds. Monika's team will also see that the "short extent of the strewnfield and the lack of primary fusion crust" on the samples "suggest a short post break-up time of flight (low altitude and/or steep entry angle) and a relatively low breakup velocity." That the Otway ice preserves this pattern means it was probably trapped in a "very slow ice motion."

Monika will deduce all this in the months and years ahead. On our first successful day at the Otway, it was simply about finding meteorites, including some big honkin' ones. I found a meteorite the size of a honeydew melon—and I chided myself for being a baby. I'm doing something incredible here, I reminded myself. Something incredible.

NOW, DRIVING ON ANOTHER Otway traverse, on my birthday traverse, with some tears behind my mask and ice and snow bristling in the air like krill in a lit ocean, I find that only part of me sees the dead beauty of the blue ice and the massif. We drive through the light and I am weary. We find meteorites, we collect them, I am efficient, I am robotic, we drive through the light and find meteorites.

I'm always prone to depression when I'm tired, and this morning was, as nearly all have been, a struggle to shake off a restless night, about which I might write, "When I sleep on my back, I skip breaths." "Get sore on side and have to switch." "Woke up saying, *My God . . .* "

As I did at LaPaz, while I'm on the snowmobile, I plan careful rearrangements of blankets, clothes, and pillow. At night, I take Benadryls and melatonin and aspirin. "That's quite a cocktail of pills," John said once. I put in earplugs against the sounds of my tentmate, the sounds of the fuel can and ice sheet expanding or contracting, the sounds of wind or silence.

"If it calms down," John said of the wind, "you'll wake up and wonder what's wrong."

When it's windy and the double tubes of the Scott tent are open and untied because John has stepped out, they swish about like hyperactive amoebae caught in a portal of light.

JOHN SENDS ME BACK to camp from the birthday traverse so I can fetch a sled and cooler for some big chondrites we've just found—the biggest we've seen, the size of basketballs. Their size startled me from my sickly self-regard.

To be alone on the snowmobile, the others disappearing behind the rise of ice, I feel an edge return, one I savor. I've been wanting some evenings to ride off alone but haven't had the guts to ask permission. Back at camp, I hitch the sled and tie down the cooler. I can't find the foam John wants to pad the rocks with, so when I return I offer my snowmobile cover instead.

John is ticked. I'd forgotten that he said I could grab the foam padding from the isopods.

"Shit," I mutter, turning to the snow, rainbow, and mountain. It doesn't help that I also lost a snowmobile hitch pin and that I ran over an ice stabilizer, those detachable crampons we put on our bunny boots when the ice is really slick.

We finish up and drive back to camp, have lunch and an afternoon break.

Knowing it's my birthday and surely knowing I'm feeling chagrined, John later takes us up the slopes to the ridge of the massif at more than 10,000 feet. We will gain the summit of that which looms over camp.

Behind the roar of my machine and the reflective silver plastic of my ski mask, which I'm wearing today in part to hide my eyes, I suddenly think of my mother, gone for a year and a half now, dead. I can't keep from crying again. We drive single file on snow thick and undulant and steep, boulders protruding from white, and at times the tears dry up and I feel an aggressive edge to the ride, enjoying in snatches the bizarreness of it all.

Suddenly, the massif flattens out. The peaks and ridges and ice rivers spread out in deadly beauty. It's like a view from Everest, it feels that high, though here the peaks are blunted by mounds of snow, passages of clouds, and feathers of light. At my feet, hoarfrost turns the snow surface into a forest of tiny crystal trees. We each wander off to take it in. I sit down on a rock near an edge beyond which is only white distance. I see no one, just the world.

At our first stop, John had said, "We're the first humans here." Petulant and ungrateful, I thought, *And the last. What a cheap sovereignty.* Then John headed to his snowmobile, saying, "Well . . . "

Monika quietly noted, "That's enough of this vista. Time for the next one."

At the top of the Otway Massif, my back to the others, I keep facing the unbelievable. Rock becomes mountain becomes snow becomes cloud. From here, I see the massive Burgess Glacier, which bisects the massif, and beyond, the rest of the Otway's mountains, including a peak named "Mom." Farther still comes the enormous stretch of the Mill Stream Glacier.

I stand up from the edge and turn around to see my red-parka companions like vermillion Siddharthas way off course for our Bodhi trees. I rub frosted steam from inside my glasses. We drive down, and it's very cold. I like John and I wish that I could lighten up.

But when we get onto blue ice—*Oh God, we're not looking for meteorites?* We are. And I'm cursing the search and the sudden overcast, I mean really bitching everything out, the way I swear at a football game but worse because this is my goddamn life now, this fucking traverse, just this cold forever, and the clouds have conspired to make me very cold and the light sucks for looking.

Then I see a black rock with a chondrule, a huge, huge chondrule, and I'm *excited* again and wave people over, only to have Gretchen bend down, look, then tell me, "It's igneous with a bit of quartz." So much for my find of the century, the biggest chondrule in the world, the Hope Diamond of chondrules, so much for that. The fifth week, she says, is the hardest.

Back at camp, the sun is out again, and I grab a mini-football from one of my duffel bags, asking, "Anyone for some toss?" Gretchen and I throw the ball in front of our two tents and talk playoffs. I feel a bit perkier, which is weird. I can't quite figure out how I keep flipping from one mood to another.

John sticks a candle in my steak this evening. "Birthday steak!" he says.

"The best birthday steak I've ever had," I say.

"In Antarctica," he says.

"In this part of Antarctica," I joke.

At eight P.M. Gretchen and Monika bring cheesecake with a match-as-candle plus a homemade card. They are all being so kind. I'm really awful to feel so anxious for home, so wired-weary.

I feel as though I've already betrayed them.

I AM TALKING TO Kathe on the satellite phone. She is wishing me a happy birthday. She is saying, "It feels like you're never coming home. Like you don't exist except as a voice on the phone."

ONE EVENING AT THE Otway, I write, "Savor: No wind!" I read the *Tao Te Ching*. As always, John keeps the radio on, so voices call from McMurdo, the field camps, the South Pole.

Endless the series of things without name/on the way back to where there is nothing.

"Moody camp, Moody camp, Mac Ops on 7995."

"South Pole, South Pole . . . "

"This is Moody Camp, Mac Ops. Go ahead."

As good sight means seeing what is very small/so strength means holding on to what is weak.

WEDNESDAY, JAN. 14, 2004, Calm, 8 degrees.

The day after my birthday. Fog toward Beardmore Glacier. We may not move camp from the Otway. In the past, I disliked tent days, but if we don't move, I tell myself that I'll nap, read, trim my beard, put my earrings in—reminder of home—call on the satellite phone, exercise, play games. Still woke up every two hours. Ralph calls from Ohio. John tells him, "Attitudes are . . . sort of ho-hum. There's only so much of this you can do."

I feel low during Scrabble. We all play desultory games. Interruption at one point: John's listening to pull-out discussion involving a science camp at

Megadunes. Gretchen glances at me with a serious look that says she's ready to go too. After a sort-of-nap, I decide to call my father and stepmother in Indiana, so I go to the poo tent and sit on the closed seat of the Human Waste bucket and we talk and I'm glad for their voices and miss them terribly. After I hang up I cry so hard my stomach hurts and it scares me.

I HAVEN'T ALWAYS FELT this way on recce.

One stunning day at the Davis Nunataks, the sun shining on ice and rock arrayed around us, I told Johnny, "The past two days I feel more like an explorer than the whole trip."

"Well," he said, "that's what we're doing." He said it kindly.

Later, we traversed a wide territory, ostensibly getting the lay of the land but mostly it felt like a joyride. We climbed up steep slopes, stood astride mountaintops, and came down a tongue of ice so steep I felt not fear but mastery and exhilaration, the swell of ice at the moraine—rocks, ice dip, ice swell, a shadow beneath the ice of submerged rocks or dirt or volcanic dust—and I thought of Georgia O'Keeffe, the voluptuous ice in her painterly shapes. As we began our trip back to camp, I saw John only as a momentary glimpse above a lip of snow, as he had gone into a tight, almost V-shaped windscoop, with smooth sides rising up like palisades. Monika and I joined in—Gretchen watched—and I zoomed up and down the sides of the bowl, laughing spontaneously. Just like The Wall of Death on shakedown!

John, with a huge smile, afterward: "That's fun, isn't it?"

"It's stupid how much fun it is!" I beamed. "I love it."

That evening in the tent, as I fixed dinner, I played bluegrass on the small speakers of my portable stereo.

"This has been my best day in Antarctica," I said.

"You weren't fighting all that weather," John noted.

After eating I hunched beside my snowmobile mirror so I could I could trim my beard. I snipped and snipped and snipped, then rubbed my hands across my face. That was me, in glacier glasses and a Bears cap, trimming my beard in a snowmobile mirror parked on the Antarctic ice. I pulled up my collar though the air was blissfully still—so much so that the sun was actually warming me—and

I put on headphones to listen to Philip Glass's "Opening" then "Islands," music so hypnotic I fell into a trance with it, and all the meteorites I'd seen, all the mountains, all the ice, solid but changing unseen by my eyes, all those rocks sitting on blue ice warming in the sun.

John had asked me this: "Did you notice there were no footsteps in that soil?"

I thought, *I am doing this, who am I to be doing this? How did I get here? I am in Antarctica. I will turn forty-one in Antarctica. I have seen meteorites before anyone else has. How is this possible? I cannot convey the epic strangeness, the sense of tranquillity, of timelessness, as the songs played, then came to an end. . . . It is a dream to be here. I've been running from the spectacle of this place. Why?*

JOHN SCHUTT FIGURES HE'S spent two years total in Scott tents.

"This," he said, "is home."

One night at the Davis Nunataks, while I sprawled in Gretchen and Monika's tent, which I often did after dinner with John, Monika said, "I don't have a home to feel homesick about."

THURSDAY, JAN. 15, 2004 Overcast, 4 degrees.

A fairly fitful night. It became especially cold, chilling my face and keeping me awake. I hate this place now. I'm going to write a to-do list again for the sake of focus. That the end of this week is the day after tomorrow is a solace, for this experience can't end soon enough. If there's a weather delay, I'll probably just crawl into my bag. I'm a failure as an explorer. I've decided to call Medical and ask for a sleeping pill and antidepressant to be sent on the next flight.

I close my eyes from ten A.M. to noon and never really sleep, then sit up, my arms on my knees, my head hanging down. My hands and arms quiver. I tell Johnny I feel shaky.

"Flu?"

"No, depression."

I find myself telling him how I feel the trip was a mistake—I feel like I'm watching myself confess, like I'm at the top of the tent's vent tube looking down—how being gone for all these weeks felt wrong from the start, because I'd only just started finding peace from change. That I've been crying. That I've been counting time and days to the point I've begun to realize it's a problem. That I haven't had a full night's sleep for a long time. Even the "good" ones are terrible.

Saying something reassuring, John leaves the tent right away and comes in with a large plastic box full of medical supplies. He rummages. Gives me sedatives. Calls Medical at McMurdo but it's lunch hour, so we wait. I say—I say that I'm ashamed that I'm not strong enough to stave this off.

John taps my leg and says, firmly, "Don't worry about it."

I reach Medical. I tell the doctor all this and the narrative, my bedlam, my bedlam. He asks if I am feeling suicidal, and I pause. My thoughts are moving slowly, like my speech. I can't say no but it's not exactly yes.

"A little."

There was a weekend after I had left my wife when, following a phone call from her—she had sobbed—I crawled into bed. I knew that if I went back she would be better. I knew that if I went back I would not be happy. I knew that if I went back I would hurt Kathe—and myself. I loved *her* not the woman I'd been with for thirteen years, more. What to do? And could I tell Kathe I felt this way? I stayed in bed for two days, thinking. The ceiling was my empty map. I knew that if I kept staring at that map I could not make it, that to be trapped beneath such emptiness was untenable. I understood then in a kind of attenuated way that one solution to end such entrapment was suicide. Later I'd learn it's called "passive suicidal ideation." The wish, without the plan.

The doctor and I agree: I should be pulled from the field.

MASSIF: A "MOUNTAINOUS MASS broken up into separate peaks and forming the backbone of a mountain range." From the French for "solid" and "massive." The classical roots lead back to Indo-European root for the word "make."

I think in all but words that I am an unmaking.

I FORCE MYSELF, UPON John's suggestion, to walk the few steps to Monika and Gretchen's tent. I can barely meet their gaze.

"You've been gone so much longer than we have."

"It's time for you to go home."

"You've been doing two jobs out here."

"You're doing the right thing."

I return to the tent in time for John's check-in, the regular evening call between LaPaz and recce. Nancy Chabot is stunned to hear my news, news that she'll have to convey to the rest of the team at LaPaz.

Friday, Jan. 16, 2004 M. sun then clouds, 10 degrees calm.

Godawful waiting to see if the plane will come. Weather is better, flyable, John says. I slept more soundly, the pill, yes, better this A.M. but still feel edgy, exhausted, depressed. If I knew we were pulling out tomorrow I could have made it, I tell myself, but the prospect of another week . . . I nearly cried when Mac Ops said the Twin Otter wasn't "off-deck."

8:41 call from Mac Ops: The Twin Otter called "KBH" is now off-deck—they're flying. I sigh a quiet "thank God" and now relief mingles with shame. Then Ralph calls. John tells him, "Chris decided to pull himself out of the field." I want to protest—not just me, the doctor too. I get on the phone and Ralph tells me, "You have to take care of yourself first."

Waiting for the plane to show up, I walk aimlessly in camp, waiting. Sit on a box. Lean on a snowmobile. Clouds hang on to dusted Otway. Overcast. Poor horizon. Poor surface.

Food no longer appeals. Gretchen says, "That's not a good sign." I crawl into the tent, not meeting any eyes and sit on a wooden box. *Please land,* I whisper.

I go outside when we hear the Twin Otter come closer now.

John: "He has his landing lights on." John had put out jerry cans and bundles of our bamboo-pole flags to mark surface and runway. KBH seems to be coming right at me. I whisper, "His flaps are down." He's low, nose up, I hear the scritching of the skis on the ice and feel my body rise up as I whisper, "Thank God." Suddenly the plane has taxied up, opened its door. Erika Eschholz, blonde, good-looking, and smiling, hunches through the now-open plane door and clambers out. She's ecstatic, for very few support staff at McMurdo ever have a chance to come out to the deep field.

"Great people, great people," I say to Erika.

"So, flu?" Erika asks me.

"Ah—no—a . . . depression thing."

I'm a gutless, sleepless loser, you beautiful blonde adventurer.

I ask John to let Medical know that I'll stop by tonight.

"Do you need more pills?" he asks.

"That would be a good idea."

We walk to a box on the cargo line.

"John," I say, "I'm so sorry."

"You're doing the right thing," he says, handing me the pills I shove in my wallet.

"Besides," he says, "look at my new roommate."

I manage a smile.

When I hug Gretchen and Monika, I say, "I'm sorry. Thank you." I shake John's hand.

Suddenly I'm getting seated. The plane taxis, bumpity-bump, across ice, I'm waving weakly, the women wave back, the plane bites into the air, and I watch the Otway.

The Otway. The same mountain I'd watched one evening of sun and little wind when I listened to Tippett's Second Symphony on my headphones when big, grand cloud shadows crossed ancient rock and when, during one crescendo, my whole being reverberated with being here, with the nameless glaciers and black splay of shadow on dark brown rock and the white-hot glow of snow and the deepening blue of the sky that colored belief and disbelief, that dared me to say that I was here and that a single life could come up against this like ice flowing into formations, the massive sweep of these mountains and time and that a single life could follow some hunters of shooting stars and become one by driving across the sky's double, picking bits of the asteroid belt before a lunch of instant soup, the most unlikely harvest, the most unlikely commute, and just as the first movement was coming to an end, Gretchen holding out the satellite phone to me to show me a brief text message from Kathe that read, "S'ton plays fetch w/ mouse like dog."

As KBH flies away from where, minutes ago, I thought I would always be—

truly I felt I'd never leave—I feel the relief I felt when I pulled out of my driveway the day I left my wife.

The Otway specimen total will be tallied at eighty-four.

I watch the massif until I can watch no more.

I AM FLYING OVER ANTARCTICA.

I am dazzled by wraiths of white at The Cloudmaker, a mountain of abject grandeur.

I am listening to the pilots, their mikes left on: "Did you see? He had tears in his eyes."

Today's cargo on flight KBH: two boxes of meteorites; me, garbage, and human waste.

Over the Queen Elizabeth Range, I see fans of crevasses gaping like the open gills of some impossible white shark. I see sharp peaks hidden by vast heaps of snow, a range that snakes beyond my understanding and perhaps beyond my desire to understand.

The air is so clear. What seems near here is almost always farther than you think.

THE VAN AT MCMURDO takes me into town, and I listen, incredulous, to Sheryl Crow's "Every Day Is a Winding Road," an early anthem for Kathe and me. The driver is playing *that* song? No fate, no signs—but *that* song?

I am surrounded by people. By people?

An official has taken charge of the isopods I had with me.

At a metal sink in the Berg Field Center, someone eyes me coldly, hating me—I'm sure, hating me—for giving another coworker, not her, a chance to go to the field. I'll wonder till I leave if I am the butt of gossip and jokes.

There, now here? Otway, now here. Not possible.

"I should have known something was wrong," Kathe says, when I call to tell her the news. But I'd kept my depression from everyone, including her. This shames me.

At McMurdo, I eat. I see the doctor, who asks me about my feelings before the ice and my feelings now, and he lets me go, down the dark narrow hallway to emerge into the bright light between the medical building and Building 155, surprised by things, a roof, a barrel, the way they shoulder into the world, the air. McMurdo doesn't feel like Antarctica anymore, not after the polar plateau.

I sleep. A lot. But I don't feel refreshed. I muster some energy to write an article for the *Antarctic Sun* about searching for meteorites, but officials above the editor censor passages about my insomnia and depression. At this, I simply shake my head.

At McMurdo, I leave notes and gifts in the ANSMET cage for the team I have let down.

Three days later—five hours of which are spent on the smelliest, hottest, most nauseating flight I've had so far, on a cargo jet this time—I'm deposited in New Zealand, where, before I fly home, I walk across Hagley Park, bewildered at the chaffinches and cirl buntings, touching and hugging trees and where, at the art museum, I stand before the paintings of Margaret Elliott, beautiful paintings of Scott tents, snow, rock, beautiful paintings that make me think, "Good riddance."

I pick up Kathe's bracelet, the one that says, *Turn your face to the sun and the shadow will fall behind you.*

Back home in Northern Utah, I listen as a therapist asks me if I think things would have turned out differently had I gone to Antarctica at another time, if I had gone later, after feeling more settled in my new life. I don't hesitate. Oh yes, I say. But I applied when I did, went when I did, and now have months—years—to reflect on what happened. I'm amazed that I lived in a tent in the middle of Antarctica for five weeks and I'm ashamed of how I left. But honestly, I'm thankful that I didn't slow the recce team and thankful that I didn't have to spend several more days at the Otway, which the team did when the weather turned to shit.

A month later, Oliver Botta writes and surprises me with his frankness and understanding. "We were all very sad to hear what happened," he says. "There is no reason for you [to] think that you let us down. What had happened to you could have happened to any of us. Just imagine it to be the same as a broken leg or any other type of accident or sickness." Oliver confesses he had "similar problems," including difficulty sleeping, which made "the days after such a night . . . just terrible. . . . I might as well have fallen into the same spiral as you did." But he didn't, even as he coped with the burden of learning that his young son believed he was never coming back. I am grateful for Oliver's kindness, which is one of several that my fellow ANSMETers show me.

I have read a good deal, discovering that sleep disturbance and depression are common in extreme environments, such as Antarctica, submarines, and long-duration spaceflights. One researcher who had been to Antarctica years before told me of the struggles with depression above and beyond the so-called "third-quarter blues" that affect so many on missions of one kind or another. Apart from causes one would suspect—such as bad news from home—depression can be triggered or accentuated by, for example, high winds. Wind-crazed pioneers in the American West would sometimes cut off their limbs.

"I think virtually everyone has gone through periods of depression while out in the field, and I mean everyone," Ralph tells me. He includes himself—and John—in this category. "Only a few scary people seem to eat the whole season up like candy; the rest of us need to work at it. To me, depression is something always there, always lurking in almost everyone; the difference is that yours was pushed off the chart by the intense isolation that is unavoidable in Antarctica."

For Ralph, the satellite phone is crucial. "I fear those calls and long desperately for them. I may get the worst news, I may get the best, but I call home and conquer that fear. At no time in life is communication something to avoid; I can't think of a single time when the best thing to say or hear is nothing at all. And because people I love are on the other side of that phone, I never ever use it to vent and I never soft-pedal my feelings." He kids, "I save the venting and soft-pedaling for my tentmate."

Ralph puts into words what I had already learned. That to speak one's feelings may be terrifying but it's infinitely preferable to the alienation that blossoms from long-held silence. With Kathe I had learned to speak. "No secrets," we say to each other, knowing the costs of having kept them before. So here was the terrible irony of my Antarctic plunge: I had learned the cost of silence and I had rejected silence in my most intimate relationship, but I hadn't yet learned to speak more honestly to those I didn't know well. And though Tim Swindle and I had stood in front of his tent at LaPaz and rejected the heroic explorers' psyches as abnormal, as extreme, some part of that aspiration to valor had kept me quiet. In short, I was afraid to admit to anyone, especially the seemingly invulnerable John Schutt, that I was doing poorly. Had I, had I done so earlier, I might have staved off the worst, I might have staved off my crack-up. But I am a slow-thinking man.

In his forgotten, often lyrical 1950 work of popular-science writing *The White Continent,* Thomas R. Henry notes that Robert Scott "was a sentimentalist who would burst into tears" upon hearing songs or stories. His rival, Roald Amundsen, "was a thwarted, unhappy man," despite having won the South Pole. Both were moody, both devoted to "adventure, wild beauty in action." But it's not all wild beauty. The explorer Richard Byrd once wrote of living in Antarctica, "You are hemmed in on every side by your own inadequacies and the crowding pressures of your associates. The ones who survive with a measure of happiness are those who can live profoundly off their intellectual resources, as hibernating animals live off their fat." Byrd wintered alone at a weather station for months before being rescued because of carbon monoxide poisoning. Me, I arrived too lean. Though of course I suffered nothing as cruel as his solitude and his sickness, many of his descriptions of time, melancholy, and difficulty resonate. Sitting by the river, beneath chokecherries nearly as red and swollen as bing cherries, I read his words: "I find I can't take my loneliness casually; it is too big. But I must not dwell on it. Otherwise I am undone." With willows swaying in the warm wind and the river rushing, with a flicker calling *waka-waka-waka* and with Kathe reading or writing inside the house, I concur with the admiral: "I don't think that a man can do without sounds and smells and voices and touch, any more than he can do without phosphorus and calcium."

Perhaps more directly relevant to my experience is a study included in the book *From Antarctica to Outer Space.* In a study called "The International Biomedical Expedition to the Antarctic: Psychological Evaluations of the Field Party," the researchers report that one member was taken out of the field. Fellow team members wrote in their journals that this individual was "preoccupied, strained, sad, tired, and lacking enthusiasm for work. He did not like the food, felt cold, slept badly; he got depressed, felt homesick, had difficulty in accepting separation from his family and friends, and reported feeling that 'he has fallen into a trap.'" I feel for that person, but admit to being comforted to read those words. I'm not the only one, I'm not the first. Unlike my ANSMET expedition, the IBEA was fraught with rancorous egos and almost adolescent hostilities. How much worse would it have been on a team like that, I wonder?

In the summer of 2007, I gave a reading from an essay based on my experience in Antarctica. Afterward, someone told me that she'd seen a newspaper article about "polar madness." I later read the study, by Lawrence Palinkas and Peter Suedfeld, which noted, among other things, typical conditions of anger, depression, self-doubt, trouble sleeping, and boredom, in varying degrees. Yup, I thought, that sounds about right.

What Ralph tells others suffering from the blues—or worse—is to take stock of where they are, to let Antarctica "in under your skin and stay there," because they probably will never be back. What the researchers studying the psychological effects of the IBEA tell readers is that "psychological selection is an absolute necessity." Perhaps so. And yet had I been screened prior to ANSMET 2003–2004—and I admit some surprise that I was not—it's likely that I would not have been allowed to go. I would not now be making art out of loss and frontiers.

In the years since, the continent has gotten under my skin. I watch, rapt, any documentary showing images of Antarctica. I seek out books, articles, websites, photographs. Knowing what I know now of the privations that necessarily attend the place, I plot ways to return. At the same time, I shake my head in disbelief at having been there once and in such good and understanding company.

All told, the two teams found 1,358 meteorites, including one Martian and a higher-than-usual proportion of achondrites, carbonaceous chondrites, and odd ordinary chondrites. It was the highest number of specimens ever recovered from the ice in a single field season.

A FEW MONTHS AFTER returning from Antarctica, I went birding on Antelope Island, which, shaped like a sea horse, rises from the south end of the Great Salt

Lake and on which sing horned larks. Their notes tinkle like cut ice. The January fog covered much of the horizon, a whiteness becoming cloud-gray. The still water shone blue, in places silver, so heavy with salt it looked almost gelatinous, and a purple band rimmed the earth. Where sky and water and fog and earth and self began or left off was hard to say. Clumps of snow edged the Davis County Causeway, and I felt for the first time that I was back, back in Antarctica, standing at the LaPaz Icefield, and I was grateful because I had things to set straight.

It was a year to the day of my confession in the tent.

From time to time, I think of what I wrote on a restaurant napkin that I mailed to Kathe from the Los Angeles Airport as I waited for my long flight to New Zealand. I wrote, "The marvels of adventure are nothing beside the clarities of home."

She pinned the napkin to a bulletin board, beside our telephone, where it remains.

Afterword

On this cloudy November morning in 2007 I find it hard to fathom that four years have passed since my journey to Antarctica, that even more have passed since I began my fascination with meteorites and meteorite hunters. I started my quest yearning for an emotional monotone of awe and adventure, which, I came to realize, was another of my dishonesties or at the least a misapprehension. We never feel just one thing, nor is the world just one thing. To attempt a journey—or a life—with only wonder in your knapsack? We need other provisions, and sometimes there is cargo we have failed to account for. Tempered by days and nights of real living, we might, on any given Tuesday morning, any given Sunday night, own up to ourselves and the world. We might admit unhappiness, confess a secret, fall in love. I first sought the fallen sky to keep me from myself. But the extraordinary is more than just the prefix.

For a long time I thought this book would end in Antarctica with triumphant discoveries of meteorites. It didn't quite work out that way. Besides, I already knew that home is more than where you start from; it's what you make, it's where you return. Eventually, I imagined the book would end with Kathe and me under the sky at Upheaval Dome, a rugged meteorite crater punched in the ground of southern Utah at Canyonlands National Park. We would go there, I thought, for the Leonids, to watch them together in our recently adopted state. A perfect narrative frame, an echo of how I had watched the Leonids with my ex-wife a decade ago, that night of fireworks in the sky and of dreams afterward trying to tell me something, and of how, while Kathe was in China watching a full-blown Leonid storm in 2001, I was on my apartment lawn in Kansas seeing nothing but clouds.

This time would be different. After seeing lives passionate and broken, after seeing lives passionate and whole, I'd watch the Leonids with the woman I had fallen in love with and from whom I had learned—along with some prompting from the biographies of star chasers—that honesty is preferable to silence, that living is preferable to complacency.

A couple of mornings ago, as we sat in the front room with coffee and our cats—Shackleton has been joined by Zinc and Burchfield—Kathe asked, "Well, what do you think?" She meant the drive to Canyonlands and our planned backpack along the rim of Upheaval Dome.

"You know," I said, "I don't really feel like going."

Kathe brought her hands together in mock supplication to the sky and said, "Thank you!"

In my other life, and it seems ridiculous to say this, I would have brooded darkly on how to admit I didn't really want to do something. The past few days I've not been brooding but I have been waffling. I'd put off a trip to Upheaval Dome precisely for this weekend, to see meteors with Kathe from the heart of a crater, to camp in one, having failed to do so in the Ries Basin. I imagined us sleeping out in the cold Utah desert, beside dramatic rocks and star-tangled junipers, then coming home to our house beside the Blacksmith Fork River, all snug in the cosmos.

But the autumn has been a busy one, and though clouds are forecast for northern Utah, meaning that here we may miss the meteor shower altogether, we each have relished the thought of a quiet, nesty weekend. Outside, there are leaves to gather and compost, weeds to pull, a fallen tree to chainsaw, and other wood to cut. Inside, there are books stacked by chairs, and a bathtub just waiting for hot water. And I need to vacuum and dust. I don't have that jury-rigged dust mop anymore but I use a feather duster for our furniture and shake it outside just the same. Puffs of dust, domestic and stellar, disperse by our front door at least twice a week.

After all, this place is no less the universe. From our backyard, looking past stands of tall grasses Kathe's planted, beyond a line of sumacs and the two tall, sprawling cottonwoods by the river, we've seen the Perseids in August and the Leonids in November more than once, watching to the calls of coyotes and owls. Friends stepping away from the circle of light at our fire pit have walked past stands of Russian sage, looked up and *ahhhed* at the passage of random meteors. Once, under a blanket, near some chokecherries, Kathe and I saw the northern lights hang like pale virga in the dark above the valley. And these past few days I've kept a telescope on the back porch to look at a diaphanous outburst of Comet Holmes, which, though tailless from our vantage, has grown a coma wider than the sun itself. It looks like a giant eye. Weeks ago, I even set up a tent in a small pasture beside our house so I could wake before dawn, drag my sleeping bag onto a path, and wait for a possible Aurigid meteor storm. Though it never came, I saw a handful of streaks, white pencil lines erasing themselves as they went. Later, I thought of that predawn vigil when I reread a scrap of paper I've kept for years, on which I had scrawled some lines from a Menomini Indian myth: "When a star falls from the sky/It leaves a fiery trail. It does not die./ Its shade goes back to its own place to shine again."

One day in early August 2007, I stood by a small juniper we had planted near the kitchen window. It was a mellow dusk, and I sat cross-legged in the gravel path next to the rocks edging a flower bed of columbine and scabiosa. A black-headed grosbeak had stunned itself on the glass, and I was sitting nearby to make sure the bird wasn't pestered or worse by a magpie or neighborhood dog. I'd checked to make sure it was alive, and nothing appeared broken, so we sat together, the bird with its eyes closed, then open, then closed, me with my legs growing cool. I watched it breathe. The sun had just gone down, and I contemplated picking up the grosbeak to warm it with my hands. I can't recall if the bird shook itself to normalcy and flew before I stood up exclaiming or if I startled it into flight. For there, about five degrees above the rounded peaks of the Bear River Range, which loom close by—a *fire in the sky*, a rippling white fireball as bright as Venus, racing on a shallow, nearly horizontal trajectory, south to north, covering 40 to 50 degrees of sky and spewing a long blue-white trail. Just before it flamed out, a puff of gray smoke arced backward.

"Oh, oh," I said, mouth agape—the first daylight fireball I'd ever seen!

The bird was gone.

I dropped a rock to mark my spot and ran inside to tell Kathe and call newspapers and meteorite dealers. She hadn't seen it from a window, and no one else I contacted had seen it either. The fireball was neither Norton nor Pasamonte, but I'd seen one in a blue sky, my own fireball from our own backyard, above our mountains. Actual matter, solid fact, the fireball felt like a nod from Sila or Tomanowos, a goodbye gesture from the sky of mythology. Instinctively, I thought of Harvey Nininger, half-imagining that he might pull up in a car with the words "American Meteorite Laboratory" painted on the side. We'd stand there by the house, and I'd tell him what I'd seen, and we'd both look out to the pasture when a meadowlark burst into the day's last song. I'd like to think he'd have told me to load up my gear and come hunting with him—after which, I'm sure, he would have detoured us to Garland, then the Drum Mountains, to look for remnants from those two Utah meteorite falls. Maybe, I sometimes think, I'll hunt for them myself.

Not long after my fireball sighting I traveled to Tucson again, this time to attend a scientific conference, a meeting of the Meteoritical Society. I was especially keen to hear the latest research on chondrule formation and, more important, to see some old friends from the ice. John Schutt was getting an award from MetSoc for all his years of service in Antarctica, and I wanted to be there for that. During the ceremony, I sat next to Monika Kress, who was at the conference to give her presentation about the Otway Massif strewnfield we'd explored. When Ralph Harvey asked the audience how many of us felt we'd entrusted our lives to John, a quarter of us stood up, including me and Monika. Ralph also noted something that had never occurred to me—that John has probably seen more meteorites for the first time than any other person in history.

"He's a one-man Apollo project," Ralph remarked.

Unsurprisingly, John deflected all such credit back to Ralph Harvey and Bill Cassidy and the "family of friends forged through the difficult conditions of working in Antarctica." He also thanked his tentmates for "forbearances of my eccentricities," at which Monika's eyes sparkled. She gave me a playful shove.

Before the ceremony ended, Monika had to leave to get to her session, so I didn't get a chance to tell her that when I saw her Otway poster before she'd arrived, I stood in front of it very quiet and very still for a very long time, feeling that peculiar mixture of pride and shame, of achievement and regret, whenever I think of my weeks on the ice. My name was listed as one of the team members who recovered meteorites from that fall; so too was my replacement's.

As things wrapped up, I fell in with John as he walked from the ballroom. "I'll never forget your kindness to me," I told him. John was taken off-guard but offered a warm response, though typically one of few words. Which was fine—it made me smile—because John, in fact, didn't have to say anything at all.

The first night of the meeting I also saw Tim Swindle, who cheerfully told me he was going back as Ralph's mid-season replacement. A third time! I envied him. I want to get back now, I told him, and he understood. "I've read my journals," Tim mulled. "I know how it's gone, but I'm addicted."

Tim also told me he was using a handful of meteorites we'd found at LaPaz for a study involving the ages of impact melts inside ordinary chondrites, in particular H chondrites, the kind with the most free metal. As Tim would later explain, he's using argon dating to see when the parent asteroid of the H chondrites was involved in collisions; this is a case when the high numbers of these relatively common meteorites is helpful, because there's a higher possibility of finding meteorites that were all part of a single collision. Tim is intrigued by evidence that the H-chondrite parent body got smacked around about 3.8 to 4 billion years ago, which is when the big lunar basins—the ones we can see with our own eyes—were created by melt-producing impacts. "And if it happened on the Moon and some asteroids, it probably happened on the Earth as well, which might be why the oldest intact rocks on Earth are about that age," he'd later tell me. The study could help us understand how impacts affect the evolution of youthful solar systems.

Though Oliver Botta wasn't at MetSoc, I remembered that he had been a coauthor on a study published the year before, which had found that some ANSMET meteorites were contaminated by an amino acid leached out of storage bags used in the field; not surprisingly, the research group suggested using another kind of bag. I recalled how we had helped Oliver collect ice samples at LaPaz for use in this very report.

In between technical talks, I saw familiar faces in the crammed corridors of the resort. There was Robert Haag and Marvin Killgore beside a display of meteorites. There was Gisela Pösges, tall and smiling as ever, who had recently fact-checked my Ries Basin material. I finally met Carolyn Shoemaker, who was as gracious as I had heard. Barb Cohen answered more of my ceaseless questions, about everything from PDFs to the Poynting-Robertson effect. In a hallway filled with chatting meteoriticists, I talked with Gary Huss, who had given me access to filing cabinets filled with his grandfather Harvey Nininger's papers. And though Ralph was tired, he was gregarious, ever ready with a laugh or wry comment. It was good to see him again, and we even snuck out of the conference one evening, along with Danish researcher Henning Haack, to catch, of course, *The Simpsons Movie.*

This wasn't my first ANSMET reunion, though. In the spring of 2004, just weeks after my return from Antarctica, I mustered the courage to attend the Lunar and Planetary Sciences Conference in Houston. The night before I left I dreamt I was at some kind of polar science camp where a huge screen revealed the sun birthing massive solar flares and one of the flares birthing a planet. Nearly hidden in a cranny were four birds' nests. Then, when I found my dream-self in a motel room filled with the ANSMET crew, they all went silent. I honestly didn't know what to expect when I actually arrived. Many of the ANSMET expedition would be at the conference, where Ralph always does a slide show and invites the prior season's participants to talk about their experiences. Those interested in going to the ice themselves come in droves.

I shouldn't have worried. Everyone there was kind to me—concerned and welcoming—and at his slide show Ralph put up a photo of me wearing those purple latex gloves and holding up a long meteorite with tongs. I told the audience that it was true I had to leave the field a few days early but only because of my stressful work—as field proctologist. Bless them all, they laughed.

One evening Nancy waved me over and said, "We were all worried about you, Chris," adding, quietly, "It was a hard day." She meant the day she went tent by tent at LaPaz to tell the others of my situation.

"I let you guys down," I said, my jaw clenched. "I let myself down."

She cooed a no-no sound, but nodded when I said I'd realized that the bigger community at LaPaz had helped keep my spirits up.

I had another reason to go that meeting in Houston. I needed to visit a concrete building at the Johnson Space Center, where some blooming crab apples and azaleas tried their best to dress up the drab, low-slung facility. It's here that the Antarctic meteorites come, to meteorite curator Kevin Righter and his staff at the Astromaterials Acquisition and Curation Office. Trucks deliver the isopods, which staffers load on carts and move into a freezer. Then the truckers go on to more usual freight, like ice cream in Colorado.

Dressed in booties, lab coat, and a hair cap that reminded me of my high school job as a fry cook, I met Kevin's two scientists, Cecilia Satterwhite and Kathleen McBride, who showed me stainless steel equipment—from big storage cabinets to sealed chambers where meteorites can be handled with those stick-your-arms-in-them-neoprene-glove-things, like the kind used for biohazards. They also made me step into an "air shower" so I'd be less dusty.

Here, the Moon rocks had been processed. Now the facility is divided into cosmic dust processing and Antarctic meteorite processing. When Cecilia or Kathleen retrieve the meteorites from the freezer, they move them to those sealed airlock cabinets, where the rocks are thawed in nitrogen to rid them of excess terrestrial water and to keep out further moisture and contaminants. The meteorites are photographed, measured, and described. Then they do "a split," knocking a chip off that can be used for further analysis if need be. After an initial classification here, most of the meteorites are sent to the Smithsonian for a more detailed classification using thin sections examined under microscopes and, sometimes, with an electron microprobe. Eventually, a newsletter about all the latest Antarctic meteorites is published so scientists worldwide can consider which meteorites to request as research specimens.

Kathleen and Cecilia can each examine ten meteorites a day, on average, as they go through the laborious initial classifications, which also involve making up new tags and numbers for the specimens. It's tedious work, leavened somewhat by the lab's boom box and by the unusual-looking finds, which they do first in order to get them into the newsletter faster. Sometimes a really odd specimen will be thin-sectioned at Johnson Space Center, and there's a set of special meteorites that will never leave here.

Kathleen put on gloves to show me one of those special meteorites—ALH 84001, the rock that for a time seemed to show signs of Martian fossils.

Probably robbed of that distinction forever now, the rock is still a star, albeit one that, as Kathleen says, weighs "less than a four-pound sack of sugar." The neoprene arms sticking from the airlock cabinets thwacked her as she walked ALH 84001 to and from its cabinet. I wasn't the only one there to see it, so the glimpse was just that, a glimpse. A chunk of rock—yes, a driveway comes to mind again—gray, friable, with an interior that looked haggard and broken. But sweet indeed. Even after all my travels with meteorites, I don't think I could have guessed this dull and ugly stone to be anything significant. But it helped spark a frenzied interest in life on other worlds, and I wondered what the astronauts might find when we finally get to Mars.

HOW WILL THEY FEEL after such a journey? Thankful? Probably. Relieved? Doubtless. Itching to go elsewhere? Hard to say. Mine has been a far, far smaller journey, or at least an earthbound one. At times, though, I felt I had landed on another planet.

I'm glad I've traveled under this sky. It seems better composed now. I'm grateful to the point of stillness and bewilderment because of what I've seen, where I've gone, and what I've gleaned. I've learned from the possessiveness that ultimately robbed people like Ellis Hughes and F. W. Cragin of happiness, though I admit I'm now loath to part with any of my tokens—the locust pod I took near the Kansas Meteorite Farm, the hammerstone and wood from Greenland, the shatter cones from Germany, the Acraman ejecta from Australia. Maybe I'll give them up someday, maybe not, but for now they crowd a shelf, they serve as reminders. On a recent trip to New York, I saw the Willamette Meteorite again—and learned that sometimes docents by the huge iron will tell the story of Ellis Hughes, the story the museum signs have effaced. One such docent spoke of how, because of the "supernova iron in your blood," it's a form of "cosmic reunion" to stand beside the Willamette, beside Tomanowos. I loved hearing that. It renewed connections.

And I've learned from the gritty hard work of the Kimberlys, trying to emulate their persistence. I've learned from the dubious ambiguities of Robert Peary, which drove him on to legacy but not to peace. I've seen how Daniel Barringer's single-mindedness, like Peary's, was both his strength and his weakness.

Obsession, they tell us, ought not become a scalpel that cuts away the rest of your life. I've relished the myths and tales of meteors that make them bodily, having become a bit more comfortable in my own skin. And from the life of Harvey Nininger, I've taken a great deal. His biography tells me there are forms of possession that lead not to miserly grasping but to connection and, yes, to wonder and knowledge. Not that Harvey didn't have his moments, but in the end, he lived the life he most needed to live, he risked and lost, he risked and triumphed, he found both legacy and calm. And early on in my pursuits, the meteorite clan that so worships Nininger showed me a kind of lusty enthusiasm that helped me to examine my insufficient compromises with appetite. Those dealers and collectors I met for the first time that warm February in Tucson, years ago, could not have known what their example would mean to me.

As my journeys brought me closer to scientists of the present—from Eugene Shoemaker to Vic Gostin, from Kath Grey to Mongo on Ice—I've struggled at times to understand the sometimes elaborate technical difficulties that come from studying meteorites, even as I've admired scientists' dedication to unraveling secrets and asking new questions. Meteorite research can often seem terribly narrow, as one scans yet another article whose upshot is, say, a subtle modification of a specimen's classification. But on such subtleties whole systems are made and understood, and from such subtleties come the bigger questions, the possible answers, from how we got here to what might happen. Speaking of subtleties, I learned as I was finishing this book that Brian May, the guitarist for the rock group Queen, recently completed a science doctorate . . . by writing a dissertation on dust in the solar system. From dust to rocks to rock stars singing about dust to rock stars writing about dust. Scientists in 2008 also tracked an asteroid in space, determined its orbit, took its spectra—then monitored its predicted fall to the ground. They found pieces of a new type of ureilite in the deserts of Sudan.

Meteorites are even helping with questions that go well beyond the solar system. They've been studied for clues to how the entire universe evolved such a harmonious set of fundamental physical values. Our universe is finely tuned among a variety of forces—from those governing the center of atoms to gravity itself—such that long-lived stars can develop, die, and seed future starbirth. Without long-lived stars, life would have little chance to develop. One such physical value is called the fine-structure constant, which sets the strength of electromagnetism. Alter it just a bit and atoms as we know them wouldn't work. Physicists are finding intriguing hints that the fine-structure constant isn't so constant after all, that over time it actually has changed. If that's true, then how one isotope decays into another would also change over time. So some researchers have studied the results of such decay in meteorites—irons such as Henbury and Hoba, among others—and have seen signs of variation, of evidence of an inconstant fine-structure constant. And that may help us reconcile the fundamental values in our universe, which seem at once so perfectly calibrated and utterly random. Personally, I'm a fan of the idea that our universe is just one in an infinity of other universes, each with its own set of "laws," all afloat in a multiverse. I relish the thought that a few iron meteorites may help us better see this grand and bizarre vision.

A 2008 discovery that some meteorites contain tiny "graphite whiskers" also has a cosmological implication. These tiny graphite particles may be a common part of space—in between stars and even in between galaxies. It turns out that graphite whiskers may be absorbing light in a way that dispenses with the need for so-called "dark energy," the mysterious force most cosmologists have invoked to explain an apparent accelerating expansion of the universe. The key here is how astronomers are reading the light from distant supernovae—a light that is deeply red-shifted, thus connoting high rates of acceleration and, therefore, the observation that the universe is expanding faster and faster as time goes on. When this was discovered, cosmologists were caught off-guard; no one expected the expansion of the universe to be speeding up. The only way to explain this was some force—as yet unknown—pushing space along. Dark energy was born. But if graphite whiskers are interacting with light such that astronomers are essentially misreading the supernovae, it may turn out the universe's expansion isn't speeding up and that dark energy doesn't exist.

Multiverse or no multiverse, dark energy or no dark energy, this particular cosmos still suffices very nicely, thank you, in ways I never would have expected. In our bathroom, we have two rings hanging over a photo frame. The picture is one I took in China of a sign that says "Poetic Home." The ring Kathe gave me is gold and engraved with trees and bears and stars. The ring I gave her is fashioned from a Gibeon iron meteorite, the lines of taenite and kamacite always crossing. Near this is the framed text of that Basho poem we love. The seasons surprise, the universe consoles and intrigues.

Nearly every morning this fall, as I've come closer to these conclusions, Shackleton the cat has explored the expanse of my desk, summiting piles of paper, traversing stacks of books, and his busy tail has swished over four meteorites. I finally bought one this year, having decided that I needed a "show-and-tell" specimen to share with audiences. Geoff Notkin, one of the cohosts of the meteorite dealers' dinner I attended in Tucson, sold me a bargain unclassified northwest Africa chondrite, a satisfyingly hefty stone the size of an apple but the color of a dark canyon stream. I worried a bit about buying a rock that was part of the flow of meteorites coming from the Sahara and other such places, but the price was right and I wasn't convinced science would miss this one stone. (Do I contradict myself? Very well, then, I contradict myself. . . .) Roughly triangular, the stone is cracked and weathered from unknown years in desert sand and sun, but I think I can still see old fusion crust in small black patches, and might those particles in one crevice be chondrules? Or beads of desert grains? Maybe someday I'd give it over to a meteoriticist for answers, maybe even for good. Geoff also included surprises, two small angular irons, a complete Campo del Cielo, regmaglypted and shaped like a lumpy foot but with edges that shine silvery from the nickel, and a coy Canyon Diablo, with a dark-brown patina and a spiffy peaked top. I hadn't known he'd give me the irons, and the evening that they all arrived in a box I held each meteorite carefully, admiring those surfaces, glad I had delayed my purchase till then. I placed them with their label cards on my desk, next to the thumbnail of Cape York I'd picked up in Greenland and next to a small fragment of a chondrite called Gold Basin, which Tim Swindle had presented to the ANSMETers at the LaPaz Christmas gift exchange. I guess I hadn't really thought of the rusted Cape York as a "meteorite," given how weathered it is, and the Gold Basin carries more power sentimentally than visually. But these two tokens gain in stature next to their three larger kin. Without really intending I have a collection. And I like that.

Recently I realized I had misremembered what I'd written on that napkin in Los Angeles; the gist was the same, but over the weeks in Antarctica I had tightened the phrase, polished it. I'd originally written, "Whatever marvels adventure brings, they are nothing beside the clarities of home." I also wrote, "How I love you. And give our Shackleton a scratch on the belly from me." In the tent at LaPaz, I'd obsessively edited my little manifesto, finally settling on "The marvels of adventure are nothing beside the clarities of home."

Now, after all this, I'm not so sure about the categorical sweep of the statement. There's room for both in a life.

Leonid weekend has passed—I write this now at the start of a new week—and though I looked out more than once Saturday night and Sunday morning I saw only clouds. As Sunday went on, the sky cleared, and the wind seemed to shake it free of our ambitions, though I know that the air is more human now than ever before. Kathe raked leaves, I trimmed branches from the poplars we planted this year, cutting myself, and after she helped me dress the wound, I went back across the river to finish pulling the latest rosettes of dyer's woad, a weed. We stowed outdoor chairs in the shed. We tied up bunches of tall grasses to keep them from breaking under snows to come. All day dippers sang from stones in the river. All day a yarrow bloomed. All day the trees were bare, and beyond those branches meteors flared hidden in the light. At day's end, we walked about the yard, tired from chores and a little stunned by the sunshine quiet and Sunday warmth. In the evening, we held each other. So how to say this? That it is here, together in the West, that we know ourselves to be home? Yes. Here, we know ourselves to be home.

Glossary of Terms

The definitions and chart that follow are necessarily quite basic. Readers should refer to the sources listed in the Notes for more expansive and detailed information.

achondrite One of the two main subcategories of stony meteorites. There are several kinds of achondrites, including howardites, eucrites, and diogenites. Achondrites have experienced melting. They are igneous products and often difficult to distinguish from Earth rocks.

asteroid A rocky body of varying size typically found in the Asteroid Belt between Mars and Jupiter. Those asteroids that cross Earth's orbit are referred to as Near Earth Objects, Earth-Crossing Objects, Near Earth Asteroids, Potentially Hazardous Asteroids, and plot devices for Hollywood movies.

bolide A very large fireball. Sometimes the term is meant to designate a meteor that explodes in mid-flight.

chondrite One of the two main subcategories of stony meteorites. The two best-known types of chondrites are ordinary and carbonaceous. The former is a common type of meteorite; the latter is rare and contains organic material. Chondrites are essentially unchanged from the formation of the solar system 4.5 to 4.6 billion years ago. They contain chondrules. Some scientists believe that without an infall of carbonaceous chondrites, life may not have developed on Earth.

chondrules Small globules of melted material found in varying degrees within many types of chondrites. The existence of a heated product in an unheated meteorite is mysterious. Without chondrules, planets may not have had a chance to form.

crater A hole in the ground blasted out by a large meteorite. You probably knew this.

fall A meteorite whose fall was witnessed.

find A meteorite that has been discovered but whose fall was not known to be witnessed.

fireball An extremely bright meteor. If observed, it will be likely to provoke an exclamation.

fusion crust The black crust that results from surface melting to a meteorite as it passes through the air.

impactor A meteorite big enough to cause significant damage.

iron One of the three main categories of meteorites. Irons form in the molten cores of asteroids. Irons are dark, heavier than Earth rocks, and can be oddly shaped. They are the most easily recognized of meteorites, though not very common.

lunar A meteorite that has been ejected off the surface of the Moon by an impact and has landed on Earth.

Martian A meteorite that has been ejected off the surface of Mars by an impact and has landed on Earth.

meteor The light of a passing extraterrestrial, nonmanufactured object through the atmosphere. A shooting star. Something to wish on, if you're so inclined.

meteorite The object from a meteor that survives passage through the atmosphere and lands on the Earth.

meteoroid A small rocky object in space that can become a meteor. Smaller than an asteroid, though this is subjective, depending on the experts consulted.

strewnfield The zone where a fall of meteorites has occurred.

stony One of the three main categories of meteorites. It includes chondrites and achondrites.

stony-iron One of the three main categories of meteorites. It includes mesosiderites and pallasites.

Meteorite Chart

Presolar nebula → Sun → solar system (outer gas giants, asteroid belt, inner rocky planets)

Moon → lunar meteorites

Mars → Martian meteorites

asteroids (heated through/differentiated) → achondrites (a form of stony meteorite)

asteroids (cores of molten iron) → iron meteorites

asteroids (from mixing zone of core and mantle/melt pockets) → stony-iron meteorites

asteroids (unprocessed, undifferentiated, or crust samples) → chondrites (a form of stony meteorite)

chondrules (formation mechanism unknown) → incorporated into chondrites (mechanism unknown)

Notes

Introduction

2 An English . . . cereal and milk: Derek Sears, "Meteorites with Your Breakfast Cereal," *Meteoritics and Planetary Science* 36 (2001): 1291.

2 Astronomers have . . . September 11 attacks: "Asteroid Memorials for September 11th," *Sky & Telescope,* January 2002: 26.

7 meteors as . . . versions of ourselves: H. M. Basurah, "Estimating a New Date for the Wabar Meteorite Impact," *Meteoritics and Planetary Science* 38, no. 7 Suppl. (2003): A155–A156. Basurah quotes an Arabic poem: "Those who saw it crouched / down with a heart flying over."

Prologue

10 There were . . . solar system today: Robert Hutchison and Andrew Graham, *Meteorites* (New York: Sterling, 1993), 52–56.

10 Supernovae . . . rips apart: Nancy Hathaway, *The Friendly Guide to the Universe* (New York: Penguin, 1995), 242–243.

10 Previously . . . marriage vow": Harry Y. McSween Jr., *Stardust to Planets: A Geological Tour of the Universe* (New York: St. Martin's Griffin, 1993), 64.

10 And some . . . on Earth: Sun Kwok, "Gems from the Stars," *Sky & Telescope,* October 2002: 38–43, 40.

10 It took . . . one supernova": McSween, *Stardust,* 66–76.

11 About 4.6 billion . . . 10 million degrees: I have relied on several accounts concerning the creation of the solar system. They include David A. Kring, *Meteorites and Their Properties,* 2nd ed. (Tucson: Lunar and Planetary Laboratory, 1998); Robert T. Dodd, *Thunderstones and Shooting Stars: The Meaning of Meteorites* (Cambridge, MA: Harvard University Press, 1986), 149, 167, 170.

11 When I first . . . by supernovae: "Chandra Spies Rare Isotopes," e-mail news item, *Sky & Telescope's News Bulletin,* September 14, 2001.

11 Not long . . . went supernova: Richard A. Kerr, "Isotopes Suggest Solar System Formed in a Rough Neighborhood," *Science* 316 (May 25, 2007): 1111.

12 Working against . . . formation: Richard M. Crutcher, "Testing Star Formation Theory," *Science* 313 (August 11, 2006): 771–772.

12 What material . . . Kring says: Kring, *Properties*, 6.

12 Electrostatic attraction . . . solar winds: David Tytell, "Building Planets in Plastic Bags," April 13, 2004. http://skyandtelescope.com/printable/news/article_1239.asp.

12 Data from . . . mixed-up cloud: Fred J. Ciesla, "Outward Transport of High-Temperature Materials Around the Midplane of the Solar Nebula," *Science* 318 (October 26, 2007): 613–15.

12 Such stuff . . . of materials: Richard A. Kerr, "Has Lazy Mixing Spoiled the Primordial Stew?," *Science* 314 (October 6, 2006): 36–37.

12 Grains, flecks . . . born: A.G.W. Cameron, "From Interstellar Gas to the Earth-Moon System," *Meteoritics & Planetary Science* 36 (2001): 9–22, 9.

12 Easily told . . . years: Ray Jayawardhana, "Planets in Production: Making New Worlds," *Sky & Telescope*, April 2003: 36–42, 39.

12 I think . . . the air": Kring, *Properties*, 6.

13 It's called "diamond dust": Vincent J. Schaefer, and John A. Day, *Atmosphere: Clouds, Rain, Snow, Storms* (New York: Houghton Mifflin, 1981), 247.

15 Earth took 10^8 years to form: Cameron, 9.

15 The gas giants . . . centers: Joshua Roth, "A Fiery Birth for Frigid Worlds," *Sky & Telescope*, July 2002: 20.

15 Inside many . . . countless collisions: Alan E. Rubin, "What Heated the Asteroids?," *Scientific American*, May 2005: 80–87, 82, 84–85, 86.

15 The cores . . . had exploded: Hutchison and Graham, 46–48.

15 It was . . . a world: McSween, *Stardust*, 66; Richard P. Binzel, "A New Century for Asteroids," *Sky & Telescope*, July 2001: 44–51.

15 Collisions between . . . rubble piles: G.S., "All in the Asteroid Family," *Sky & Telescope*, February 2002: 24; R.A.K., "Roughed Up and Far from Home," *Science* 311 (March 31, 2006): 1859.

16 Researchers in 2001 . . . a planet: A. Morbidelli, J.M. Petit, B. Gladman, and J. Chambers, "A Plausible Cause of the Late Heavy Bombardment," *Meteoritics & Planetary Science* 36 (2001): 371–80, 371.

16 Meteorites are . . . presolar nebula: Multiple sources provide overviews of meteorite basics.

17 As meteoriticist . . . baby pictures": Dodd, 34.

17 Finally, the . . . hyphen suggests: O. Richard Norton, *Rocks from Space: Meteorites and Meteorite Hunters* (Missoula, MT: Mountain Press, 1994), 157–73. Hereinafter, Norton, *Rocks*.

17 Authors David Coleman . . . and rubble—chondrites: David L. Coleman and Sarah K. Kennedy, *The Pocket Guide to Asteroids* (Dubuque: Jensan Scientifics, 2000), 11–12.

17 "Melted globules . . . fiery rain": Quoted in McSween, *Stardust*, 52–53.

18 Cosmochemist . . . 72!: Derek Sears, "Chondrules and Chondrites," *Meteorite*, May 2007: 7–11.

18 Within the . . . its problems: Sears, "Chondrules," 9; Kurt Liffman, "Workshop on Chondrites and the Protoplanetary Disk Kaua'i, Hawai'i, 2004," *Meteoritics & Planetary Science* 41, no. 1 (2006): 3–6; Harold Connolly Jr. and S. J. Desch, "On the Origin of the 'Kleine Kügelchen' Called Chondrules," *Chemie der Erde/Geochemistry* 64 (2004): 95–125.

18 Despite Sorby's . . . impact events: Connolly and Desch, 113.

18 A study . . . and melted: Sonnet L'Abbe, "Discovery of 'Young' Material in Meteorites Defies Linear Theory of Solar System's Origin," August 19, 2005. http://www.spacedaily.com/news/meteor-05d.htm.

18 Another research . . . made of: Phonsie J. Hevey and Ian S. Sanders, "A Model for Planetesimal Meltdown by 26Al and its Implications for Meteorite Parent Bodies," *Meteoritics & Planetary Science* 41, no. 1 (2006): 95–106, 104–5.

18 Just how weird . . . solar system: Connolly and Desch, 112, 120.

19 And there's more . . . of glue: Mario Trieloff and Herbert Palme, "The Origin of Solids in the Early Solar System," in *Planet Formation: Theory, Observation, and Experiments*, ed. Hubert Klahr and Wolfgang Brandner (Cambridge, MA: Cambridge University Press, 2006), 64–89.

19 On this world . . . creation's body: J. Kelly Beatty, "Fountains of Chondrules from the Sun's Cloudy Birth," *Sky & Telescope*, October 2001: 18.

20 A scientist . . . dust particles": Quoted in Kenneth F. Weaver, "Meteorites: Invaders from Space," *National Geographic* 170, no. 3 (September 1986): 390–418, 413.

20 Researchers such . . . star systems: "Scientists Get First Close Look at Stardust." http://www.sciencedaily.com/releases/2003/02/030228073248.htm.

20 Scientist Monika Kress . . . the beach": Monika Kress, "Collecting Cosmic Dust," *Mercury*, November–December 2001: 24–30, 25.

20 In the . . . them up: Susan Taylor, James H. Lever, Ralph P. Harvey, and John Govoni, "Collecting Micrometeorites from the South Pole Water Well," *CREEL (Cold Regions Research and Engineering Laboratory) Report 97–1*, May 1997, U.S. Army Corps of Engineers.

20 Kress notes . . . the sea: Kress, 27.

21 Researchers who . . . as alien: Kress, 28, 29.

21 Like some . . . building blocks: Kress, 26.

21 Meteoroids hit . . . times larger: Hutchison and Graham, 8. For entry velocities, Mike D. Reynolds, *Falling Stars: A Guide to Meteors and Meteorites*, (Mechanicsburg, PA: Stackpole, 2001), 2.

21 Even a . . . huge fireball: Coleman and Kennedy, 5–6.

21 For a time . . . a sunspot: Norton, *Rocks*, 14–15; Phillis Engelbert and Diane L. Dupuis, *The Handy Space Answer Book* (Canton, MI: Visible Ink Press, 2005), 144 for sunspot temperature.

21 You can . . . for calcium: Tom Wiest, "Reach for the Stars," *Backpacker*, December 1999, 31–32, 31.

21 The glow . . . it passes: Coleman and Kennedy, 5.

21 Eventually all . . . slower speeds: Norton, *Rocks*, 52.

21 Perhaps counterintuitively . . . surface area: Norton, *Rocks*, 15.

22 About 40,000 . . . each year: Gisela Winckler and Hubertus Fischer, "30,000 Years of Cosmic Dust in Antarctic Ice," *Science* 313 (July 28, 2006): 491.

22 Or it's more . . . each day: Norton, *Rocks*, 45.

22 Robert Dodd . . . feet deep: Dodd, 7.

22 A French meteorite . . . ever recovered: Alain Carion, *Meteorites*, Trans. Anne Black (Paris: Alain Carion, no date), 7–8.

22 Dodd says . . . being found: Dodd, 8, 15.

22 A high number . . . particular spot": Dodd, 8.

22 Meteoriticist Ralph . . . different falls: Ralph Harvey, phone interview, June 29, 2007.

22 David Kring . . . been located: Kring, *Properties*, 20.

24 With ideal . . . 24 hours: Norton, *Rocks*, 15.

24 Our sky . . . my skin": Quoted in Kenneth J. Atchity, *The Classical Greek Reader* (New York and Oxford: Oxford University Press, 1996), 60.

24 Bright meteors . . . electric field": Colin Keay, "Electrophonic Sounds from Large Meteor Fireballs," *Meteorite!*, August 1998, 8–10, 10; Graham W. Wolf, "Electrophonic Fireball Sounds: A History," *Meteorite!*, November 1997, 8–9.

24 As Ralph . . . a hum.": Harvey, June 29, 2007.

24 There is . . . had supposed: Wolf, 8–9; Keay, 8–10.

25 Heated, impacted . . . wedding bands: Coleman and Kennedy, 30.

25 Historian . . . planet Earth": Jeffrey Burton Russell, Foreword to Joseph A. Amato, *Dust: A History of the Small and the Invisible* (Berkeley: University of California Press, 2000), xi.

25 And a few . . . on Earth: Michelle Thaller, "Cosmic Dust Bunnies," *Christian Science Monitor*, April 18, 2003. http://www.csmonitor.com/2003/0418/p25s02-stss.html.

Book I: Distances Measured in Various Units

Chapter I.1. The First Asteroid

29 On January 1, 1801 . . . something better": Quoted in Clifford J. Cunningham, *The First Asteroid: Ceres 1801–2001* (Surfside, FL: Star Lab Press, 2001), 36. For additional details of the discovery, John Wagoner, "Celebrating Ceres at 200," *S&T's News Bulletin for January 5, 2001*, and same-title sidebar, by J. Robinson Leif, *Sky & Telescope*, July 2001: 49.

30 For a time . . . random desire: Cunningham, 49.

30 Later, Ceres . . . asteroid belt: Diameter and relative size listed in multiple sources, including, for example, http://www.seds.org/nineplanets/nineplanets/asteroids.html.

30 Ceres is just . . . ice in its mantle: http://hubblesite.org/newscenter/archive/releases/2005/27/text.

30 In 1995 . . . the feature Piazzi: J. K. B., "The Face of Ceres," *Sky & Telescope,* January 2002, 21.

30 In 2011 . . . asteroid, Vesta: C. M. Pieters et al., "Vesta: Exploration, Predictions, and Surprises," *Abstracts of the 65th Annual Meeting, The Meteoritical Society,* A117; Guy Webster, "JPL Asteroid Mission Gets Thumbs Up from NASA," news release, December 21, 2001, Jet Propulsion Laboratory.

30 After the Ceres . . . been catalogued: Arthur Berry, *A Short History of Astronomy: From Earliest Times Through the Nineteenth Century,* reprint of 1898 ed. (New York: Dover, 1962), 376–77.

30 By 1923 . . . named "Piazzia": Hathaway, 121.

30 Now, using sophisticated . . . such information: Richard P. Binzel, "A New Century for Asteroids," *Sky & Telescope,* July 2001: 44–51; G. B., "An Explosion of Asteroids," *Sky & Telescope,* October 2002, 95.

31 A few . . . of the skies": Quoted in Hathaway, 121. Pages 120–21 include some information about the predictions of a planet between Mars and Jupiter, Franz Xaver von Zach, Piazzi's discovery, the naming of the asteroid, and other early discoveries.

31 Piazzi might . . . first to do so: Cunningham, 66.

32 I pull out . . . on the other: Vagn F. Buchwald, *Handbook of Iron Meteorites: Their History, Distribution, Composition and Structure,* vol. 1–3 (Berkeley: University of California Press, 1975), 1317.

33 A 1922 . . . able to ignore": "Meteorites in Oregon," *The Morning Oregonian,* April 26, 1922: 8.

33 He was . . . something new: Cunningham, 35.

Chapter I.2. Ellis Hughes's 15-Ton Caper

34 Its acre . . . bound for the stars: "Facts About the Architecture of the Rose Center for Earth and Space at the American Museum of Natural History," fact sheet, AMNH Communications Department, February 2000; site visit, New York City, January 2–5, 2001.

37 When struck . . . a meteor": J. Hugh Pruett, "Ellis Hughes: He Won Fame by Losing a Meteor," *The Sunday Oregonian,* October 23, 1938: 6. Pruett says that two University of Oregon students interviewed Hughes in 1938, gleaning the details of the initial discovery, including the conversation between Hughes and Dale.

37 His wife . . . in Oregon: Joanne L. Jardee, headline cut off on provided copy, *Old Stuff,* Fall 1978: 12, 14.

37 They lived . . . by radio: Marilyn Miller and Marian Faux, eds., *The New York Public Library American History Desk Reference* (New York: Macmillan, 1997), 374.

37 The men . . . dead cows: Quotations from and information about the trials comes from the following: *In the Supreme Court of the State of Oregon, March Term 1905, Oregon Iron & Steel Company, Respondent, vs. Ellis Hughes, Appellant, Appellant's Abstract of Record in the Circuit Court of the State of Oregon for Clackamas County,* 33–34. Hereinafter Trial transcripts.

37 According to . . . capital": Douglas J. Preston, *Dinosaurs in the Attic: An Excursion into the American Museum of Natural History* (New York: St. Martin's, 1986), 204–6, 204.

37 Reportedly, the . . . were rebuffed: Trial transcripts, 14.

37 In any case . . . her husband: Preston, 204.

37 "It would probably . . . next day": Quoted in Pruett, 6.

38 So throughout spring . . . mile away: Erwin F. Lange, *The Willamette Meteorite 1902–1962* (West Linn, OR: West Linn Fair Board, 1962), not paginated; Jerry Easterling, "Willamette Meteorite," *Oregon Territory,* November 13, 1977, 4–5.

38 He wove . . . from space: Lange.

38 Clearing brush . . . cone-shaped bottom: Lange.

38 "Excellent luck . . . own hands": Pruett, 6.

38 Using some . . . Egyptian pyramid: Easterling, 4–5; Lange.

38 He thought . . . 5 tons: Trial transcripts, 35.

38 Day after day . . . the meteorite: Trial transcripts, 17.

38 In a 1960s . . . shovels are for": Jerry Tippens, "State Retains Claim to Title of Meteorite Capital of U.S.," *Oregon Journal,* May 7, 1963, 17.

39 "Slowly the wagon . . . and reanchored": Lange.

39 Soon enough . . . from space: Pruett, 6; Mary Goodall, *Oregon's Iron Dream* (Portland: Binfords & Mort, no date), 13.

39 Many of them . . . fallen then: "Is a Real Meteor," *Oregon City Enterprise,* November 6, 1903: front page.

39 Visitors could . . . the city: Pruett, 6; Marlene Wallin, e-mails from the Oregon Historical Society, citing newspaper ads, November 14, 2006, November 21, 2006.

39 According to . . . warp it": Quoted in typescript based on 1938 Pruett article, but "burlap" quote not in the original 1938 Pruett article.

39 A San Francisco . . . closely guarded": "Enormous and Valuable Meteor Is Found by Chance in Oregon," *San Francisco Examiner,* November 5, 1903, no page, "Meteors and Meteorites," vertical file, Oregon Historical Society Research Library.

40 In fact . . . strange monster": A. W. Miller, "Solid Bodies Which Whiz Through Space Are Puzzle to Scientists," *The Sunday Oregonian,* June 2, 1912: 6, reprint of Miller's paper to Oregon Academy of Sciences, Corvallis.

40 "With the official . . . prove interesting": "Is a Real Meteor."

40 The paper reported . . . its discoverer: "Is a Real Meteor."

40 Among the guests . . . offered $100: Pruett, 6; Lange.

40 Then company . . . own property": Beth Ryan, "Date Mystery Still Surrounds West Linn's Famous Meteorite," *The Oregonian,* July 1, 1973: 6.

40 Several thousand . . . now Canada: Richard Pugh and John Eliot Allen, "Origin of the Willamette Meteorite: An Alternate Hypothesis," *Oregon Geology* 48, no. 7 (July 1986): 79–80, 85, 80; John Eliot Allen, Marjorie Burns, and Sam. C. Sargent, *Cataclysms on the Columbia: A Layman's Guide to the Features Produced by the Catastrophic Bretz Floods in the Pacific Northwest* (Portland: Timber Press, 1986), 182.

40 Henry Ward . . . largely inaccessible": Henry A. Ward, "The Willamette Meteorite," *Proceedings of the Rochester Academy of Sciences* 4 (March 24, 1904): 137–48.

41 Rooting around . . . the world: Ward, 139, 137, 139.

41 Of the cavities . . . small bath-tubs!": Ward, 144.

41 Researchers writing . . . of the rock: Pugh and Allen, 79–80; Buchwald, 1318.

41 It is the largest . . . in the world: Pugh and Allen, 79.

41 At least twice . . . cosmic shock": Buchwald, 1319.

41 No other . . . melted twice: Courtenay Thompson, "Tribes Claim Willamette Meteorite," *The Oregonian,* November 17, 1999. http://www.oregonlive.com/news/99/11/st111704.html.

41 The Willamette . . . on the Earth: "The Willamette Meteorite." http://www.amnh.org/rose/meteorite.html.

41 All this makes . . . are irons: Label text, "The Willamette Meteorite," American Museum of Natural History.

42 When cut . . . Widmanstätten: Norton, *Rocks,* 214–33, for discussion of irons.

42 The cooling . . . 200 degrees Fahrenheit: Dodd, 132–133.

42 Deep inside . . . medium octahedrite IIIAB: Pugh and Allen, 79.

42 The relative . . . the interior: Hutchison and Graham, 34–36.

42 On November 27, 1903 . . . with Hughes": Ward, 139.

43 Of course it's impossible . . . April 28, 1904: Oregon Climate Data Service, Oregon State University.

43 The transcripts . . . the compass," he testified: Trial transcripts, 36.

43 When he first . . . his house: Trial transcripts, 15.

43 He also lied . . . Iron Company's": Trial transcripts, 16, 27.

43 After company counsel . . . to take it": Trial transcripts, 19–20.

43 Wood continued . . . Ellis replied: Trial transcripts, 28.

43 When Wood . . . I did": Trial transcripts, 29.

43 After this bizarre . . . moon creature: Trial transcripts, 37–43.

44 After Susap . . . was magic: Trial transcripts, 43–44.

44 The testimony . . . contradictory evidence": Trial transcripts, 9–10, 12.

44 That same day . . . Oregon Iron and Steel: Circuit Court Journal, April 28, 1904.

44 In late November . . . their place: Circuit Court Journal, November 21, 1904.

44 The Latourettes . . . with the meteorite: Goodall, 13.

45 A newspaper article . . . worth $10,000: "Fell from Sky—Meteorite's Possession Is Fought in Court," "Meteors and Meteorites," vertical file, Oregon Historical Society Research Library.

45 Meanwhile, the rock . . . of course: Pruett, 6.

45 At least . . . over the years: Buchwald, 1317.

45 The case . . . no excavation: *In the Supreme Court of the State of Oregon, Oregon Iron & Steel, Respondent, vs. Ellis Hughes, Appellant, Appellant's Brief,* 1–8.

46 Oregon Iron and Steel's angry brief . . . recognized today: *In the Supreme Court of the State of Oregon, Oregon Iron & Steel, Respondent, vs. Ellis Hughes, Appellant, Brief for Respondent,* 1–11

46 So on Monday . . . fantastic things": *Oregon Reports,* vol. 47, Oregon Iron vs. Hughes, 313–322.

46 a hot, dry day in Salem: Oregon Climate Data Service, Oregon State University.

46 For Ellis . . . matter drop: Jardee, 12, 14.

46 Quickly, Oregon . . . a barge: Easterling, 4–5.

46 Edward . . . the barge: Tippens, 17.

Chapter I.3. Tomanowos

47 She was . . . for $16,000: Document noting this from H. C. Bumpus Collection, Box 47, Archives, Department of Library Services—Special Collections, American Museum of Natural History. Hereinafter, Bumpus, AMNH. Details of ceremony in multiple sources.

47 Henry Ward . . . entire specimen: Henry A. Ward, to H. C. Bumpus, February 28, 1904, and February 3, 1905, Bumpus, AMNH.

47 Through a Portland . . . the company: William MacMaster to Cuyler, Morgan & Company, January 20, 1906, Bumpus, AMNH.

47 A month later . . . Mrs. Dodge: Lange.

47 While Oregonians . . . the museum: E. O. Hovey to H. C. Bumpus, February 6, 1906, Bumpus, AMNH. Multiple letters and internal museum documents were generated as this issue reached resolution some three years later.

48 The giant rock . . . to New York: H. C. Bumpus to William MacMaster, February 22, 1906, Bumpus, AMNH.

48 The meteorite . . . the box: No author, typed museum note, March 27, 1906, Bumpus, AMNH.

48 A receipt . . . Meteorite (Boxed)": Bumpus, AMNH.

48 In New York . . . April 14, 1906": Anonymous, copy of agreement with F. Schillinger, handwritten note concerning arrival, April 2, 1906, Bumpus, AMNH.

48 Director H. C. Bumpus . . . be displayed: Bumpus to Mrs. Dodge, April 17, 1906, Bumpus, AMNH.

48 Later that year . . . the firm accepted: W. M. Ladd to H. C. Bumpus, November 1, 1906, Bumpus, AMNH.

48 When the city . . . you need'": Ben Fritchie, e-mail to author, September 9, 2001.

49 A few years . . . quickly rebuffed: Lange.

49 In 1990 . . . Johnny Carson: Sources include Linda McCarthy, "Students Want Willamette Meteorite Returned to Oregon," *The Oregonian,* May 30, 1990: D2; Linda McCarthy, "Students Want Oregon Meteorite Returned," *The Oregonian* March 12, 1991: C1, C7; Help End Willamette Meteorite's Absence Committee newsletter 1, no. 4 (May 3, 1991), "Timeline-HEWMAC."

49 Meanwhile, an attorney . . . presumed ownership: James J. Hennigan to Annie Campbell, Stephanie Corey, and HEWMAC, April 3, 1991.

49 One of the teachers . . . with us": Bill Vanderheide, e-mail to author's research assistant, January 12, 2003.

50 Nine years . . . be returned: http://www.usgennet.org/alhnorus/ahorclak/MeteorHome.html.

50 The group . . . hundreds of photographs: http://www.turtletrack.org/Issues/CO02262000/CO02262000_Meteorite.htm.

50 Heavy Head . . . they freaked": Quoted in John Jurgensen, "Indians Want Meteorite," Associated Press, February 19, 2000. http://abcnews.go.com/sections/us/DailyNews/planetarium000219.html.

50 The tribe's bid . . . Grand Ronde Reservation: "The First People of Clackamas County." http://www.usgennet.org/alhnorus/ahorclak/indians.html.

50 In the 1950s . . . 5-acre cemetery: Brochure, "The Confederated Tribes of Grand Ronde Community of Oregon," no date.

50 After years . . . is now thriving: http://www.turtletrack.org/Issues/CO02262000/CO02262000_Meteorite.htm.

50 According to a National . . . of the rock: Courtenay Thompson, "Tribe's Rock of Ages Will Test Federal Repatriation Laws," *The Oregonian,* November 21, 1991: B1, B3.

51 Use of . . . passed down: Courtenay Thompson, "Tribes Claim Willamette Meteorite," *The Oregonian,* November 17, 1999. http://www.oregonlive.com/news/99/11/st111704.html.

51 An anthropologist . . . the tribes": Deward Walker quoted in Thompson, "Tribe's Rock of Ages," B3.
51 Indeed, the Comanches . . . to be meteorites: Dorcas S. Miller, *Stars of the First People: Native American Star Myths and Constellations* (Boulder, CO: Pruett, 1997), 243.
51 Meteorites have . . . Mississippi, and Arizona: John G. Burke, *Cosmic Debris: Meteorites in History* (Berkeley: University of California Press, 1896), 223–25.
51 For the Pawnee . . . the Creator: Ray A. Williamson, *Living the Sky: The Cosmos of the American Indian* (Norman: University of Oklahoma Press, 1984), 230–31.
51 Sometimes, however . . . meteor storm: Miller, 236.
51 Despite the rich . . . the meteorite: "The Story of a Meteorite," *New York Times,* March 3, 2000, A20.
51 Neil deGrasse Tyson . . . the meteorite: Benjamin Weiser, "Museum Sues in Bid to Keep Meteorite." http://nytimes.com/learning/students/pop/022900sci-planetarium-meteor.html.
51 And that's when . . . once and for all: Quoted in Weiser.
51 Tribal chairwoman . . . a blue pearl": Gary Zarr, interview with author, January 5, 2001.
52 The tribe and museum . . . people together": Quoted in Peter Sleeth, "Tribe Decides Meteorite's Site Right," *The Oregonian,*; June 23, 2000, C1, C8.
52 There were still . . . be here": Chris Mercier, "Like a Movie Script, Tomanowos Has a Great Story," *Smoke Signals,* September 1, 2001: 6.
52 The site . . . poke around: Site visit, Portland area, September 5, 2001.
52 On July 8, 1910 . . . Phebe's suit: Clackamas County Court Records, #10139.
53 A few months . . . from her ex: Clackamas County Court Records, #11820.
53 Three years . . . Hughes died: Mountain View (Oregon City, Oregon) Cemetery records.
53 In a 1938 . . . material possession": Pruett, 6.
53 At least Ellis . . . the 1920s: Herbert L. Hergert, "Early Iron Industry in Oregon," *Reed College Bulletin* 26, no. 2 (1948): 1–40, 37.
53 Standing by . . . meteorite caper: "Ellis Hughes Funeral Set," *The Oregonian,* December 5, 1942: 8.
54 For a time . . . like his dad: Jardee, 12, 14.
54 After leaving . . . need healing": Site visit, Portland area, September 5, 2001.

Book II. What Breaks Out Entire

59 The fireball . . . every direction": S. E. Bemis, Howard Graves, and Henry Barber. Quoted in H. H. Nininger, *Our Stone-Pelted Planet: A Book About Meteors and Meteorites* (Boston: Houghton-Mifflin, 1933), 120–22. This material was quoted in other venues as well, such as the *American Journal of Science and Arts.* Hereinafter, *Stone.*
59 Next came . . . he was standing": Bemis et al., 120–22.
59 No one died . . . *400 pounds:* Bemis et al., 120–22.
60 Even the smaller . . . space rock dissipated: Harvey H. Nininger, *Find a Falling Star* (New York: Ericksson, 1972), 194–95. Hereinafter, *Find.*
60 Estherville was . . . metallic nuggets": *Find,* 194.
60 On their exterior . . . some Esthervillians recorded: Bemis et al., 120–22.
60 "Rough and . . . the stones: J. Lawrence Smith, "Study of the Emmet County Meteorite," *American Journal of Science and Arts,* 3rd series, 19, no. 114 (June 1880): 459–63, 460.
60 After the fall . . . into symbol: "Meteor Marker Dedicated Tues," *The Estherville Enterprise,* October 14, 1929, A1.
60 Estherville was . . . *Our Stone-Pelted Planet:* Nininger, *Stone,* 122.
60 It would be . . . thirty years old: H. H. Nininger, *Out of the Sky: An Introduction to Meteoritics* (New York: Dover, 1959), 200 (Hereinafter, *Out*); Ellis L. Peck, *Space Rocks and Buffalo Grass* (Mt. Clemens, MI: Peach Enterprises, 1979), 14; Irven L. Corder, "A History of Kiowa County," *Kiowa County Signal* 83.47 (March 22, 1961): 2; Kiowa County—History, Clippings, 978.1 -K62 Clipp., Kansas State Historical Society. The article consists of brief oral history of Mary Kimberly Evans Kendall's.
61 One sweltering day . . . one of them'": Irven L. Corder, "A History of Kiowa County," *Kiowa County Signal* 83, no. 50 (April 12, 1961): 2, Kiowa County—History, Clippings, 978.1 -K62 Clipp., Kansas State Historical Society. This article continues the oral history.
61 Frank had to . . . blade resharpened: Corder, April 12, 1961.

61 "Frank, do you . . . a meteorite": *Find*, 81.

62 In another version . . . the rest of my life": *Stone*, 67.

62 The meteorite . . . about hitting it": Corder, April 12, 1961.

62 Liza probably knew . . . his discovery: Various sources, including *Find*, 81, and George F. Kunz, "On the Group of Meteorites Recently Discovered in Brenham Township, Kiowa County, Kan.," *Science* 15, no. 384 (June 13, 1890): 359–62, 360.

62 One of those . . . these stones meteorites: Kunz, 359–62.

62 Naturalist Henry Ward . . . didn't care: Burke, 218.

62 in fact . . . growing seasons: "Historical Items of Interest of Kiowa County and Greensburg," *Greensburg News*, June 21, 1934, photocopy, no page provided, Kiowa County Clips, Kansas State Historical Society.

63 They weighed . . . stable roof: A variety of sources list the uses to which the meteorites were put, including Kunz, 360.

63 Over on Jud . . . pigpen fence: Kunz, 359–62.

63 Sometimes the men . . . sporting contests: *Find*, 81.

63 Eliza couldn't . . . lunar volcanoes: Peck, 16.

63 "Her pile . . . the region": Edwin Way Teale, *Journey into Summer* (New York: Dodd, Mead, 1960), 254.

63 A granddaughter . . . use it": Bulah Kimberly Carson, "Frank Kimberly," in *A History of Kiowa County 1880–1980* (Lubbock, TX: Taylor Publishing for Kiowa County [Kansas] Libraries, 1979, 387–88.

63 The day Cragin . . . suit him: Peck, 12.

64 So from Peck . . . nothing: Peck, 12.

64 Off the train . . . hotel reservation: Peck, 14.

64 From atop . . . and wagon: Peck, 16.

64 "The buggy . . . upon greeting: Peck, 17.

64 "How do . . . Cragin's horse: Peck, 17.

64 Soon enough . . . I knew it!": Peck, 17–18.

65 The well-known . . . a cool $3,800: http://www.meteoriteman.com/sales.htm#esquel.

65 Their appearance . . . olivine zone: Hutchison and Graham, 37–38, and Harry McSween, *Meteorites and Their Parent Bodies*, 2nd ed. (Cambridge, MA: Cambridge University Press, 1999), 219.

66 The metal webbing . . . interiors of irons: Hutchison and Graham, 37–38.

66 Pallasites get . . . came from: Burke, 29–30.

66 The other type . . . crust materials: Edward R. D. Scott, Henning Haack, and Stanley G. Love, "Formation of Mesosiderites by Fragmentation and Reaccretion of a Large Differentiated Asteroid," *Meteoritics and Planetary Science* 36 (2001): 869–81, 870.

66 Within a single . . . later conglomeration: Hutchison and Graham, 37–38.

66 Having matched . . . once volcanic: McSween, *Meteorites and Their Parent Bodies*, 224–25; Hutchison and Graham, 37–38.

66 A 2001 . . . 90 miles' diameter: Scott et al., 869.

66 As for the pallasites . . . iron meteorites: Robert N. Clayton, "Meteorites and their Parent Asteroids," *Science* 313 (September 22, 2006): 1743–44.

67 It would be decades . . . blazing crash: http://www.lpi.usra.edu/features/heggy.

67 It's thought that . . . by dirt: http://brenhammeteoritecompany.com/science.htm.

67 Ellis Peck found . . . specimen he could: Peck, 19.

67 Eliza told . . . "Moon Rock": Peck, 19; F. W. Cragin to Mary Ann Cragin (mother), May 6, 1890, F. W. Cragin Papers, MS 0362, Box 1, Folder 3, Tutt Library Special Collections and Archives, Colorado College.

67 Cragin and the men . . . the Kimberlys celebrated: Peck, 20–21.

68 Despite how . . . masses": E. H. S. Bailey, "On the Minerals Contained in a Kiowa County (Kansas) Meteorite, *Science*, October 10, 1890: 5–6, photocopy from K552.6, Clip V. 1 ns, Kansas State Historical Society.

68 If so . . . scientific facts": Ellis Peck to Harvey Nininger, August 6, 1973, "BRENHAM Find Pallasite," Papers of Harvey Harlow Nininger, Department of Archives and Special Collections, Arizona State University Libraries, Tempe, AZ. Hereinafter, Papers of HHN.

68 It's after I find . . . Mary said: Corder, April 12, 1961.

68 She remembered how . . . another load of water": Corder, April 12, 1961.

69 In a typescript . . . various specimens: Harvey Nininger, "Chasing Meteorites," typescript, "BRENHAM Find Pallasite" file, Papers of HHN.

69 Long after . . . costing $400: F. W. Cragin to F. D. Kimberly, March 13, 1890, F. W. Cragin Papers, MS 0362, Box 1, Folder 4, Tutt Library Special Collections and Archives, Colorado College.

70 He paid Kimberly . . . "a memento": F. W. Cragin, to Mary Ann Cragin (mother), May 6, 1890.

70 Eventually I ask . . . where he worked: William A. Fischer, "Professor Francis W. 'Cragstone' Cragin and the Kiowa County, Kansas Meteorite 1890," unpublished typescript, no page numbers, F. W. Cragin Papers, MS 0362, Box 1, Folder 5, Tutt Library Special Collections and Archives, Colorado College.

70 And a further . . . called it: R. T. Walker, "Prof. Francis W. Cragin: An Appreciation," unpublished typescript, no page numbers, F. W. Cragin Papers, MS 0362, Box 1, Folder 2, Tutt Library Special Collections and Archives, Colorado College.

71 Then there's Harvey . . . daughter's, Mary: *Find,* 81.

71 He said Eliza . . . five years: *Find,* 82; *Out,* 200.

72 The March 19, 1890 . . . he concluded: "Topeka Society of Natural History," *Topeka Daily Capital.* March 19, 1890; 4. "A Rare Lot of Meteorites," *Topaka Daily Capital,* March 19, 1890, 8.

72 That summer . . . the risk: F. W. Cragin to Washburn College Trustees, August 2, 1890, F. W. Cragin Papers, "1890–1891 Natural History Dept.," Box 2.6, Archives, Mabee Library, Washburn University.

73 The geologist . . . after Cragin: F. H. Snow, "Kiowa County, Kan., Meteorites," *Science* 15, no. 379 (May 9, 1890): 290. Snow sets out the general historical points and provides a timeline of visits in this letter.

73 It seems that . . . the discovery: Kunz, 359–62.

73 Hay eventually . . . the fall site: Robert Hay, "The Kiowa Co. (Kansas) Meteorites," *The American Journal of Science,* 3rd Series, 43, no. 253 (January 1892): 80.

73 The third man . . . March 29: Snow, 290.

73 Snow obtained . . . a mile long: Snow, 290.

73 He also reported . . . comfortable circumstances": Snow, 290.

73 Indeed, Eliza Kimberly . . . Kiowa County": Teale, 254.

73 Though Snow merely . . . anxious search": Snow, 290.

73 According to Harvey Nininger . . . to Snow: *Find,* 83.

73 According to Teale . . . for $500: Teale, 254.

73 *A Kansas City Star* . . . Prof. Snow": "A Heavenly Kansas Crop," *Kansas City Star,* April 9, 1905, photocopy, no page provided, clippings files, Papers of F. H. Snow, University of Kansas Library.

74 This collecting . . . coal cinder: Peck, 63.

74 As the money . . . he said: *Find,* 82.

74 The frenzy . . . wouldn't sell: "A Heavenly Kansas Crop."

74 The one original . . . the pickle jar): F. H. Snow, Field notebook, 1890, Special Collections, University of Kansas.

74 Then again . . . far sighted": Teale, 256.

75 I reach Kiowa County . . . in meteorites: Site visit, Kiowa County and Greensburg, Kansas, March 29, 2002.

76 Teale watched . . . diffident rabbit: Teale, 261–63.

77 Eliza and Frank . . . early 1930s: Peck, 75, 74.

77 Others walked . . . an immediate sale: http://kansascity.com/mld/kansascity/news/local/14358544.htm.
http://www.cnn.com/2006/TECH/space/10/16/meteorite.kansas.ap/index.htm.
http://kansas.com/mld/kansas/13139061.htm.
Kevin Murphy, "From Cosmos to Kansas," *Kansas City Star,* December 12, 2005: A1, A6.
http://www.topix.com/content/kri/2007/10/kansas-meteorite-goes-unsold-at-auction.

78 Months after . . . the skin: Bill and Ellie Hansen, e-mail to author, January 23, 2002.

78 And then I remember . . . jeweler's tools: *Find,* 83–84.

78 Most of what I've learned . . . not very successfully: Dorothy Price Shaw, "The Cragin Collection." *The Colorado Magazine* 25, no. 4 (July 1948): 166–78. All Cragin details, unless otherwise indicated, come hereafter from Shaw, including Cragin's daughter's quote.

79 Another author . . . historical details": Terry R. Koenig, "F. W. Cragin and His Famous Collection," *Wagon Tracks: Santa Fe Trail Association Quarterly* 6, no. 1 (November 1991): 11–12. 11.

Book III. Higher Latitudes

Chapter III.1. Please Bring Your Wu Wei to the Upright and Locked Position

83 "Yet as you get . . . already arrived": Alan Watts, *What Is Tao?* (Novato, CA: New World Library, 2000), 87–88.

84 "splendid distances": Robert E. Peary, "The Value of Arctic Exploration," *The National Geographic Magazine* 14, no. 12 (December 1903): 429–36, 429.

84 "Be rather the Mungo Park . . . by the ice": Henry David Thoreau, *Walden* (New York: Signet, 1999), 254.

87 So it was . . . speeding trucks: Site visit, Cresson, Pennsylvania, August 2, 2001.

89 Greenland? . . . on the planet: "Greenland—A Crash Course," Arctic Adventure fact sheet.

89 Living on the fringes . . . the Danish government: Deanna Swaney, *Iceland, Greenland and the Faroe Islands*, 3rd ed. (Hawthorn, Victoria: Lonely Planet, 1997), 356–58.

89 Somehow, though . . . marine mammals: Sam Hall, *The Fourth World: The Heritage of the Arctic and Its Destruction* (New York: Vintage, 1988), 132, 135.

89 Today pollution . . . 56,000 inhabitants: "Greenland—A Crash Course," Arctic Adventure fact sheet.

89 Greenland is a nation . . . to survival: Swaney, 381.

89 On May 6, 1856 . . . business people: I have relied on four books for information regarding Robert Peary's life: Robert M. Bryce, *Cook & Peary: The Polar Controversy Resolved* (Mechanicsburg: Stackpole, 1997), Kenn Harper, *Give Me My Father's Body: The Life of Minik, the New York Eskimo* (New York: Washington Square Press, 2000), Wally Herbert, *The Noose of Laurels: Robert E. Peary and the Race to the North Pole* (New York: Atheneum, 1989), and John Edward Weems, *Peary: The Explorer and the Man* (Boston: Houghton Mifflin, 1967).

89 According to biographer . . . the other side": Weems, 4–6.

90 They next moved . . . Elisha Kent Kane: Weems, 6–11.

90 In college, Peary . . . joined the Navy: Weems, 23, 37, 21, 25, 33–34, 36–67.

90 He also thought . . . "contented": Weems, 71.

90 In 1886, Peary . . . thirty years old: Weems, 76–83.

90 Peary thought . . . before the world": Quoted in Weems, 91.

90 Into this grand . . . the honeymoon: Weems, 94.

90 On his next Greenland . . . an African-American: Weems, 106–27.

90 While his men . . . swapped wives: Harper, 18.

90 Blue-eyed, tall . . . devoid of emotion": Harper, 2.

90 In the spring of 1892 . . . raised more money: Weems, 128.

91 But in 1894 . . . frostbite and snow blindness: Weems, 138–42.

91 Back in camp . . . "seal and walrus": Bryce, 133.

91 So when Peary's . . . while he stayed on: Bryce, 133.

91 Meanwhile, Jo . . . for good: Bryce, 135–36.

Chapter III.2. Thule and the *Barb*

93 Thule . . . rusted mica: Site visits, Dallas, August 14, 2002, and Thule Air Force Base, Greenland Coast, meteorite sites, August 15–23, 2002.

93 Built after the Second World War . . . 12th Space Warning Squadron: http://www.peregrinefund.org/archived_conserve. "Arctic Program Greenland Conservation Projects-Archive."

94 Across the bay . . . around it: N. O. Christensen and Hans Ebbesen, *Thule: In Days of Old* (Charlottenlund: Arktisk Institut, 1985), 6.

98 In the summer . . . grave I'd visited: Rolf Gilberg, "Polar Eskimo," *Arctic*, series ed. David Damas, *Handbook of North American Indians*, vol. 5 (Washington, DC: Smithsonian Institution Press, 1984), 577–94, 577. Multiple sources outline the story of Ross and his discovery of iron among the Inuit.

98 On May 16, 1894 . . . but three irons: Robert Peary, *Northward Over the "Great Ice,"* vol. 2 (New York: Frederick A. Stokes, 1908). Unless otherwise indicated, all quotes from Peary regarding discovery and retrieval of the meteorites are from this work. Discovery narrative, 127–151, 155, regarding departure of the *Falcon* without Lee, Henson, or Peary; retrieval narrative, 553–618.

99 Conditions were so . . . lead Peary again: Weems, 142–68, esp. 143–44.

99 It was called Woman . . . god Tornarsuk: Peary, 559.

100 While the tale . . . into a shaman: John MacDonald, *The Arctic Sky: Inuit Astronomy, Star Lore, and Legend* (Toronto and Iqaluit: Royal Ontario Museum and Nunavut Research Institute, 1998), 142–44.

100 Peary paused . . . home camp: Weems, 142–68, esp. 145.

Chapter III.3. The Isthmus and Meteorite Island

107 Now, at the northern . . . surpassingly fine": Josephine Diebitsch-Peary, *My Arctic Journal: A Year Among Ice-fields and Eskimos* (New York: Contemporary, 1897), 29.

109 Researchers Vagn Buchwald . . . about the time of Christ: Vagn Fabritius Buchwald and Gert Mosdal, *Meteoritic Iron, Telluric Iron and Wrought Iron in Greenland* (Copenhagen: Meddelelser om Gronland, *Man & Society* 9, 1985), 3–49, 3.

109 Thus the Inuit . . . on the coast: Richard Vaughan, *Northwest Greenland: A History* (Orono: University of Maine Press, 1991), 108.

109 The Cape York fall . . . half-mile long: Henning Haack, e-mail to author, September 12, 2007.

110 It's by studying . . . been on Earth: Various sources discuss the "ages" of meteorites, including F. Heide and F. Wlotzka, *Meteorites: Messengers from Space* (Berlin: Springer-Verlag, 1995); Hutchison and Graham, 40–44.

110 Exposure to conditions . . . sources of ordinary chondrites: "Space Weathering" Cements Asteroid-Meteorite Link," *Sky & Telescope*, December 2000, 24–25.

110 As to terrestrial exposure . . . thousands of years: Hutchison and Graham, 40–44.

111 The P-Funders soon find . . . the meteorites: Vaughan, 104.

112 I'm alone . . . even Peary himself: Hall, 233.

113 According to Kenn Harper . . . I have failed": Quoted in Harper, 13.

114 In the archives . . . Meteorite Island in 1896: William Smith, *Log Book, The S.S. Hope of Greenock, From St. John's and bound for Greenland Via Sydney*, 1896, Special Collections, Archives, Department of Library Services, American Museum of Natural History. Quotes from Smith in text from this log.

114 During the next days . . . Voker, Haul!": Quoted in Vaughan, 103.

117 But I remember . . . belong to Greenland": Holger Pedersen, e-mail to author, July 17, 2002.

121 Everywhere grow . . . names of grasses: E. C. Pielou, *A Naturalist's Guide to the Arctic* (Chicago: University of Chicago Press, 1994), 179.

121 "In the darkness . . . that split the sky . . . : Eleanor Lerman, *The Mystery of Meteors* (Louisville: Sarabande Books, 2001), 3–4, 3.

122 The psychiatrist . . . to be unfulfilled: Melvin Konner, *Why the Reckless Survive and Other Secrets of Human Nature* (New York: Viking, 1990), 15.

122 For Konner . . . fully every day": Konner, 139.

123 The compass . . . September 30, 1897: Harper, 21.

123 Peary had telegraphed . . . name of science: Harper, 21–22.

123 For years . . . *Scientific American:* Walter L. Beasley, "A New Home for the Peary Meteorite," *Scientific American*, December 24, 1904: 461.

123 Throngs watched . . . of the museum: Photographs, AMNH archives; Preston, 40–41; museum records regarding arrival time and date.

123 In a deceptive . . . buy the stones: Josephine Peary to Henry Osborn, president, AMNH, March 15, 1908, from the Archives, Dept. of Library Services, Special Collections, American Museum of Natural History, Box 124.

123 Meanwhile, Peary . . . effort for the Pole: Harper, 12.

123 Eventually, Mrs. Peary . . . gave them to the museum: Vaughan, 106.

124 The sale price was . . . "long-time fraud": Harper, 70–72.

124 As to the feat . . . single-mindedness and resolve": Vaughan, 105.

124 Other meteorites . . . of Nuuk: Vaughan, 108.

124 Peary earned . . . wasted": Herbert, 106–8.

124 Further . . . and her tent": Harper, 12.

124 As to supplies . . . claimed the Pole: Harper, 168.

125 But the most . . . in their tupiks: Harper, 25–42, 85–145.

125 From time to time . . . unhappiness just as complete": Hans Lynge, "Ersinngitsup Piumasaa, The Will of the Invisible," excerpt, in *From the Writings of the Greenlanders Kalaallit atuakkiaannit*, comp., trans., and ed. Michael Fortescue (Anchorage: University of Alaska Press, 1990), 144.

BOOK IV. THE WEATHER OF BELIEF

131 Beneath a sky that once rained bacon . . . over there: Ursula B Marvin, "Meteorites, the Moon and the History of Geology," *Journal of Geological Education* 34, no. 3 (May 1986): 140–65. 140, for list of Pliny's dropping sky bits; Burke, 16, 218; site visit, Ensisheim, December 17–19, 2002.

131 At Ensisheim . . . New World: Ursula B. Marvin, "The Meteorite of Ensisheim: 1492 to 1992," *Meteoritics* 27, no. 1 (March 1992): 28–70, 51.

133 He was, it seems . . . the fall: M. De Dree, "On the Stones Said to Have Fallen at Ensisheim, in the Neighbourhood of Agen, and at Other Places," *The Philosophical Magazine* 16, no. 64 (September 1803): 289–305, 290.

133 He was not . . . sheep: Marvin, "Meteorite of Ensisheim," 62.

133 Sometime between . . . with sound: *Story of a Meteorite* (Ensisheim: Saint George's Brotherhood of the Meteorites' Guardian and the History Society, no date), booklet, 5; De Dree, 291.

133 The aerial explosion . . . fire burbles: Marvin, "Meteorite of Ensisheim," 39.

133 Along the flight . . . prayed for safety: Marvin, "Meteorite of Ensisheim," 52–53, 63.

133 Slowed by air . . . 200 miles an hour: Dodd, 4.

133 A 1513 illustration . . . to paradise: Marvin, "Meteorite of Ensisheim," 48.

134 Another drawing . . . the prospect: Ingrid D. Rowland, "A Contemporary Account of the Ensisheim Meteorite, 1492," *Meteoritics* 25 (1990): 19–22, 19, 21.

134 There had . . . to find it": *Story of a Meteorite*, 5.

134 That they did . . . closer to God: Alexandra M. Witze, "The Great Stone of Ensisheim Turns 500," *Sky & Telescope* (November 1992): 502–3.

134 When the German . . . of people: *Story of a Meteorite*, 7; Marvin, "Meteorite of Ensisheim," 37.

134 Brant wrote . . . for the French: Sebastian Brant, "Of the Aerolite Fallen in Front of Ensisheim in 1492," quoted in *Story of a Meteorite*, 6; Marvin, "Meteorite of Ensisheim," 35.

134 A cardinal . . . venereal disease: Marvin, "Meteorite of Ensisheim," 50.

134 Three weeks . . . of the rock: Marvin, "Meteorite of Ensisheim," 29.

135 This rock . . . . from destruction": Marvin, "Meteorite of Ensisheim," 46.

135 The meteorite stayed . . . remaining fragment: *Story of a Meteorite*, 7, 9.

136 In ancient European . . . partly understood: Dodd, 51; Irene Seco Serra, "Gods Who Fell from the Sky," *Meteorite!* 6, vol. 2 (May 2000): 25–28; Philip M. Bagnall, *The Meteorite and Tektite Collectors Handbook: A Practical Guide to Their Acquisition, Preservation and Display* (Richmond, VA: Willmann-Bell, 1991), 2; Arthur Bernard Cook, *Zeus: A Study in Ancient Religions* (London: Cambridge University Press, 1940), 939–42; Burke, 221.

136 Diogenes of . . . for him: Bagnall, 1.

137 Egyptian hieroglyphs . . . in pyramids: Bagnall, 1–2.

137 In Mecca . . . it's terrestrial: William K. Hartmann, "Sociometeoritics," *Meteoritics and Planetary Science* 36, no. 10 (October 2001): 1294–95, 1294.

137 The Hittites . . . the earth: Judith Kingston Bjorkman, *Meteors and Meteorites in the Ancient Near East*, Publication No. 12 (Tempe, AZ: Center for Meteorite Studies, 1973), 110.

137 In Asia . . . meteorite falls: Bagnall, 1.

137 The Japanese goddess . . . Philip Bagnall: Bagnall, 2.

137 The world's oldest . . . at a temple: Marvin, "Meteorites, the Moon," 140.

137 Different cultures . . . wish upon a falling star": Burke, 215.

137 In the Middle Ages . . . good eyes: Maria Leach, ed., *Funk & Wagnalls Standard Dictionary of Folklore, Mythology and Legend* (New York: Funk & Wagnalls, 1950), 1009.

137 A Swiss . . . illness: Burke, 215.

137 The Baronga . . . by yourself": Quoted in Burke, 217.

137 In Siberia . . . worms: Tamra Andrews, *Legends of Earth, Sea and Sky: An Encyclopedia of Nature Myths* (Santa Barbara, CA: ABC-CLIO, 1998), 143.

137 The Dobu . . . sneezes blood": Carlos Trenary, "Universal Meteor Metaphors and Their Occurrence in Mesoamerican Astronomy," *Archaeoastronomy* 10 (1987–88): 98–116, 102.

137 And meteor storms . . . the world: Trenary, 109.

137 Nigerian natives . . . British troops: Burke, 226.

137 Various tribal . . . speaking of them: Burke, 219.

137 Still, from . . . beads to blades: Lincoln LaPaz, *Topics in Meteoritics: Hunting Meteorites: Their Recovery, Use, and Abuse from Paleolithic to Present* (Albuquerque: University of New Mexico Press, 1969), 79–80.
137 I think . . . or bats: Trenary, 104, 112; Burke, 217.
137 Just as the Inuit . . . other societies: Trenary, 99.
138 In one Chinese . . . have gas: Trenary, 102.
138 A Baja California . . . Carlos Trenary: Trenary, 103.
138 Trenary says . . . exhalations and semen": Trenary, 103.
138 The perceived . . . Trenary writes: Trenary, 103.
138 The Tucano . . . of the sun": Trenary, 102.
138 Historian John Burke . . . passionate cries: Burke, 225.
138 In one version . . . a scholar: Bjorkman, 117.
138 If the sky . . . we despise: Burke, 216.
138 Fate breaks . . . a meteor: Burke, 216.
138 The Hindus . . . near at hand: *Funk & Wagnalls Folklore,* 1009.
138 Among some Mayans . . . bad, really: A. F. Aveni, ed., *World Archaeoastronomy* (Cambridge: Cambridge University Press, 1989), 289–99.
138 The Irish believed . . . Saint Lawrence: Burke, 217.
138 Then again . . . life and death: Bjorkman, 104–5.
139 For all the wonderful . . . many centuries: Burke, 6–8.
139 William Fulke . . . instead of meadows?: Theodore Hornberger, ed., *A Goodly Gallerye: William Fulke's Book of Meteors (1563)* (Philadelphia: American Philosophical Society, 1979).
139 Ernst Chladni . . . solar system: Burke, 41–42.
139 Chladni's euphonium . . . producing sound: http://totse.com/en/fringe/fringe_science/chladni1.html.
140 And the once-prevalent . . . great distances: Burke, 13–16.
140 So it was . . . musical planets): Hathaway, 155.
140 Inspired by his own . . . beyond the atmosphere: Ursula B. Marvin, "Ernst Florens Friedrich Chladni (1756–1827) and the Origins of Modern Meteorite Research," *Meteoritics & Planetary Science* 31 (1996): 545–88, 547; Burke, 42.
140 Chladni went . . . lawyer's mind: Marvin, "Chladni," 547.
140 A handful . . . from the sky: Burke, 24, 30.
140 Part of the difficulty . . . of stones?: Burke, 23–24.
141 Two years after . . . evidence of heat: Burke, 42–45; Marvin, "Chladni," 549.
141 After all, the complexity . . . ancient times: Hathaway, 14.
141 Emboldened . . . do this, however.): Marvin, "Chladni," 547.
141 German reviews were disapproving: Marvin, "Chladni," 558.
141 Even Lichtenberg . . . publication: Ron Cowen, "After the Fall," *Science News* 148, no. 16 (October 14, 1995): 248–49.
141 Critics resorted . . . of lightning: Marvin, "Chladni." 569–570.
142 But the critics . . . India, 1798: Marvin, "Meteorites, the Moon," 142.
142 Chemists began . . . of the Earth: Marvin, "Chladni," 558; Cowen, 248.
142 In England . . . examining them: Marvin, "Chladni," 561–62; Dodd, 40–41.
142 Howard, with the French . . . their findings: Marvin, "Chladni," 567–68; D. W. Sears, "Sketches in the History of Meteorites I: The Birth of the Science," *Meteoritics* 10, no. 3 (September 30, 1975): 215–25, 219, 223.
142 One commentator . . . August 1802: Quoted in Burke, 53.
143 *The owner was . . . at Fontenil:* J. B. Biot, *Relation d'un voyage fait dans le département de l'orne, pour constater la réalité d'un météore observé à l'Aigle le 26 floréal an 11,* Trans. Kathe Lison, Paris: Baudouin, Imprimeur de l'Institut National, 1803), 24–25; site visit, L'Aigle, December 16–17, 2002.
143 On the day . . . of meteorites: C. Biot, "Account of a Fire-ball Which Fell in the Neighbourhood of Laigle: In a Letter to the French Minister of the Interior," *The Philosophical Magazine* 16, no. 63 (1803): 224–28, 225. All subsequent Biot quotes from this article.
143 Those who saw . . . been fiction: Biot, 226.
143 Indeed, the falling . . . branches: Biot, 226.
143 The fireball . . . eldritch balloon: Biot, 227.

143 Though Biot . . . near Paris: Biot, 225.
143 The largest specimen . . . a meteorite shower: Biot, 228.
148 In a stunning . . . moral world": Quoted in Marvin, "Chladni," 566.
148 For him . . . random universe: Marvin, "Chladni," 572.
148 The same year . . . or space: Sears, "Sketches," 216.
148 By 1810 . . . astronomical textbooks": Sears, "Sketches," 215.
148 Only in 1812 . . . appear: Marvin, "Chladni," 577.
148 Advances in microscope . . . structures: Brian Mason, *Meteorites* (New York: John Wiley & Sons, 1962), 7.
148 Many puzzles . . . the object itself: Marvin, "Chladni," 550.
148 In America . . . Ensisheim: Marvin, "Meteorites, the Moon," 146.
149 Eventually . . . Brenham pallasite): Peck, 76.
149 The lunar-volcano . . . bits to Earth: Marvin, "Chladni," 579.
149 The debate . . . the solar system: Marvin, "Chladni," 545.
149 But not until 1964 . . . by only one year: Tom Gehrels, "History of Asteroid Research and Spacewatch." http://spacewatch.lpl.arizona.edu/history_text.html.
149 In Paris . . . its metal flash: Site visit, Paris, December 23, 2002.
149 Jean-Baptiste Biot . . . by man": Biot, 228.

Book V. Mr. Barringer's Big Idea

153 Sometimes coming events cast their shadows before: Harold J. Abrahams, ed., *Heroic Efforts at Meteor Crater, Arizona: Selected Correspondence Between Daniel Moreau Barringer and Elihu Thomson* (Rutherford, NJ: Fairleigh Dickinson University Press, 1983), 211.
154 "The view . . . never to be forgotten": George P. Merrill, "The Meteor Crater of Canyon Diablo, Arizona; Its History, Origin, and Associated Meteoric Irons," *Smithsonian Miscellaneous Collections* 50 (1908): 461–98, 464.
154 Those who knew . . . been collected there: Janet Gillette, e-mail to author, June 6, 2003.
154 The U.S. Geological Survey's . . . at 4:30 P.M.: G. K. Gilbert, Notebook 51, 1891, 2. Accession No. 3448, Stack 150/20/35/2, Box 34, Department of the Interior, United States Geological Survey, Geologic Records. Photocopy of notebook provided by the National Archives and Records Administration.
154 Half an inch . . . have been bracing: "Meteorological Journal," U.S. Naval Observatory, Washington, DC, October 21–22, 1891.
154 At forty-eight . . . top geologist: Kathleen Mark, *Meteorite Craters* (Tucson: University of Arizona Press, 1995), 27, 30.
154 Historian William Graves Hoyt . . . yet produced": William Graves Hoyt, *Coon Mountain Controversies: Meteor Crater and the Development of Impact Theory* (Tucson: University of Arizona Press, 1987), 37.
154 A student of . . . Western surveys: Mark, 27; Hoyt, 36.
154 Gilbert wrote . . . where I live: Hoyt, 37.
155 It would take . . . to Flagstaff: Gilbert, Monday, October 26, 1891, Notebook 51, 2.
155 Just two months earlier . . . formed the hole: Hoyt, 31–32.
155 Desk-bound . . . wasn't so sure: Hoyt, 40.
155 On Sunday . . . (=Coon Butte)": Gilbert, November 1, 1891, Notebook 51, 9.
155 The weather was dry and pleasant: "Voluntary Observer's Meteorological Record," Department of Agriculture, Weather Bureau, Holbrook, Arizona, November 1891.
155 As they set to . . . cosmic or subterranean: Hoyt, 41.
155 Whatever had formed . . . substance beneath": Gilbert, November 4, 1891, Notebook 51, 13–14.
155 The U.S. Geological Survey . . . rim and pit: "Voluntary Observer's Meteorological Record," Department of Agriculture, Weather Bureau, Holbrook, Arizona, November 1891.
155 Gilbert wrote . . . one should too: Gilbert, November 12, 1891, Notebook 51, 36–37.
156 Yet they could find . . . bottom of the hole": Gilbert, November 14, 1891, Notebook 51, 53.
156 Working through math . . . through the crust": Gilbert, November 8, 1891, Notebook 51, 24.
156 Curiously, on Thursday . . . nor its presence": Gilbert, November 12, 1891, Notebook 51, 35.
156 Gilbert also calculated . . . exist in the rim": Mark, 27.

156 After noting . . . of weathering: Gilbert, Wednesday, November 11, 1891, Notebook 51, 34.
156 "It follows . . . of the crater": Gilbert, November 14, 1891. Notebook 51. 53.
156 His man . . . pocket of water: Gilbert, November 14, 1891, Notebook 51, 54.
157 The Smithsonian's head curator . . . a steam explosion": Mark, 30.
157 In an ironic twist . . . tops of mountains: Marvin, "Meteorites, the Moon," 151–52, 154; Mark 28–29; Hoyt, 54–72.
157 Like a few . . . lunar craters: Marvin, "Meteorites, the Moon," 153–55.
157 He posited . . . that formation: Mark, 28.
157 Gilbert was . . . to find out: Mark, 28.
158 Objects arriving . . . a low-angle impact: Hoyt, 60–61.
158 Later scientists . . . proven wrong: Mark, 28–29.
160 How much . . . illegal logging: Brandon Barringer to H. H. Nininger, April 10, 1956, Papers of HHN.
160 According to a family . . . San Xavier Hotel: *Heroic Efforts,* 38; Nancy Southgate and Felicity Barringer, *A Grand Obsession: Daniel Moreau Barringer and His Crater* (Flagstaff: Barringer Crater Company, 2002), 3.
161 For a time . . . Army scout: Brandon Barringer, "Daniel Moreau Barringer (1860–1929) and His Crater (The Beginning of the Crater Branch of Meteoritics," *Meteoritics* 2, no. 3 (December 1964): 183–99, 190. Hereinafter, Brandon.
161 Then, in 1886 . . . interest Gilbert: Hoyt, 32–33.
161 As to Volz . . . the world: Hoyt, 34.
161 Barringer seems . . . in my mind": *Heroic Efforts,* 38.
161 Born in 1860 . . . in Philadelphia: Southgate and Barringer, 7.
162 Barringer was kicked . . . Harvard and Virginia: Southgate and Barringer, 8–9, 18.
162 His affinity . . . from mines: Southgate and Barringer, 18.
162 This cigar-smoking . . . sight unseen: Southgate and Barringer, 19, 35–37; Hoyt, 74; Brandon, 184–86.
162 With partner . . . at the crater: Brandon, 186.
162 Everyone expected . . . confidence!)": Brandon, 186.
162 Drills were sunk . . . the shafts: Brandon, 186–187; Hoyt, 78–79, on workers' attitudes.
162 Drill holes . . . three times that: Southgate and Barringer, 21.
162 But they revealed . . . rock, and water: See, for a summary, Dean Smith, *The Meteor Crater Story* (Meteor Crater Enterprises, 1996), 22.
162 The meteorites weren't . . . to Coon Butte: Brandon, 186–87.
163 Meanwhile, in 1905 . . . $24,000 total: Brandon, 187.
163 By 1908 . . . bleeding money: Southgate and Barringer, 22.
163 Barringer began . . . company operations: Hoyt, 110.
163 Preparations . . . got under way: Hoyt, 86.
163 To compound . . . not yet want: Hoyt, 87.
163 Benjamin Tilghman and . . . attacked Gilbert: D. M. Barringer, "Meteor Crater in Northern Central Arizona," read before the National Academy of Sciences, November 16, 1909, 1–24, exclusive of plates.
163 ("Science and . . . Wolfgang Elston): Wolfgang E. Elston, "How Did Impact Processes on Earth and Moon Become Respectable in Geological Thought?," *Earth Sciences History* 9, no. 1 (1990): 82–87, 85.
164 The inventor . . . negative reaction: Abrahams, introduction to *Heroic Efforts,* no page listed; Southgate and Barringer, 30.
164 In their various . . . reversed strata: Barringer, 1909, 5.
164 Ejecta could . . . massive boulders: Barringer, 1909, 5.
164 They pointed to . . . the crater rim: Barringer, 1909, 3; Mark, 32.
164 The fact that . . . had occurred there: Barringer, 1909, 2; Mark, 33–34.
164 Further, rocks . . . microscopically: Barringer, 1909, 6–7; Mark, 32.
164 Exposure to . . . a single hollow: Mark, 32.
164 Gilbert had . . . any careful geologist.'": Quoted in Hoyt, 91.
164 The rusty shale . . . single body: Barringer, 1909, 9–12.
164 If the meteorite . . . unmoved magnet: Mark, 32; Hoyt, 96.
164 Gilbert had also . . . the rim: Mark, 32.
164 He had failed . . . crater floor: Brandon, 190.

165 And Tilghman. . . slowed it down: Brandon, 192.
165 Lacking a mathematical . . . such estimates: Barringer, 1909, 4.
165 The Smithsonian's George . . . fullest investigation": Merrill, 462.
165 That one word . . . were right: Mark, 34, 36.
165 Among other points . . . have produced: Mark, 35.
166 He did say . . . of slags": Merrill, 494.
166 Barringer despised this idea: Barringer, 1909, 15.
166 Products that . . . utter absurdity": *Heroic Efforts,* 18.
166 For example, after . . . the allegations: Hoyt, 147–50.
166 And he and . . . massive meteorite: Hoyt, 132–33.
166 Unbeknownst . . . over Siberia: Gabi Mocatta and Dmitry M. Yurkovsky, "The Tunguska Phenomenon: 90 Years of Investigations," *Meteorite!* 4, no. 4 (November 1998): 24–25, 24.
166 Throughout the region . . . rubber bands: Roy A. Gallant, *Meteorite Hunter: The Search for Siberian Meteorite Craters* (New York: McGraw-Hill, 2002), 3, 5.
166 The explosion . . . Hiroshima bomb: Mocatta and Yurkovsky, 24.
166 A slight change . . . Barringer's ideas: Gallant, 53.
167 No one . . . northern lights: Gallant, 17.
167 In 1908 . . . Meteor Crater: Southgate and Barringer, 22.
167 Within a few . . . $100 million: Southgate and Barringer, 31.
167 Later . . . $700 million: Hoyt, 163.
167 Benjamin Tilghman . . . could be found: Brandon, 187.
167 Barringer and Tilghman . . . months later: Southgate and Barringer, 33–34; Hoyt, 103–4.
167 In order to . . . living quarters: Southgate and Barringer, 37.
167 The crater . . . not him alone: *Heroic Efforts,* 48.
167 Elihu Thomson . . . each opportunity": *Heroic Efforts,* 24.
167 Thomson said . . . other evening": *Heroic Efforts,* 87.
167 After canvassing . . . back the drilling: Brandon, 187–88.
167 Something else . . . impact arguments: Brandon, 192; Hoyt, 39, 101.
168 A drilling supervisor . . . the meteor": C. W. Plumb, "Extracts from the Weekly Reports by C. W. Plumb on the Progress of Drill—Hole Number One, at Meteor Crater," typescript manuscript, 1, hand-dated November 29, 1926 (March 26, 1922, entry), 1217 Barringer. Copied from the Archives, Dept. of Library Services, Special Collections, American Museum of Natural History.
168 U.S. Smelting . . . $200,000: *Heroic Efforts,* 272; Brandon, 188–89.
168 A mining . . . stepped in: Brandon, 189.
168 Then, after . . . he thought so: Hoyt, 220.
168 In a 1924 paper . . . round hole: Hoyt, 185.
168 Barringer hated . . . meteorite swarm: Hoyt, 240–41.
168 Colvocoresses decided . . . large meteorite: Mark, 37.
168 Colvocoresses wrote . . . if not more: George M. Colvocoresses, "Report on Meteor Crater as a Mining Prospect," concluding summary attached to main report, typescript manuscript, Section 4A, 3, November 1926, 1217 Barringer, copies from the Archives, Department of Library Services, Special Collections, American Museum of Natural History.
168 In one investment . . . in the world": Daniel Moreau Barringer to Henry Fairfield Osborn, November 23, 1926, 1217 Barringer B, copied from the Archives, Department of Library Services, Special Collections, American Museum of Natural History.
169 Osborn declined . . . writing him: Osborn to Barringer, November 26, 1926, and Barringer to Osborn, November 29, 1926, 1217 Barringer B, copied from the Archives, Department of Library Services, Special Collections, American Museum of Natural History.
169 Despite the poor . . . funds: *Heroic Efforts,* 274.
169 In 1928, drilling began anew: Southgate and Barringer, 43.
169 Barringer had . . . Barringer's name: Brandon, 194–s95; Hoyt, 250–57.
169 As new shafts . . . could ever be found: Brandon, 190; Hoyt, 257–59, 260–63.
169 In 1929 . . . friendly assessment: Hoyt, 259–60.

169 He didn't get . . . left to mine: *Heroic Efforts,* 306, 308–9; Hoyt, 264–71.
169 William Graves Hoyt . . . drilling at once: Hoyt, 273.
170 A few thought . . . preventing vaporization: *Heroic Efforts,* 188, 200–201, 221, 236, 245; Hoyt, 279.
170 But even Thomson . . . Moulton's numbers: Hoyt, 294.
170 Moulton had noted . . . a crater: Elston, 85.
170 In his second . . . 1,300 calculations: Elston, 84; Hoyt, 291.
170 He even offered . . . was mistaken: Mark, 39.
170 It wasn't . . . Colvocoresses's estimates: Hoyt, 292.
170 (Scientists now . . . with the Earth.): Norton, *Rocks,* 53.
170 So the meteorite . . . small pieces: Elston, 83.
170 Despairing, the frantic . . . a project backer: Hoyt, 287–89, 294–95.
170 In the autumn . . . $5 million: Southgate and Barringer, 50–51.
171 To Barringer's chagrin . . . after all: Hoyt, 275.
171 For a time . . . "Star dust": Elston, 84.
171 Barringer was . . . age sixty-nine: Hoyt, 296.
171 In the family . . . ambitions": Quoted in Southgate and Barringer, 76.
171 In the years . . . in the area: David Kring, "Calamity at Meteor Crater," *Sky & Telescope,* November 1999: 48–53, 49, 50–51.
171 When the fireball . . . David Kring: Kring, "Calamity," 51.
171 At 45,000 . . . 150 feet: Smith, 9–10.
171 took just a half-minute . . . planet's atmosphere: Kring, "Calamity," 51.
171 The explosive force . . . global extinctions: Smith, 10.
171 Still, following . . . pick up, it did: Kring, "Calamity," 52–53.
172 According to one . . . through the air: Smith, 12.
172 What was left? . . . iron meteorites: Kring, "Calamity," 53.
172 Recent research . . . isn't finished: N. Artemieva and E. Pierazzo, "The Canyon Diablo Impact Event: The Projectile Fate"; M. Poelchau, T. Kenkmann, and D. A. Kring, "Structural Crater Rim Analysis at Meteor Crater." Both papers presented August 13, 2007, Meteoritical Society, Tucson, AZ.
172 Writers seeking . . . the crater floor: Smith, 10.
173 A booklet . . . Forest Ray Moulton: Smith, 24.

Book VI. Harvey Nininger Sees the Light

177 Do with . . . needs doing: *Find,* 11.
177 On a hot day . . . "The Castle": Cindy Gottsch (Northwestern Oklahoma State University), e-mails to author, December 2004. Provided historical information on the State Normal and photos of the exterior and interior of "The Castle." Climate information from August and September 1907, reports of Oklahoma and Indian Territory Section of the Climatological Service of the Weather Bureau.
177 Harvey Harlow Nininger . . . a library either: Harvey Harlow Nininger, Oral History, March 22, 1982, Colorado Museum of Natural History, now known as Denver Museum of Nature & Science; hereinafter, DMNS. All oral histories that follow are from this series of three interviews of Nininger conducted by museum staff and volunteers.
177 Harvey had read . . . ever read: H. H. Nininger, "It Wasn't Always Meteorites: The Rest of the Nininger Saga" (typescript, privately distributed, July 1983), 69, 72–73; hereafter, IWAM.
177 "Feverishly . . . decades later: IWAM, 72.
177 Harvey walked . . . from the shelves: IWAM, 76.
178 The boy . . . kind of hunt: IWAM, 1.
178 Harvey pulled out Ralph Waldo Emerson's *Essays:* IWAM, 72.
178 In a reverie . . . from another": Ralph Waldo Emerson, *Self-Reliance and Other Essays* (New York: Dover, 1993), 20.
178 Shame had been . . . his best: IWAM, 3, 10, 11, 31, 36.
178 But he was . . . master fear: IWAM, 5–6, 8–9, 45–46.
178 Among people . . . blacked out": IWAM, 53–54.

178 Harvey Nininger realized . . . my character": IWAM, 51–54.

178 That stubbornness . . . after the fact": Oral History, March 22, 1982, DMNS.

178 Despite such . . . Alva: "The Evolution of One Man's Philosophy," Papers of HHN.

179 At the State Normal . . . "Observe Nature": IWAM, 87.

179 This boy . . . a new continent": IWAM, 62–64, 72.

Chapter VI.1. Epiphany on Euclid Street

180 In McPherson . . . the evening air: *Find,* 12.

180 Harvey Harlow Nininger . . . Craik's home: Susan Taylor (librarian, McPherson College, Kansas), e-mail to author, August 5, 2003.

180 Friday, November 9, 1923 . . . blankets: Brian Fuchs (High Plains Regional Climate Center), e-mail to author, July 14, 2003.

180 The mercury . . . November: Fuchs.

181 "Suddenly . . . than before": Quoted in *Find,* 13.

181 "One moment . . . threads": Rainer Maria Rilke, "Sunset," in *Selected Poems of Rainer Maria Rilke,* trans. Robert Bly (New York: Harper & Row, 1981), 85.

181 Three months earlier . . . he said: *Find,* 11.

181 He told his wife . . . about meteorites: Oral History, March 22, 1982. DMNS.

181 Harvey Nininger remembered . . . ghosts and dragons": *Find,* 11–12, 11.

181 As a Brethren . . . out-of-doors: Carl F. Bowman, *Brethren Society: The Cultural Transformation of a "Peculiar People"* (Baltimore: Johns Hopkins University Press, 1995), 6.

181 The meteor presented . . . finish dessert: Taylor, August 5, 2003.

181 Craik could not . . . 1010 Euclid: *Find,* 13, 15. Addresses from McPherson city directories at McPherson Museum.

181 In his autobiography . . . ever been done?'": *Find,* 13

182 182 Given this future . . . a local settler: Taylor, August 5, 2003.

182 At five-foot-five . . . straight back": "Autobiography," manuscript, Papers of HHN; Doris Banks (Nininger's daughter), interview with author, June 19, 2004.

182 He also sometimes . . . has called it: Barbara Buskirk and Dorothy Richards (friends of Nininger's), interview with author, May 6, 2002.

183 After he made . . . winter sledding: IWAM, 190–91.

183 Over the years . . . Fluffy: Banks, June 19, 2004.

183 Home movies . . . the arms: Additional scene detail from home movies provided me by Gary Huss (Nininger's grandson).

183 On November 10, 1923 . . . angle of descent: *Find,* 13.

183 Harvey and physics . . . Kansas and Oklahoma: All reports cited here from "Meteor Nov. 9, 1923," Papers of HHN.

184 When he interviewed . . . had seen the meteor: Oral History, March 22, 1982. DMNS.

184 In his talk . . . Morris quit: Nininger, "My Introduction to Meteorites," taped talk provided by Peggy Schaller (Nininger's granddaughter), no date.

184 If the fireball . . . space rocks: *Find,* 15.

184 That season in 1921 . . . he said: Oral History, April 12, 1982, DMNS; Nininger, "Should a Doctor Tell," January 1963 typescript letter to *Reader's Digest,* Papers of HHN.

185 It had happened . . . credit to him: IWAM, 126–31, dialogue verbatim from this source.

186 To track down . . . borrowed funds": *Find,* 16–18.

186 A year later . . . for $690: Nininger to E. O. Hovey, December 6, 1924, Archives, Department of Library Services–Special Collections, American Museum of Natural History, 446.

186 In 1923 . . . $1.50: McPherson Public Library, copies of newspaper advertisements, *Weekly Republican,* November 1923.

187 At the time . . . a year: *Find,* 44.

187 Soon after obtaining . . . F. H. Snow: *Find,* 18–19.

187 Harvey continued . . . scientific truths: *Find,* 19–20.

187 He could not . . . bought two: *Find,* 33.

187 But some six years . . . trying years": Nininger, "Meteorites in the Antarctic," taped talk, no date. Provided by Schaller.

187 He had trouble sleeping: "Meteorites in the Antarctic."

187 Addie worried . . . their family: IWAM, 303.

187 When his . . . or robin": IWAM, 306.

188 To allow . . . some strain: *Find,* 20.

188 Privately . . . even biology: *Find,* 37.

188 While Harvey kept "talking, talking, talking meteorites": IWAM, 304.

188 he continued . . . Summer School: IWAM, 226.

188 Freedom's name . . . I'll pick cotton": IWAM, 231, 229, dialogue verbatim from this source.

188 Addie—by the time . . . went traveling": Addie Nininger, "Addie's Story: The Diaries of Addie D. Nininger," typescript manuscript, 35, introduced and compiled by Margaret Nininger Huss (daughter) and provided to author by the family (Westminster, CO: 1992). Hereinafter, "Addie's Story."

188 This was . . . in a tent: Peggy Schaller, interview with author, September 21, 2003.

188 For Harvey . . . losing": IWAM, 230.

188 So they . . . recreational vehicle: Details of the Henry trip, IWAM, 231–301, as well as "Addie's Story," 35–63.

189 But they did not . . . again scoffed: Details of the Natural History Trek, IWAM, 306–51, as well as "Addie's Story," 64–142.

190 No meteorite collectors . . . 750-mile trip: *Find,* 23.

190 Though in debt . . . Alex Richards: *Find,* 24; "A Couple of Detours," typescripts, 2, in file "Detours: Mexico and Paragoul," 2, Papers of HHN.

190 The dark-haired . . . did not: Details of Toluca trip: *Find,* 23–30.

190 He was handy . . . in appearance": "Detours," 2.

190 With its 80-gallon . . . for reverse: "Detours," 2.

191 He would pretend . . . of his *patrón*": Barbara Buskirk, interview with author, May 15, 2002.

191 Taking supplies . . . the trip: *Find,* 24.

191 The two . . . potholes: *Find*: 25; "Roads in Mexico Like Alaskan Snake Story, There Simply Aren't Any," *The Spectator,* March 18, 1930, no page given on copy.

191 Alex and Prof . . . robbery: "Nininger Has Thrilling Experience in Bandit Country in Old Mexico in the Dead of Night," *The Spectator,* January 21, 1930: 1; *Find,* 25

191 Once, three men . . . for murderers: *Find,* 25–26.

192 "And that's what . . . Barbara says: Buskirk and Dorothy Richards, interview with author, May 15, 2002.

192 Harvey Nininger's granddaughter . . . where he was": Schaller, interview.

192 Grandson Gary Huss . . . anyone depressed": Gary Huss, e-mail to author, July 16, 2004.

192 Later Harvey . . . such chances": "Detours," 3.

192 One of those . . . to Barbara: Buskirk and Richards, May 6, 2002.

193 Needing to protect . . . alarm if startled: Alex Richards, "The Hawk That Went Exploring," *The Open Road for Boys,* September, no year, page numbers cut off. Copy provided by Barbara Buskirk.

193 One night . . . the story: Buskirk and Richards, May 15, 2002.

193 After twenty-one days . . . Frederick Mullerried: "Nininger Making Study of Mexican Meteorites," *The Spectator,* November 19, 1929; 2; *Find,* 26.

193 The November 5, 1929 . . . with bandits": "Nininger Well Pleased with His Discoveries," *The Spectator,* November 5, 1929: 1.

193 For the next . . . organizing meteorites: "Nininger Making Study of Mexican Meteorites," *The Spectator,* November 19, 1929: 2; *Find,* 26.

194 194 Staying at the YMCA: "Detours," 4.

194 Probably they talked . . . never looked back": Richards, "The Hawk That Went Exploring."

194 Soon Harvey . . . go along: *Find,* 27.

194 To buy . . . of theft: *Find,* 28.

194 The going . . . and meadows: Details of Toluca taken from Addie Nininger's 1952 portrait of the village. The images she saw then would have been substantially the same in 1929. "Addie's Story," 150–52.

194 Among the men . . . 20-pound iron: *Find,* 28–29.

194 Afterward . . . long time: "Detours," 7.
194 "When we returned . . . from Toluca: "Detours," 7; *Find*, 29.
194 The original . . . Jiquipilco.): William D. Panczner, "When Is a Toluca Meteorite Not from Toluca, Mexico?," *Meteorite*, November 2003: 8–12, 11.
195 According to Barbara . . . area natives: Buskirk, May 6, 2002. Date of this event is unclear. It might not have happened on this trip, though Barbara Buskirk believes it did. Bruce Bair, "Meteor Man: A Lifetime Spent Chasing Objects from Space," *The Hays Daily News*, August 16, 1987: 1, 11.
195 When Prof . . . wet nurse: Buskirk, May 15, 2002.
195 In Mexico City . . . December 3, 1929: *Find*, 29. "The Nininger Party to Be Home December 24," *The Spectator*, December 10, 1929: no page on copy.
195 Prof and Alex . . . his hopes: Banks, June 19, 2004.
195 Then, a boy . . . and study: *Find*, 30–31.
195 Then came . . . and South. Details of this fall are taken from *Find*, 37–41; Nininger, "The Story of the Paragould, Arkansas, Meteorite," no date; taped talk provided by Schaller, "A Couple of Detours," Papers of HHN; *Sky*, 19—21; and various newspaper clippings, some without dates or publications listed, from the files of the Kansas State Historical Society, K552.6, v. 1.
197 If you are . . . pieces of the sun": Dodd, 32.
197 Despite being . . . chondrules blur together): Ralph Harvey, Phone interview, July 3, 2007.
197 And it turns out . . . *chemically* primitive": Dodd, 35–40.
198 Paragould is . . . ordinary chondrites: Multiple sources, including Michael R. Jensen, William B. Jensen, and Anne M. Black, *Meteorites from A to Z* (Michael R. Jensen, 2001), 84, 2.
198 Only in the . . . visibility of chondrules: Burke, 138–44, 283–301.
198 Certainly by 1930 . . . after Paragould: *Stone*, 33.
198 As to the origins . . . current concepts": McSween, *Meteorites and Their Parent Bodies*, 53.
198 Harvey Nininger . . . on the Earth: *Stone*, 17–20.
199 The Paragould chondrite . . . as yet known.): Multiple popular sources.
199 Naturally, Harvey . . . financial break: Clarita Nunez (Field Museum), e-mail to author, February 18, 2005.
199 In 1930 . . . security with routine": IWAM, 228, 184, 307.
199 One might suspect . . . 1912: "Addie's Story," 13.
199 On Valentine's Day . . . strong to resist": DMNS Archives Box IA, Nininger File Folder 1–20.
199 Harvey Nininger already . . . Audubon Society: Nininger, "Degrees," taped talk provided by Schaller, no date; "Recommendations," Elmer Craik to T. Gilbert Pearson, January 5, 1925, Papers of HHN.
199 "I am thinking . . . move to Denver": DMNS Archives, Box 1A, Nininger File Folder 1-20.
199 He knew his . . . anything else": Nininger, "My Introduction to Meteorites," taped talk.
200 Harvey says . . . than continue to teach: Nininger, "My Independence," manuscript, Papers of HHN.
200 Without waiting . . . his job: Untitled file, Papers of HHN.
200 Doris Banks . . . curiosity: Banks, June 19, 2004.
200 Concrete's replaced . . . face of success: Site visit, McPherson, Kansas, March 30, 2002.
201 "Not for me, ways of routine," Harvey had written: IWAM, 147.

Chapter VI.2. Never Done
202 In his autobiography . . . McPherson College: *Find*, 41–42.
203 Harvey followed . . . 1317 East 18th: Nininger to Figgins, no date, October 1930, DMNS Archives, Box IA, Nininger File Folder 1–20.
203 Atop . . . rocking chair: Banks, June 19, 2004.
203 Harvey does not . . . he believed: IWAM, 166.
203 I wonder if . . . of catalpas: IWAM, 2.
203 Earlier in the year . . . his autobiography: *Find*, 42.
203 But his correspondence . . . moving the rocks: Nininger, September 12, 1930, manuscript, with typescript of handwritten document, DMNS Archives, Box IA, Nininger File Folder 1–20.
203 This was a complete . . . the meteorites: Figgins to Nininger, March 31, 1930. DMNS Archives, Box IA, Nininger File Folder 1–20.

203 When he quit . . . $600: *Find*, 44, 42.

203 "I guess they . . . once said, laughing: Oral History, March 29, 1982, DMNS.

203 That the two . . . for food: Nininger to Figgins, September 17, 1924, DMNS Archives, Box IA, Nininger File Folder I–20.

203 Figgins replied . . . of return": Figgins to Nininger, September 22, 1924, DMNS Archives, Box IA, Nininger File Folder I–20.

204 By fall 1930 . . . for Harvey: Oral History, April 12, 1982, DMNS.

204 A year later . . . position of each: "Wards," letters of 1932 especially, Papers of HHN.

204 To make matters . . . Nininger specimens: *Find*, 44.

204 The recent move . . . unexpected blow": Nininger to Figgins, January 25, 1932, DMNS Archives, Box IA, Nininger File Folder I–20.

204 To earn . . . a grievous disappointment": Nininger to Figgins, January 25, 1932, DMNS Archives, Box IA, Nininger File Folder I–20.

204 He apparently . . . Port Orford, Oregon, meteorite: *Find*, 62.

205 Harvey was able . . . purchasing specimens: *Find*, 56–57.

205 A 1931 financial . . . meteorite curator: J. D. Figgins, to John C. Merriam, February 5, 1934, "Recommendations," Papers of HHN.

205 Simply gathering . . . eclipse in it": "Baumgardt, Mars M," Papers of HHN.

205 And when a scientific . . . not money": "American Journal of Science," Papers of HHN, April 1935 letters.

205 Harvey might . . . in 1933: "Personal." Elmer Craik to Nininger, July 5, 1933, Papers of HHN.

205 In his autobiography . . . accurate: *Find*, 45.

205 In April 1935 . . . university again declined: "Denver University," Papers of HHN.

206 In a note . . . needed them": "Hard Times," Papers of HHN.

206 Letters from the mid-30s . . . to pass: Roy S. Clarke and Howard Plotkin, "Frederick C. Leonard (1896–1960): First UCLA Astronomer and Founding Father of the Meteoritical Society," paper, 65th Meteoritical Society Meeting, Los Angeles, CA, July 25, 2002.

206 And as early as 1936 . . . awards: "F.C. Leonard 1930- Business and 'Confidential' and Proposed Position at UCLA," Papers of HHN.

206 "There were periods . . . life insurance: *Find*, 45–46.

206 Harvey also gardened . . . newspaper route: Banks, June 19, 2004.

206 "Once," Harvey wrote . . . the budget": IWAM, 352.

206 Mending . . . interested him: Banks June 19, 2004.

206 "He could . . . his rocks: Gary Huss, interview with author, June 13, 2004.

207 Granddaughter . . . each other": Schaller, September 21, 2003.

207 Harvey once . . . that I quit": "Degrees," taped talk.

207 "The sleepless hours . . . many!": "Top Crest Writing Tablet," untitled and unfilled, Papers of HHN.

207 Whether Harvey . . . faith: Schaller, September 21, 2003.

207 Harvey typically . . . a week: *Find*, 142.

207 but it was Addie's . . . first: Schaller, September 21, 2003.

207 Daughter Doris Banks . . . one-track mind:" Banks, June 19, 2004.

207 Addie's diaries . . . several times: "Addie's Story," 6, 12, 14, 18, 19, 24, 28.

207 Addie also was . . . major crisis: "Addie's Story," 14, 19, 20, 21.

207 Harvey had gone . . . may cost:" "Addie's Story," 20–21.

207 Doris says . . . very ill": Banks, June 19, 2004.

207 Still, in the . . . Scouts: IWAM, 352.

208 Doris recalls . . . in winter: Banks, June 19, 2004.

208 "To keep . . . car's floorboard: Banks, June 19, 2004.

208 "Have you noticed . . . the next find": Nininger, "Lecture to the Colorado Mineral Society," taped talk May 2, 1975. Provided by Schaller.

208 In Denver . . . along the way: Oral History, April 12, 1982, DMNS.

208 He'd put on . . . he got there: *Find*, 47.

208 The man who . . . get his share: *Find*, 46–47.

208 Harvey even . . . had the key: *Find*, 48.

209 "He was a big . . . off to college: Banks, June 19, 2004.

209 By truck or train . . . Huizopa iron: *Find*, 48–49.

209 On a Canadian . . . to science: *Find*, 53–55.

209 Harvey also continued . . . interest in it": Letters in "Prof. Dr. F. K. G. Mullerried 1931–1934–1936," Papers of HHN.

209 Harvey sometimes employed middlemen too: Letters in "Prof. Dr. F. K. G. Mullerried 1931–1934–1936," and "Toluca," Specimen Collection Files, Papers of HHN.

209 Harvey's success . . . Believe it or Not!": *Find*, 97–98; "Autobiography," Papers of HHN.

209 Such publicity . . . same academics: *Find*, 97.

209 He must take . . . regular doctorate: "Autobiography" and "How I Feel at 80," Papers of HHN.

210 Harvey's otherwise . . . friendly and supportive: Moulton to Nininger, March 13, 1942; "F. R. Moulton," Papers of HHN.

210 In the Nininger . . . thing to do": Nininger to Henry Hunter, October 4, 1934; "A. R. Allen," Papers of HHN.

210 Of course . . . of slag: Letters to Nininger in DMNS Archives about fireball sightings and mail-in samples, DMNS Archives, Box IA, Nininger File Folder 1–20.

210 Then there were . . . at Earthlings: "Miscell. Writings Various Authors," Papers of HHN.

210 Harvey even . . . lightning bolts: Nininger, "Pasamonte 45 Years Later," taped talk, no date. Provided by Schaller.

210 Saturnians aside . . . Kansas alone: *Find*, 108.

211 The finds . . . 300 stones: *Find*, 109–110.

211 Harvey also . . . or discovery": *Find*, 107.

211 A sky flash . . . turned out to be a meteorite: *Find*, 110–111.

211 And one in . . . good rate: *Find*, 126.

211 Such chases . . . was near: *Find*, 63–65, 79–80.

211 Harvey happened . . . sailed overhead?: *Find*, 65–70, 101, 80.

211 Pasamonte, Harvey . . . cosmic dust: *Find*, 67.

211 The Pasamonte meteorite . . . asteroid in 2010: Heide and Wlotzka, 131–133; Dodd, 139–141, 52–53; Michael J. Drake, "The Eucrite/Vesta Story," *Meteoritics & Planetary Science* 36 (2001): 501–513, 501, 512; Donald D. Bogard and Daniel H. Garrison, "39AR-40AR Ages of Eucrites and Thermal History of Asteroid 4 Vesta," *Meteoritics & Planetary Science* 38, no. 5 (2003): 669–710, 669.

212 When not chasing . . . advantage of this: *Find*, 57, 59–60.

213 A custodian . . . in gutters: *Find*, 57–59.

213 At the heart . . . country stoicism: *Find*, 110, 56.

213 An additional boost . . . Verne novel": "Adventures in Writing" and "Our Stone-Pelted Planet (Appreciation)," Papers of HHN; photocopy clip of "About Meteors," anon., *New York Times Book Review*, April 16, 1933, no page on copy.

213 Part of the book . . . Brenham pallasites: *Find*, 83.

213 When Harvey visited . . . its nature: *Find*, 86–87.

213 After Eliza and Frank . . . road scrapers": *Find*, 88.

213 Letters between . . . crush on Bob: "Brenham," Specimen Collection Files, Papers of HHN.

213 In all . . . major find: *Out*, 201.

213 Harvey's work there . . . strikes the earth": Nininger to Moulton, November 25, 1936, "Geological Society of America J. Lawrence Smith Fund," Papers of HHN.

214 Two years later . . . to $50,000 a year: Nininger to Moulton, January 22, 1938, "Geological Society of America J. Lawrence Smith Fund," Papers of HHN.

214 Moulton was impressed . . . had long ago": Moulton to Nininger, March 19, 1938, "Geological Society of America J. Lawrence Smith Fund," Papers of HHN.

214 Not everyone . . . to run it: Nininger to *Science* editors, June 20, 1933, "Correspondence with Journals," Papers of HHN.

214 Harvey's self-assuredness . . . been suspected": Nininger to Figgins, June 16, 1933, DMNS Archives, Box IA, Nininger File Folder 1–20.

214 Of these ideas . . . and *Meteoritics:* H. H. Nininger, *The Published Papers of H.H. Nininger: Biology and Meteoritics.*

George A. ed. Boyd (Tempe: Center for Meteorite Studies, Arizona State University, 1971), hereafter, *Published Papers.*

214 He chronicled . . . magnetic device": *Find,* 130; multiple papers, *Published Papers.*

215 He stuck . . . of fountains: Oral History, March 29, 1982, DMNS; Banks, June 19, 2004.

215 Harvey paid attention . . . like Moulton: *Find,* 131–135.

215 He wrote that . . . from meteorites: H. H. Nininger, "It Pays to Keep Your Eyes Open," *Published Papers,* 390–92, 390.

215 More important . . . new to science: Zeke Scher, "How Harvey Nininger's World Was Created," *Empire, Sunday Magazine of the Denver Post* (May 22, 1975): 12–13, 17–18, 19–21 [likely]. Last pages not included in DMNS files.

215 Two years later . . . desert heat: Richard O. Norton, "Personal Recollections of Frederick C. Leonard III Part II," *Meteorite!* 2, no. 4 (November 1996): 20–23. http://www.meteor.co.nz/nov96_1.html. 1–5. 3.

216 Harvey writes . . . U.S. government: *Find,* 119–21.

216 It's a good story . . . recovery": O. Richard Norton, "Goose Lake Meteorite—The Rest of the Story," *Meteorite!,* February 1999: 30–32, and May 1999: 30–33.

216 A file of "Unfinished Papers" . . . such a deal: "Unfinished Papers," untitled manuscript, first page missing, Papers of HHN.

217 Searching for meteorites . . . ordinary rock": *Find,* 23.

217 It meant searching . . . [and] ditches": Nininger, "My Introduction to Meteorites," taped talk.

217 Looking for . . . pursue them": Nininger, "Pasamonte 45 Years Later," taped talk.

218 On a December 1933 . . . Harvey wrote: *Find,* 92–96.

218 Two years later . . . on the pillows": *Find,* 114–15 concerning Hugoton. Additional unpublished details drawn from these files in the Papers of HHN: "Birth of the Meteoritical Society," a document included called "Hugoton" and "Unfinished Papers"; Oral History, April 12, 1982. DMNS.

220 Years later . . . ephemeral": IWAM, 286.

220 Hugoton . . . nerves tingling": *Find,* 113.

220 So addictive . . . a month later: "Misc. Writings" and "Nininger Family," Papers of HHN.

221 (The First World War . . . epidemic.): "Degrees," taped talk; IWAM, 164–168.

221 The Second World War . . . meteorite-hunting: *Find,* 145–146, 142.

221 Harvey could . . . university level: *Find,* 131.

221 Much good . . . on meteorites: Ursula Marvin, "The Meteoritical Society: 1933–1993," *Meteoritics* 28 (1993): 261–314, 267–68.

221 Now calling . . . collection anywhere: *Find,* 45; Scher, 14.

222 Perhaps when . . . going down: Nininger, "Memories and Dreams," taped talk no date. Provided by Schaller.

Chapter VI.3. Strongly Spent

223 Harvey and Addie looked . . . Harvey's own museum: IWAM, 298. Published details of the move are found in *Find,* 152–55. Other details from "Ariz. Crater Exploration Arrangements with Barringer," Papers of HHN, which includes handwritten manuscript, "Arizona's Meteorite Crater and Me," no date. Doris Banks also provided notes and cards.

223 The Niningers could . . . bolts and nuts": *Find,* 153.

224 In England . . . the first time: Neil Bone, *Meteors* (Cambridge, MA: Sky Publishing, 1993), 105.

224 Harvey called . . . within [our] view": *Find,* 154.

224 After days . . . centerpiece: Details of life there and customers from *Find,* 152–62.

224 The sixty visitors . . . blacksmith": *Stones from the Skies,* brochure for the American Meteorite Museum. Provided by Doris Banks.

224 A later brochure . . . the moon!": *A Unique Experience,* brochure. Provided by Doris Banks.

224 Tourists could . . . 33,000 visitors: *Find,* 159; Addie to Margaret, November 19, 1946, untitled file, Gary Huss.

225 But what was . . . pay a fee: *Find,* 155, 158, 161.

225 Tourists were . . . tourist trap: *Find,* 150.

225 Controversy still . . . up the stairs: Doris Banks, interview, June 19, 2004.

225 Regardless . . . to operate: *Find,* 166.

225 In between . . . way inside: *Find*, 168, 151, 157–58.

225 The Niningers . . . they'd planted: *Find*, 156; Addie to Margaret, December 4, 1946, untitled file, Gary Huss.

225 Supplies were . . . the work: Addie to Margaret, December 4, 1946, untitled file, Gary Huss.

225 Mornings could . . . exercise: "Religion in My Life," Papers of HHN, Addie to Margaret, July 13, 1947, untitled file, Gary Huss.

225 Some mornings . . . prophets: IWAM, 297.

225 And he . . . epic ground: IWAM, 298.

225 He had been . . . stopped. Nininger stopped: *Find*, 176; *Sky*, 214.

226 He still . . . specimens too: *Find*, 148–49.

226 Now, living . . . hurt you": *Find*, 151.

226 Surrounding Meteor Crater . . . mining rights: Smith, *Meteor Crater Story*, 42.

226 The war . . . the picture": Nininger to Brandon Barringer, May 2, 1940, "Barringer, Brandon 1934–1944," Papers of HHN.

226 A few months . . . crater rim: Nininger to Barringer, January 13, 1941, "Barringer, Brandon 1934–1944," Papers of HHN.

226 Brandon politely declined: Barringer to HHN, January 16, 1941, "Barringer, Brandon 1934–1944." Papers of HHN.

226 So Harvey . . . an arrangement: Barringer to Nininger, March 16, 1942, "Barringer, Brandon 1934–1944," Papers of HHN.

226 But Harvey . . . stone building: Nininger to Warren Tremaine, May 18, 1940; Nininger to Tremaine Cattle Company, March 26, 1942; Nininger to Tremaine Brothers, April 25, 1942; Nininger to Warren Tremaine, February 28, 1944; Tremaine to Nininger, May 9, 1942. All in "Tremaine," Papers of HHN. The last letter invites Nininger to work with the Tremaines after the war so both may benefit from some arrangement.

226 (Harvey spoke . . . the crater.): Nininger to Gillespie, May 10, 1946, "Gillespie," Papers of HHN.

227 In one letter . . . ended up: Nininger to Brandon, April 22, 1942, "Barringer, Brandon, 1934–1944," Papers of HHN.

227 At times . . . an emergency": Nininger to Carl Tremaine, April 29, 1942, "Tremaine," Papers of HHN.

227 Not that . . . the property": Brandon to Nininger, November 6, 1946. "Barringer," Papers of HHN.

227 In early 1947 . . . crater itself: Nininger to Brandon, January 3, 1947, "Barringer," Papers of HHN.

227 He also had . . . absolutely false": Nininger to Ernest Chilson, August 3, 1947, "Tremaine," Papers of HHN.

227 Chilson, not . . . his finds: Chilson to Nininger, August 15, 1947, "Tremaine," Papers of HHN.

227 Despite a more . . . state property, he claimed: Nininger to Chilson, September 24, 1947, "Tremaine," Papers of HHN.

228 A "shocked" Brandon . . . than competitive": Brandon to Nininger, no date, "Barringer," Papers of HHN.

228 Brandon even . . . misstate a fact": Barringer to Johnson, August 11, 1947, copy to HHN, "Barringer," Papers of HHN.

228 As to the collecting . . . with impunity: Nininger to Brandon, January 3, 1947, "Barringer," Papers of HHN.

228 Indeed, a series . . . his finds: Brandon to Nininger, October 6, 1947; D. Moreau ("Reau") Barringer to Nininger, October 16, 1947; "Ariz. Crater Explorations Arrangements with Barringer," Papers of HHN.

228 With these various . . . toward Earth: Details for this extended scene are drawn from a variety of sources, including photographs, promotional brochures, my interview with Doris Banks, and especially Addie Nininger's 1947 letters to her children. Untitled file, Gary Huss home.

229 Spewing fire . . . customer's hair: Brodie Farquhar, "Fallen Star Hunted and Found Near Norton," *The Oberlin Herald*, March 25, 1998, Section B front page.

229 Horses . . . panic: Glenda Hahn, interview, August 16, 2005.

229 Harold and Glenda . . . on fire": Hahn, interviews July 25, 2005, and August 16, 2005.

229 Dale Leidig . . . and dust": Dale Leidig, interview, August 2, 2005.

229 A B-29 . . . an hour: "Sky Blast Here Was a Meteorite?", *Norton Daily Telegram*, February 19, 1948, front page, photostat of typescript, K552.6 Misc., Kansas State Historical Society, copy of front page article of this issue, Norton County Library; *Out*, 32–34.

229 Somewhere near . . . some mass: "Weather Today," *Norton Daily Telegram*, February 23, 1948, front page; Michael Mansur, "One Winter Day, Fire Split the Sky," *Kansas City Star*, July 4, 1993, A1.

229 Before exploding . . . busy telegraphs: "Sky Blast Here Was a Meteorite?," *Out*, 33.

230 M. R. Krehbiel . . . out of control": Quoted in "Editor Gives Description of Meteor Blast," *Wichita Evening Eagle*, February 19, 1948: 19, K552.6 v. 1, Kansas State Historical Society. Clip copy.

230 The *Norton Daily Telegram* . . . strewnfield might be: "Aid Is Invited Locating Fragments from 'Fire Ball,'" *Norton Daily Telegram*, February 19, 1948, photostat of typescript, K552.6 Misc., Kansas State Historical Society, photocopy of article from front page of the paper, Norton County Library.

230 After being alerted . . . was no: "Arizona Scientist on Way to Norton," *Norton Daily Telegram*, February 19, 1948: front page, photostat of typescript, K552.6 Misc., Kansas State Historical Society, copy of article, Norton County Library; Nininger, "9," unpublished chapter, deleted from *Find a Falling Star*, Undated typescript provided to me by Doris Banks.

230 On February 20 . . . snowbirds": "Scientist Asks Reports from Here on Meteror [*sic*] Show," *Norton Daily Telegram*, February 20, 1948: front page, photostat of typescript, K552.6 Misc., Kansas State Historical Society, copy, Norton County Library.

230 Harvey arrived . . . be worth: "The Blowoff Corner," *Norton Daily Telegram*, February 21, 1948, photostat of typescript, K552.6 Misc., Kansas State Historical Society; "A Noted Scientist Seeks Fragments of Big Meteorite," *Norton Daily Telegram*, February 23, 1948, photostat of typescript, K552.6 Misc., Kansas State Historical Society, copy Norton County Library.

231 On Monday . . . north of Norton": "A Noted Scientist Seeks Fragments of Big Meteorite."

231 Later, he'd say . . . large fragments": Nininger, "9."

231 The weather . . . late-winter: "Weather Today," *Norton Daily Telegram*, February 23, 1948: front page.

231 A snowstorm . . . February 24: Nininger, "9"; "'Double Hunt' Is Proposed," *Norton Daily Telegram*, February 24, 1948: front page.

231 Perhaps they . . . in California: "The Blowoff Corner," *Norton Daily Telegram*, February 25, 1948, photostat of typescript, K552.6 Misc., Kansas State Historical Society.

231 Harvey's "preliminary investigation" . . . the strewnfield: "'Double' Hunt Is Proposed."

231 Meanwhile, Lincoln LaPaz . . . seems inaccurate.): Lincoln LaPaz, *Space Nomads: Meteorites in Sky, Field and Laboratory* (New York: Holiday House, 1961), 26.

231 Decades later . . . this assertion: Lincoln LaPaz, "Commentary," January 22, 1977:8. Provided by LaPaz son-in-law Harry Baldwin.

232 Why did LaPaz . . . were located: Lincoln LaPaz, "The Achondritic Shower of February 18, 1948," *Publications of the Astronomical Society of the Pacific* 61, no. 359 (April 1949): 63–73, 64.

232 A month after . . . no meteorites: LaPaz, "Achondritic Shower," 65.

232 Harvey knew . . . nor samples: Nininger, "9."

232 At least . . . from anyone: D. Moreau ("Reau") Barringer to Nininger, April 29, 1948, "Arizona Crater Explorations Arrangements with Barringers," Papers of HHN.

232 Brandon Barringer . . . Theodore Johnson: Brandon to Nininger, March 29, 1948, "Barringer," Papers of HHN.

232 Once again . . . in the area: Nininger to Brandon, April 27, 1948, "Barringer," Papers of HHN.

232 Brandon responded . . . anywhere you please": Brandon to Nininger, April 29, 1948, "Barringer," Papers of HHN.

233 This was not . . . a regular accounting": Brandon to Nininger, January 7, 1948, "Barringer," Papers of HHN.

233 Concerned that . . . us and the Tremaines": Nininger to Brandon, May 11, 1948, "Barringer," Papers of HHN.

233 Harvey clarified . . . you and the Tremaines": Nininger to Brandon, May 11, 1948, "Barringer," Papers of HHN.

233 Brandon's response . . . your studies": Brandon to Nininger, May 12, 1948, "Barringer," Papers of HHN. This may be a transposed date for the letter; it might have been written on May 21, 1948.

233 But perhaps . . . the cause: Nininger to Harlow Shapley, March 29, 1948, "Shapley, Dr. Harlow," Papers of HHN.

234 Three months earlier . . . the land: Nininger to Brandon, January 4, 1948, "Barringer," Papers of HHN.

234 Brandon replied . . . the state: Brandon to Nininger, January 7, 1948, "Barringer," Papers of HHN.

234 In June 1948, Harvey dropped a bombshell: Nininger to Brandon, June 26, 1948, "Barringer," Papers of HHN.

234 When Harvey read . . . end this way": Brandon to Nininger, July 6, 1948, "Barringer," Papers of HHN.

235 Amazingly, Harvey . . . these doings": Nininger to Brandon, July 12, 1948, "Barringer," Papers of HHN.

235 Brandon Barringer privately . . . his achievements": Barringer to William Gruenwald, January 27, 1949, Baldwin file.

235 To Lincoln LaPaz . . . may fall": Brandon to LaPaz, July 9, 1948, Baldwin file.

235 Letters indicate . . . the family: See letters between Nininger and Reau Barringer, between August 31, 1948, and June 24, 1949, "Barringer," Papers of HHN.

235 (A letter . . . he should have.): J. Paul Barringer to Nininger, September 9, 1977, "Arizona Crater Exploration Arrangements with Barringers," Papers of HHN.

235 And for the rest . . . Meteor Crater property: Smith, *Meteor Crater Story*, 40. Nininger noted permission to search on state lands in a letter to the Tremaines, April 1, 1949, "Tremaine," Papers of HHN. The letter noted that any holes would be filled in once exploration was complete.

235 At least he . . . been banished: Johnson to Nininger, August 12, 1948, "Tremaine," Papers of HHN.

236 Months earlier . . . Norton bolide: Nininger, "9."

236 The first such . . . April 27: LaPaz, "Achondritic Shower," 65, 67; "Meteorite Fragments Found Near Norton," May 1, 1948: front page, "Meteor Feb. 18 – 1948," Papers of HHN.

236 He had waited . . . but wasn't: Mansur, 3.

236 The meteorite . . . diamonds": "Meteorite Fragments Found."

236 In his several . . . our support": Nininger, "9."

236 He could not . . . nothing more: Nininger, "9."

236 Fragile, friable . . . original chondritic state: *Rocks from Space*, 203–4.

236 And today we know . . . hundreds": Ralph Harvey, e-mail to author, August 2, 2005.

237 In 1948 . . . according to LaPaz: LaPaz, "The Achondritic Shower," 65.

238 But they may . . . soldier his life": Nininger, "9."

238 "That man" . . . meteorite specimens: G. Jeffrey Taylor, "History of the Institute of Meteoritics," undated typescript manuscript provided by Wolfgang Elston; Marvin, "Meteoritical Society," 269.

238 Indeed, researchers . . . 60 million years: Burke, 299–300.

238 This suggests . . . with Jupiter's gravity: Tim McCoy, e-mail to author, August 8, 2005.

238 "Some collectors of . . . his mother's housekeeper.): "Wichitan Possesses Polished Meteorite Called 'Sky Jewel,'" *Wichita Evening Eagle*, November 24, 1952, photocopy from K 552.6 v. 1, Kansas State Historical Society.

238 After leaving Norton . . . living in California: Hahn, July 25, 2005; Nininger, "9."

239 Harvey could have . . . and rightly so": Nininger, "A Bit of History," no date, typescript, Papers of HHN.

239 Harvey told . . . he harvested": Nininger, "9."

239 Throughout this second . . . locate specimens: Nininger, "Getting Matters Straight," "Unfinished Papers," Papers of HHN.

239 Harvey apparently . . . Nininger's materials: LaPaz, "The Achondritic Shower."

239 The University of New Mexico . . . delight of Lincoln LaPaz: LaPaz, "The Achondritic Shower," 65, 69; "Norton Meteorite of Military Value," *Norton Daily Telegram*, May 4, 1948: front page.

239 (The McKinley stone . . . collection years later.): Horton Newsom, e-mail to author, August 8, 2005; James Wray to Bertrand Schultz, April 27, 1967, Files of the Nebraska State Museum. Wray: "I shall notify Lincoln of the imminent completion of the negotiations so that he and I may begin figuring out how to get a nice slice off of *his* McKinley stone" [emphasis mine].

239 The find occurred . . . the *Daily Telegram* reported: "McKinley Ranch Meteorite Is 100-Pound 'Granddaddy'; May Be Even Larger Ones?" *Norton Daily Telegram*, May 3, 1948: front page; "Meteor Feb. 18 – 1948," Papers of HHN.

239 LaPaz paid $500 . . . the high school: Copy of bill of sale and bank check, May 4, 1948, Baldwin; "Exhibit Meteorite at Lecture Tonight in High School," *Norton Daily Telegram*, May 5, 1948, photocopy with no page citation provided by same.

239 Harvey Nininger left Furnas . . . been no reply: Nininger to LaPaz, May 14, 1948, Baldwin.

240 So how to sort . . . should be searched: LaPaz, "Commentary," typescript. January 27, 1977, Baldwin.

240 "It would be a pleasure" . . . to include meteorites: LaPaz to Nininger, August 17, 1938, "Recommendations," Papers of HHN.

240 Harvey might have . . . set up the talk: LaPaz to Nininger, August 17, 1938, "Recommendations," Papers of HHN.

240 According to . . . many issues: None of the materials provided by Baldwin address these issues. University of New Mexico staff could locate no relevant files.

240 One thing is clear . . . the nation's first: Taylor, "History of the Institute."

240 Harvey would suspect . . . New Mexico: Nininger, "A Bit."

240 One public difficulty . . . to the occult: Marvin, "Meteoritical Society," 268–69.

241 LaPaz took aim . . . nickel-iron meteorite": Marvin. "Meteoritical Society," 269, 271.

241 LaPaz was also . . . them for sale": Marvin, "Meteoritical Society," 273.

241 Never mind . . . from Harvey: "University of N.M., Albuquerque," Papers of HHN.

241 Despite these and . . . physical safety: Nininger, "A Bit," Papers of HHN.

241 "[LaPaz] has always . . . under his jurisdiction": Nininger to Gillespie, February 28, 1946, "Gillespie," Papers of HHN.

241 Frederick Leonard . . . embraced LaPaz: Nininger to Harlow Shapley, December 29, 1947, "Shapley, Dr. Harlow," Papers of HHN.

242 A former student . . . being pedantic": O. Richard Norton, "Personal Recollections of Frederick C. Leonard," *Meteorite!*, August 1996. http://www.meteor.co.nz/aug96_03.html.

242 Certainly Leonard's . . . writing at times: "Leonard, Frederick C. 1946–," Papers of HHN.

242 LaPaz could deploy . . . "cerebration": Lincoln LaPaz, "Some Aspects of Meteorites," University of New Mexico Fifth Annual Research Lecture, May 2, 1958, 185.

242 Bill Cassidy . . . working privately: William Cassidy, e-mail to author, no date recorded.

242 When I ask . . . to be unpredictable." Wolfgang Elston, e-mail to author, September 8, 2004.

242 The couple . . . beginning to wonder": Krehbiel to Nininger, June 3, 1948, "Norton County Fall Aubrite (enstatite ach)," Papers of HHN.

242 Unbeknownst . . . keep threshing: "Locate Second Big Fragment of Norton County Meteorite," *Norton Daily Telegram*, August 18, 1948: front page; "Meteor Feb. 18 – 1948," Papers of HHN.

242 Then Harold . . . six-foot hole": H. H. Nininger, "Tracing the Norton, Kansas, Meteorite Fall," *Sky & Telescope* 7, no. 12 (October 1948): 293–95; in *Published Papers*, 559–61.

243 But–and this . . . Hahn's field: Nininger, "A Bit."

243 There was more . . . of any finds: Nininger, "A Bit"; LaPaz, "The Achondritic Shower," 69. Farquhar, B1.

243 The Niningers arrived . . . the Hahns: "Huge Meteor Fragment Found by Harold Fahn" [sic], *Norton County Champion*, August 19, 1948: front page; Hahn, August 16, 2005.

243 Lincoln LaPaz would . . . had": LaPaz to Schultz, October 1, 1948, Files of the Nebraska State Museum.

243 The fact that . . . LaPaz mentions: Hahn, July 28, 2005, and August 16, 2005.

243 The *Norton County Champion* . . . saved for science: From "Meteor Feb. 18 – 1948," Papers of HHN.

243 Meanwhile, the *Daily Telegram* . . . Wednesday, August 18: "Locate Second Big Fragment of Norton County Meteorite," *Norton Daily Telegram*. August 18, 1948: front page; "Meteor Feb. 18 – 1948," Papers of HHN.

243 The *Omaha Morning World-Herald* . . . LaPaz wasn't mentioned: "Big Meteorite Piece Found," *Omaha Morning World-Herald*, August 20, 1948, front page: from "Meteor Feb. 18 – 1948," Papers of HHN.

244 The hole was . . . with its pond: Hahn, July 25, 2005.

244 After the one-ton . . . before Lincoln LaPaz: Hahn, July 25, 2005.

244 "We really liked him . . . about that meteorite": Hahn, July 25, 2005.

244 Glenda didn't go . . . notes and photographs: Hahn, July 25, 2005.

244 Harvey and Harold . . . fusion crust: Nininger, "A Bit."

244 "I don't know . . . fragments in Kleenex": Nininger, "Getting Matters Straight," "Unfinished Papers," Papers of HHN.

244 An hour later . . . with LaPaz: Nininger, "A Bit."

244 According to Harvey . . . going to be called: Nininger, "A Bit."

245 Climbing out . . . toss each aside: Nininger, "A Bit."

245 Harvey said . . . underfoot, Harvey alleged: Nininger, "The Norton Meteorite," "About Life," Papers of HHN.

245 The fate of . . . the main mass": McCoy, August 8, 2005.

245 Glenda Hahn . . . the hole: Hahn, July 25, 2005.

245 Soon enough . . . been taken away?": B. F. Butler to Nininger, no date. "Norton County Fall Aubrite (enstatite ach)," Papers of HHN.

245 LaPaz, Leonard . . . the Niningers: Nininger, "9."

245 There don't seem . . . the legal assistant": Nininger, "Misc. Writings," Papers of HHN.

246 The final bid . . . $250 respectively: LaPaz to Schultz, September 20, 1948, Files of Nebraska State Museum.

246 A letter from . . . for the universities": LaPaz to Schultz, September 12, 1948, Files of Nebraska State Museum.

246 After he won . . . after he bought it": Hahn, July 25, 2005.

246 LaPaz told Schultz . . . Nininger's fabrication": LaPaz to Schultz, October 1, 1948, Files of the Nebraska State Museum.

247 After Harvey left . . . the small crater: Hahn, July 25, 2005, and August 16, 2005.

247 Harvey, then sixty-one . . . to make the survey": Nininger, "Misc. Writings," Papers of HHN.

247 Harvey called . . . to show for it: Nininger, "A Bit."

247 Harvey admitted . . . doesn't always win": Nininger, "Misc. Writings" and "A Bit of History," Papers of HHN.

247 In the end . . . you can't stop": Addie to Margaret, November 25, 1947, untitled file, Gary Huss home.

247 Sentiments in the . . . specimens are obtained": R. D. Bower to Nininger, August 31, 1948, "Norton County Fall Aubrite (enstatite ach)," Papers of HHN.

247 The director . . . are obtained" : Krehbiel to Nininger, June 3, 1948, "Norton County Fall Aubrite (enstatite ach)," Papers of HHN.

247 In September . . . come forth yet": Krehbiel to Nininger, September 16, 1948, "Norton County Fall Aubrite (enstatite ach)," Papers of HHN.

247 The next month . . . its surface features: Nininger to LaPaz, October 25, 1948, "Norton County Fall Aubrite (enstatite ach)," Papers of HHN.

247 But he . . . made jewelry: Nininger, "University of N. M., Albuquerque," Papers of HHN.

247 LaPaz privately . . . request: LaPaz to Dorrit Hoffleit, December 1, 1948, Baldwin.

247 After the Norton . . . old Doc Splurge and Jabber": Nininger to Frank Cross, November 5, 1949, "Cross, Frank," Papers of HHN.

248 He even journaled . . . for the better: Nininger, "Cross, Frank," Papers of HHN.

248 Harvey would . . . until 1963: Marvin, "Meteoritical Society," 273–74, 281.

248 Some acrimony . . . suspect Harvey Nininger: LaPaz to Schultz, February 5, 1958, Files of the Nebraska State Museum.

248 And despite . . . a long time": Forrest Warren to Nininger, October 4, 1948, "Norton County Fall Aubrite (enstatite ach)," Papers of HHN.

248 Meanwhile, according to . . . him [Nininger]": Schultz to LaPaz, September 14, 1948, Files of the Nebraska State Museum.

248 LaPaz even . . . seeking specimens: Statements dated August 1948, Baldwin.

248 LaPaz still believed . . . suggested for searching: LaPaz to Hoffleit, December 1, 1948, Baldwin.

248 Amazingly, the Norton . . . falsehoods and innuendos": LaPaz to Schultz, October 1, 1948, Files of the Nebraska State Museum.

248 In a long . . . thoroughly demoralizing": LaPaz, "Commentary," Baldwin. Also in "Correspondence Norton Controversy 1977–1980," Papers of HHN.

249 Decades didn't soften . . . nerve to write that": Nininger, "Who Discovered the Norton County Fall?," taped talk provided by Schaller, no date.

249 While Harvey had . . . mother's funeral: "Meteor Expert Chased Balloons," *Wichita Evening Eagle*, August 3, 1948, photocopy from Kansas State Historical Society, K552.6, v. 1.

249 That summer . . . contracted polio: LaPaz to Schultz, October 9, 1948, Files of the Nebraska State Museum.

249 Today the area . . . meteorite was found: Horace Collins, site visit, Norton County, Kansas, and nearby Nebraska area, August 16, 2005.

250 A year after . . . breakdown: Addie to Margaret, July 22, 1949, untitled file, Gary Huss home.

250 But in the months . . . vaporized impactor: *Find,* 177–178.

250 He spent . . . at a time": *Find,* 174.

250 Armed with magnets . . . a magnet and sieves: *Find,* 179-180.

251 There is still . . . in studies today: Hoyt, 362–63. I. Leya, et al., "Pre-Atmospheric Depths and Thermal Histories of Canyon Diablo Spheroids," *Meteoritics & Planetary Science* 37 (2002): 1015–25.

251 He also found . . . had previously thought": Hoyt, 342.

251 Harvey concluded . . . field work": Nininger, undated release on AMM letterhead, "(Meteor) Crater Explorations," Papers of HHN.

251 In a book on . . . large impacts: H. H. Nininger, *Arizona's Meteorite Crater: Past, Present, Future* (Denver: American Meteorite Laboratory, 1956), 50.

251 Meanwhile, a new . . . Route 66: *Find,* 167.

251 Not that Harvey . . . away once again: *Find,* 172–174.

251 In 1953 . . . town of Sedona: *Find,* 188.

251 From there . . . unprepared": Nininger, "Advertising - Am. Met. Museum," Papers of HHN.

251 Harvey and Addie . . . $275,000 respectively: *Find,* 206, 220.

251 With one banker . . . had been in years": IWAM, 399.

251 In 1959 . . . still lives: *Find,* 217.

252 From Sedona . . . hunt meteorites: IWAM, 426.

252 In his old age . . . new to science: IWAM, 394.

252 He'd revolutionized . . . how they formed: *Find,* 190.

252 For some . . . his science": McCoy, August 8, 2005.

252 But in 1967 . . . highly technical: Marvin, "Meteoritical Society," 285.

252 As early as 1937 . . . suites of minerals: Marvin, "Meteoritical Society," 274–75, 278; Gretchen Benedix, e-mail to author, August 30, 2005.

253 About the time . . . that is, asteroids: Dodd, 69–71.

253 The well-known . . . until much later: Gary Huss, e-mail to author, September 30, 2007.

253 "I can read . . . to him: *Find,* 201.

253 Harvey Nininger was right . . . the subject:" *Find,* 244.

254 "Nature is good . . . in 1971: Nininger, "Xmas Greetings 1971," Papers of HHN.

254 Even in his . . . in the day: Nininger, "Life at 80," Papers of HHN; Scher, no page number on copy.

254 On one of the . . . almost wept": Nininger. "Unfinished Writings." Papers of HHN.

254 The two men . . . Harvey again: Nininger, "Leonard, Frederick C. 1946–," Papers of HHN.

254 Harvey Nininger . . . injured his head: Schaller, September 21, 2003.

254 "My only quarrel . . . once wrote: IWAM, 419.

254 And Addie? . . . dear Harvey: Schaller, September 21, 2003.

255 "What a great boon . . . gloriously red": IWAM, 357.

Book VII: A Serious Case of the I Wants

259 Gary is giving . . . April snow. All quotes from Gary Curtiss and Matt Morgan from interviews with author, April 6, 2003.

264 Exuberance, psychologist . . . energy and delight": Kay Redfield Jamison, *Exuberance: The Passion for Life* (New York: Knopf, 2004), 25.

264 Jamison believes . . . sexual energy: Jamison, 22–24.

264 Exuberance . . . recapture the joy": Jamison, 32.

264 Psychologist John D. Gartner . . . describing them": John D. Gartner, *The Hypomanic Edge: The Link Between (A Little) Craziness and (A Lot of) Success in America* (New York: Simon & Schuster, 2005), 2.

264 They take risks . . . often oversexed: Gartner, 4–5.
264 Hypomanics . . . epiphany: Gartner, 154.
264 their ability . . . *leap on it*": Gartner, 142.
264 Anticipation . . . in the brain: Jamison, 113–115
264 Strongly associated . . . feels good: Jamison 113, 148–149, 116–117.
265 Dopamine is . . . Berns writes: Gregory Berns, *Satisfaction: The Science of Finding True Fulfillment* (New York: Henry Holt, 2005), 5.
265 "The release . . . motivational system": Berns, 15.
265 Yes, it does feel . . . each time: Shari Caudron, *Who Are You People?: A Personal Journey into the Heart of Fanatical Passion in America* (Fort Lee, NJ: Barricade, 2006), 179–80.
265 Of course . . . novelty": Berns, 12–15.
265 In what measure . . . you to do": Berns, xiv.
265 Having to work . . . memory": Berns, 142–47, 143.
265 In combination . . . Berns calls it: Berns, 147.
265 The more skilled . . . we seek: Berns, 148.
265 "But I'm . . . never quit": Harvey Nininger, typescript, "Misc. Writings," Papers of HHN.
265 And if you . . . beget meaning: Kay Redfield Jamison, *Touched with Fire: Manic-Depressive Illness and the Artistic Temperament* (New York: The Free Press), 1993, 104–5, 110.
265 As Gregory Berns . . . same thing": Berns, xvi.
266 Dick Pugh . . . but for safety: Dick Pugh, phone interview, December 4, 2000.
266 In a matter . . . education for all": Sean D. Hamill, "'Park Forest Meteorite Fall': One Year, Many Deals Later," *Chicago Tribune*, March 26, 2004, Section C (Tempo Section): 1–2.
267 So while . . . blue-collar area: Hamill, 2.
268 "Heed glides . . . the floor": Jamison, 246.
268 In Tucson . . . signs are posted: Site visit, Tucson, February 8–10, 2001.
268 There are . . . gem polishers: Kevin Krajick, "Mining for Meteorites," *Smithsonian* 29, no. 12 (March 1999): 90–100, 93.
269 Norton himself . . . from Mars: O. Richard Norton, "Tucson '98," *Meteorite!* 4, no. 2 (May 1998): 30–32, 31, 32.
271 Ureilites . . . impact melting: Alan E. Rubin, "Shock, Post-Shock Annealing, and Post-Annealing Shock in Ureilites," *Meteoritics & Planetary Science* 41, no. 1 (2006): 125–33, 125.
272 When I return . . . escape from: Site visit, Tucson, February 7–8, 2003.
273 As one . . . a meteorite": Jim Philips,"Collecting Meteorites, Some Thoughts," *Meteorite!* 3, no. 4 (November 1997): 32–33, 32.
273 Freud thought . . . collectors (of erotica): Geoff Nicholson, *Sex Collectors: The Secret World of Consumers, Connoisseurs, Curators, Creators, Dealers, Bibliographers, and Accumulators of "Erotica."* (New York: Simon & Schuster, 2006), 17.
273 Be that . . . they receive: Nicholson, 45.
273 Some collect . . . at bay": Nicholson, 83.
273 The very act . . . lovers and collections: Nicholson, 75.
273 "Every collector . . . Don Juan": Quoted in Nicholson, 76.
274 The comparison . . . to attain: Nicholson, 77.
274 Education professor . . . stories, history": Martin Horejsi, e-mail to author, November 9, 2005.
275 For Horejsi . . . solar system": Horejsi, November 9, 2005.
275 Australian dealer . . . breccia instead?'": Jeff Kuyken, e-mail to author, November 5, 2005.
275 This savvy . . . a meteorite feeding frenzy": O. Richard Norton, "Tucson Meteorite Show '99." *Meteorite!* 5, no. 2 (May 1999): 16–18. 16.
275 There's another . . . he says: Robert Hagg, phone interview, October 11, 2002.
276 In fact . . . for them: http://www.space.com/scienceastronomy/astronomy/meteorite_hunters_01128.html.
276 He was . . . strangers: Krajick, 96.
276 In one . . . sky *is* falling": Robert Haag, *The Robert A. Haag Collection Field Guide of Meteorites* (Tucson: Robert A. Haag, 1997), inside front, 3.

276 Haag began . . . definitely better": Haag, 2–3.
276 This is selling specimens . . . the camp?": Haag, 8–9, 15.
276 This is using . . . system works": Haag, interview with author, October 12, 2002. All subsequent Haag quotes from this interview.
276 Mentored by . . . man possessed": Norton, *Rocks*, 289–90.
276 After selling . . . precious stone: Norton, *Rocks*, 293.
277 Then he . . . Glenn Huss: Haag.
277 Haag was . . . out of the business: Haag.
278 His wife . . . own joke: Heidi Haag, interview with author, February 8, 2003.
279 A story . . . was dropped: *New Scientist* 134, no. 1821 (May 16, 1992): 6.
279 Dealer Jeff Kuyken . . . material overseas: Kuyken, November 5, 2005.
279 Haag also . . . cool million: Norton, *Rocks*, 302–3.
279 Haag was . . . the press: Abram Katz, "Local Man Charged with Stealing Prized Meteorite in Brazil." *New Haven Register*, July 1, 1997: A1; Abram Katz, "Bethany Man Freed in Theft of Meteorite." *New Haven Register*, September 13, 1997: A1.
280 While some . . . was recovered: Repost on http://www7.pair.com/arthur/meteor/archive/archive2/July97/msg00015.html.
280 After the Brazil . . . in Africa": Brian Homewood, "The Hottest Rock in the World," August 2, 1997. http://www.newscientist.com/article/mg15520934.500-the-hottest-rock-in-the-world.html.
280 When I asked . . . small samples": John Wasson, e-mail to author, October 19, 2005.
280 Criminals even . . . were apprehended: Ronnie Mckenzie, "Comments on the Article Titled 'From Grootfontein to Windhoek: Meteorites of Namibia,'" *Meteorite* (August 2007): 6–8.
280 One way . . . labeling practices": http://www.imca.cc/.
281 Alan Rubin . . . Reed, and Killgore: Alan Rubin, e-mail to author, October 4, 2005.
281 Conflict revolves . . . the fragments: Jeff Grossman, e-mail to author, October 5, 2005.
281 Difficulties with . . . Tim Swindle: Tim Swindle, "Desert Meteorite Workshop," *Meteorite* (November 2006): 20–21.
281 Some researchers . . . search": Grossman, October 5, 2005.
282 Jeff Kuyken . . . in the market": Kuyken, November 5, 2005.
282 Scientists aren't . . . with dealers: http://www.oregonlive.com/editorials/oregonian/index.ssf?xml/ . . . /1013259343246 10101.xm.
282 Matt Morgan . . . Grand Ronde: wysiwyg://62/http://www.nytimes.com/20.../national/14METE.html.
282 Despite occasional . . . Ibitira eucrite": Alan E. Rubin, book review of *The Robert Haag Collection of Meteorites*, *Meteoritics* 38, no. 4 (2003): 663.
282 Who knew? . . . read it!": Rubin, 663.

Book VIII. Church of the Sky

287 "And her fragile bones: how simply, how completely, Janis would disappear. And all of us.": Robert Olen Butler, "Doomsday Meteor Is Coming," in *Tabloid Dreams: Stories*, 133–45. (New York: Henry Holt, 1996), 139.
287 Or, rather . . . in Arizona: "Tracing the Ries-Meteorite in the City of Nördlingen." trans. Christopher Hikle, brochure, 2000.
287 A satellite photograph . . . the crater's shape: Gisela Pösges and Michael Schieber, *The Ries-Crater Museum Nördlingen*, trans. Peter Scherer and Albert Pösges, Academy Bulletin Nr. 253, Bavarian Academy for Teacher Training Dillingen (Munich: Dr. Friedrich Pfeil, 1997), 73.
288 Tucked inside . . . for the basin: Julius Kavasch, *The Ries Meteorite Crater: A Geological Guide*, trans. by Beryl Höfling (Donauwörth: Ludwig Auer 1986), 24–25, 30.
288 Long before . . . dragons: Wolfgang Kootz, *Nördlingen in the Region of Ries on the Romantic Road* (Dielheim, Germany: Kraichgau, no date), 3, 12.
288 In 1960 . . . a hunch: David H. Levy, *Shoemaker by Levy: The Man Who Made an Impact* (Princeton, NJ: Princeton University Press, 2000), 83; Carolyn Shoemaker, e-mail to author, March 12, 2002.

288 He was stopping . . . Arizona": Paper reprinted in G. J. H. McCall, *Meteorite Craters. Benchmark Papers in Geology 36* (Stroudsburg, PA: Dowden, Hutchinson and Ross, 1977), 170–186.

288 Carolyn says . . . first time": Carolyn Shoemaker, e-mail to author, March 8, 2002.

288 It was . . . weeks before: Carolyn Shoemaker, March 12, 2002; Levy, 83.

288 The 1950s . . . for scientists: Levy, 59.

288 Even some . . . suspects: Levy, 60–61.

288 Harvey Nininger . . . at the time: "Cataclysm and Evolution," originally in *Popular Astronomy* 50, 5, 270–272, May 1942, in *Published Papers*, 488–490.

288 In addition . . . Alfred Wegener: Levy, 53.

289 At the time . . . Rocky Mountain West: Levy, 33, 35, 73, 39.

289 But it was . . . reputation: Levy, 78–79.

289 After reading . . . by volcanism: Levy, 73–74.

289 There was more . . . explosion outward: Eugene Shoemaker, "An Account of Some of the Circumstances Surrounding the Recognition of the Impact Origin of the Ries Crater," typescript. 1984, p. 2. Provided to the author by Carolyn Shoemaker.

290 On June 20 . . . didn't mention Nininger: Levy, 81–83.

290 A month after . . . he needed": Carolyn Shoemaker, March 8, 2002.

290 Found them . . . were impact rocks!": Quoted in Levy, 83.

290 He saw . . . impact ejecta": Eugene Shoemaker, "An Account of," 10.

290 Carolyn says . . . in town: Carolyn Shoemaker, March 8, 2002.

290 The next day . . . Carolyn says: Carolyn Shoemaker, March 8, 2002.

290 Of this epiphanic . . . to realize it": Eugene Shoemaker, "An Account of," 5.

291 He was right . . . a major find": Levy, 83.

291 By the early . . . worldwide: http://www.unb.ca/passc/ImpactDatabase/CINameSort2.htm.

291 Not long after . . . impact events: Bevan M. French, "The Importance of Being Cratered: The New Role of Meteorite Impact as a Normal Geological Process," *Meteoritics & Planetary Science* 39, no. 2 (2004): 169–197, 171.

291 Eugene Shoemaker would . . . crashed into the Moon: Levy, 33, 85–266.

292 The Ries Basin is about . . . the basin floor: Pösges and Schieber, 8.

293 Inside the outer . . . during the impact: Kavasch, 22.

293 Before the actual . . . the Tethys Sea: Pösges and Schieber, 34–35, 32–33.

293 Then it came . . . 300 miles away: Concise overviews of the event can be found in "Tracing the Ries-Meteorite in the City of Nördlingen," a brochure, and the "Ries" entry in Paul Hodge, *Meteorite Craters and Impact Structures of the Earth* (Cambridge, MA: Cambridge University Press, 1994), 91–94.

294 Within one . . . Earth atmospheres: Pösges and Schieber, 38, 41; Kavasch, 34, 38.

294 The meteorite and surrounding . . . from a scalp: Pösges and Schieber, 41–42; Kavasch, 34–36.

295 Still, not a . . . in historic times: Heide and Wlotzka, 79.

295 Philip M. Bagnall . . . Swedish sailors: Philip M. Bagnall, "Starstruck," *Meteorite!*, February 1995: 18–19, 18.

295 In 1929 . . . a meteorite: Coleman and Kennedy, 42.

295 A few years . . . by meteorites: Garry J. Tee, "A Shepherd and His Flock, Killed by Meteorites in 1725," *Meteorite!*, November 1995: 23–24.

295 In 1827 . . . from space: "Meteorites That Have Struck People." http://imca.repetti.net/metinfo/metstruck.html.

296 The afternoon of . . . beetles: Hutchison and Graham, 10.

296 On July 14, 1847 . . . not injured: "Meteorites That Have Struck People." http://imca.repetti.net/metinfo/metstruck.html.

296 (Braunau . . . upon landing): Heide and Wlotzka, 24.

296 In 1946 . . . a meteorite: Coleman and Kennedy, 43.

296 In 1992 . . . not seriously hurt: Walter Branch, Ph.D. "Chronological Listing of Meteorites That Have Struck Humans, Animals and Man-Made Objects (HAMS)." http://imca.repetti.net/metinfo/metstruck.html; Heide and Wlotzka, 80.

296 Only one . . . 10 pounds: Bagnall, 18.

296 The wife of . . . like a kiss: "A Big Bruiser from the Sky: Meteorite Injures Alabama Woman," *Life* December 13, 1954: 26–27.

296 A sickly . . . the stone: Hal Povenmire, "Sylacauga, Alabama Meteorite Revisited, *Meteorite,* May 2003: 26–28. 27.

296 Things went poorly . . . divorce: Ian Frazier, "On Impact." *The New Yorker,* July 9, 2007, and July 16, 2007. 72–79. 78.

296 One of the few . . . the vehicle: Multiple sources tell this story. Interesting information on the fate of the car since the impact can be found at http://www.nyrockman.com/peekskill.htm.

297 In Wethersfield . . . 1982: Michael R. Jensen, William B. Jensen, and Anne M. Black, *Meteorites from A–Z* (Michael R. Jensen: 2001, 119; Kenneth F. Weaver, "Meteorites: Invaders from Space," *National Geographic* 170, no. 3 (September 1986): 390–418, 394.

297 In 1984 . . . a meteorite: http://imca.repetti.net/metinfo/metstruck.html.

297 The mailbox . . . auction: http://www.iht.com/articles/ap/2007/10/29/america/NA-GEN-US-Meteorite-Auction.php.

297 As to animals . . . 1972: http://www.schoolersinc.com/meteorites_p_25.htm.

297 A long-told . . . untrue: Kevin Kichinka, "Nakhla, Part II," *Meteorite!* August 1998: 14–17, 16.

297 The odds . . . 7,000 years or so": Ben Fenton, "What Chance of Being Hit by a Meteorite? Don't Ask a Scientist." http://www.telegraph.co.uk/news/main.jhtml?xml=/news/2006/07/31/nmeteor31.xml&pPage=/core/Matt/pcMatt.

297 The chances . . . to one: "Meteors, Meteorites and Meteor Showers." http://www.bbc.co.uk/dna/h2g2/A480520.

297 By contrast . . . a car: "Go for It and Enjoy," *Hope Health Newsletter* 25, no. 10 (October 2005): 7.

297 We know . . . Heidelberg: Curtis Peebles, *Asteroids: A History* (Washington, DC: Smithsonian Institution Press, 2000), 57.

298 He named . . . Apollo: Peebles, 57.

298 The mile-wide . . . astronomically speaking: Peebles, 57.

298 The historic . . . glowing: Site visit, Heidelberg, March 19, 2002.

298 A handful of . . . Reinmuth's Apollo: Peebles, 56–58.

298 (A Near Earth Asteroid . . . our planet): Timothy Ferris, *Seeing in the Dark: How Backyard Stargazers Are Probing Deep Space and Guarding Earth from Interplanetary Peril* (New York: Simon & Schuster, 2002), 169.

298 One of them . . . Earth: Peebles, 58.

299 Something called . . . cooler side: William Bottke, "JAM II (Joint Astrodynamicists & Meteoriticists) Meeting.," *Meteorite!* 4, no. 4 (November 1998): 36–37; Clark Chapman, "The Long Trip to Earth," *Nature* 407, no. 6804 (October 5, 2000): 573–576; Peebles, 70.

299 A more nuanced . . . rubble: David P. Rubincam and Stephen J. Paddack, "As Tiny Worlds Turn," *Science* 316 (April 13, 2007): 211–12.

299 A few years ago . . . on entry: Peebles, 78–80.

299 Another . . . passed by: "A Tiny Asteroid Whizzes By." http://skyandtelescope.com/printable/news/article_1071.asp.

299 A year later . . . an asteroid: "Asteroid Heading Past Earth Was Closest One Ever Recorded," *Houston Chronicle,* March 19, 2004: 4A.

299 By summer 2007 . . . Asteroids: Near Earth Object Program, "NEO Groups." http://neo.jpl.nasa.gov/neo/groups.html.

299 Astronomers are also . . . been discovered: Alan Fitzsimmons, "Ice Among the Rocks," *Science* 312 (April 28, 2006): 535–36.

299 In terms of . . . per million years: J. Kelly Beatty, "Gauging the Impact," *Sky & Telescope,* October 1999: 32–33.

300 The scientific community . . . researchers suggest: Beatty, "Gauging," 32–33.

300 When objects . . . the planet: Beatty, "Gauging," 33.

300 Scientists estimate . . . 10,000 megatons of TNT: Multiple sources. Coleman and Kennedy, 18–23.

300 In 2001 . . . nada: Siim Veski, Atko Heinsalu, Kalle Kirsimäe, Anneli Poska, and Leili Saarse, "Ecological Catastrophe in Connection with the Impact of the Kaali Meteorite About 800–400 B.C. on the Island of Saaremaa, Estonia," *Meteoritics & Planetary Science* 36 (2001): 1367–75.

300 Every five . . . peasants: Philip A. Bland and Natalya A. Artemieva, "The Rate of Small Impacts on Earth," *Meteoritics & Planetary Science* 41, no. 4 (2006): 607–31, 607.

300 A rock . . . 90 million years: Coleman and Kennedy, 19, 22–23.

300 Some 3.5 billion . . . the dinosaurs: Kenneth Chang, "Scientists Find Signs Big Meteor Hit Earth 3.5 Billion Years Ago," August 23, 2002. http://select.nytimes.com/.

300 Other researchers . . . and earthquakes: "Earth's Surface Transformed by Massive Asteroids." http://www.spacedaily.com/news/early-earth-05h.html.

301 Cosmic impacts . . . Chesapeake Bay: Seung Ryeol Lee, J. Wright Horton Jr., and Richard J. Walker, "Confirmation of a Meteoritic Component in Impact-Melt Rocks of the Chesapeake Bay Impact Structure, Virginia, USA—Evidence from Osmium Isotopic and PGE Systematics," *Meteoritics & Planetary Science* 41, no. 6 (2006): 819–33, 833.

301 Scientists also . . . up to see: Richard A. Kerr, "A Big Splat in the Asteroid Belt Doomed Earth's Dinosaurs," *Science* 317, no. 5843 (September 7, 2007): 1310.

301 If you want . . . are likely to change: "Asteroid 1950 DA." http://neo.jpl.nasa.gov/1950da/.

301 Within my lifetime . . . constellation Cancer: http://neo.jpl.nasa.gov/news/news149.html.

301 Recently, scientists . . . mile-wide crater: "NASA Solves Half-Century Old Moon Mystery." http://www.spaceflightnow.com/news/n0302/22Moon.

302 To keep track . . . precise orbits: Stuart J. Goldman, "Rocks from and in Space," *Sky & Telescope*, January 2001: 84.

302 We're not . . . history of impacts: J. Kelly Beatty, "NEAR Falls for Eros," *Sky & Telescope*, May 2001: 34–37.

302 Eros also . . . in litter: Richard A. Kerr, "Peeling Back One More Layer of Asteroid Mystery," *Science* 313 (July 14, 2006): 158–59. Multiple authors, Special Section, "The Falcon Has Landed: Hayabusa at Asteroid Itokawa," *Science* 312 (June 2, 2006): 1327–53.

302 The following . . . in 2015: http://dawn.jpl.nasa.gov/mission/timeline.asp.

303 Today I have . . . discovery: Site visit, Ries Basin, March 15–19, 2002; field day with Gisela March 18, 2002.

Book IX. Life Work

Chapter IX.1. The Resurrection of Acraman

311 Just a century . . . the lake endured: Pösges and Schieber, 48–56.

311 The most profound . . . refuges for microbial life: Gordon Osinski, "Shocked into Life," *New Scientist*, September 13, 2003: 40–43.

312 Haughton Crater is . . . impact event: Pascal Lee and Gordon Osinski, "The Haughton-Mars Project: Overview of Science Investigations at the Haughton Impact Structure and Surrounding Terrains, and Relevance to Planetary Studies," *Meteoritics and Planetary Science* 40, no. 12 (December 2005): 1755–58.

312 Several thousand years . . . five dozen sites: Gordon R. Osinski, Pascal Lee, John Parnell, John G. Spray, and Martin Baron, "A Case Study of Impact-Induced Hydrothermal Activity: The Haughton Impact Structure, Devon Island, Canadian High Arctic," *Meteoritics and Planetary Science* 40, no. 12 (December 2005): 1859–77, 1860.

312 It's even possible . . . on Mars: Justin J. Hagerty and Horton E. Newsom, "Hydrothermal Alteration at the Lonar Lake Impact Structure, India: Implications for Impact Cratering on Mars," *Meteoritics & Planetary Science* 38, no. 3 (2003): 365–81, 365.

312 "The main heat source . . . the rim of the crater: Osinski, "Shocked," 42.

312 This "vapor-dominated regime" was "short-lived": Osinski et al., "A Case Study," 1872.

312 The hot springs . . . years: Osinski et al., "A Case Study," 1875.

312 As at the Ries, a lake developed: Osinski, "Shocked," 40

312 Fossils from . . . Oz notes: Osinski, "Shocked," 42–43.

313 And microbes are . . . planet's surface: Andrew Glikson, "Early Asteroid Impacts and the Origin of Terrestrial Life," *Meteorite!*, November 2000: 8–14, 9.

313 Charles Cockell . . . truly biologic process": Charles S. Cockell, Pascal Lee, Paul Broady, Darlene S.S. Lim, Gordon R. Osinski, John Parnell, Christian Koeberl, Lauri Pesonen, and Johanna Salminen. "Effects of Asteroid and Comet Impacts on Habitats for Lithophytic Organisms—A Synthesis," *Meteoritics and Planetary Science* 40, no. 12 (December 2005): 1901–14. 1908, 1901, 1912.

313 This is a radical . . . on the earth": Richard A. F. Grieve, "Impact Cratering on the Earth," *Scientific American* 262, no. 4 (April 1990): 66–73, 73.

313 To understand . . . Slimeworld": Gabrielle Walker, *Snowball Earth: The Story of the Great Global Catastrophe That Spawned Life as We Know It* (New York: Crown, 2003), 12.

314 Prokaryotes . . . business would swoon: Patricia Barnes-Svarney, ed, *The New York Public Library Science Desk Reference* (New York: Macmillan, 1995), 98–99.

314 About 1.5 billion . . . food to energy: Eldon D. Enger, *Concepts in Biology* (Dubuque: Wm. C. Brown, 1985), 69–91, 404–7; entries in such reference works as *The Oxford Dictionary of Science.*

314 But as Stephen . . . multicellular *animals*: Stephen Jay Gould, *Wonderful Life: The Burgess Shale and the Nature of History* (New York: Norton, 1989), 58.

314 The first such . . . 100 million years: Gould, 59; Walker, 215; *Brachina Gorge Geological Trail,* no date, brochure, Mines and Energy, South Australia.

314 On a November afternoon . . . tickle my back: Site visit, South Australia, November 18–21 November 2003.

315 *Dickinsonia costata* . . . head and tail": Walker, 208.

315 Lungless, this worm breathed through its skin: Patricia Vickers Rich et al. *The Fossil Book: A Record of Prehistoric Life* (Mineola, NY: Dover, 1996), 96.

315 Nonetheless, *Dickinsonia* . . . animals: Rich, 92–93.

315 With the rank . . . dumbstruck: Sue Barker, Murray McCaskill, and Brian Ward, *Explore the Flinders Ranges* (Adelaide: Royal Geographical Society of Australasia, South Australian Branch, 2000), 91.

316 Lorraine walks me . . . world's first reefs: Vickers, 114, 117; *Corridors Through Time: The Geology of the Flinders Ranges National Park,* no date, brochure, Department of Mines and Energy.

317 Others believe . . . complex creatures: Walker, 211; Gould, 311–314.

317 Researchers such as Richard . . . Precambrian: Richard Fortey, *Trilobite! Eyewitness to Evolution* (New York: Knopf, 2000).

318 While most of . . . water to land: Rich, 369–71.

318 Also among . . . Pikaia: Gould, 208, 321–23.

318 "In a geological . . . curious ways": Gould, 64.

319 The Moon loomed . . . hours shorter: George Williams and Phillip Schmidt, "Proterozoic Equatorial Glaciation: Has 'Snowball Earth' a Snowball's Chance?," *The Australian Geologist* 117 (December 31, 2000): 21–25. 22.

320 Williams . . . at the equator: Walker, 90.

320 But here's . . . icy cold: Walker, 20; Paul Hoffman, and Daniel P. Schrag, "Snowball Earth," *Scientific American* 282 (January 2000): 68–75, 69–70.

320 There was more . . . you need seasons: Williams and Schmidt, "Proterozoic Equatorial Glaciation," 22–23.

320 Williams champions . . . present orientation: Walker, 165–67; Williams and Schmidt, "Proterozoic Equatorial Glaciation," 25.

320 There were other . . . Barbados: Walker, 67–70; Hoffman and Schrag, "Snowball Earth," 73.

320 So how to . . . Snowball Earth": Walker, 83–100.

320 Since Kirschvink . . . a thaw: Hoffman and Schrag, "Snowball Earth," 72–73.

320 The whole process . . . as now: Hoffman and Schrag, "Snowball Earth," 75.

321 As to evidence . . . in temperatures. Walker, 168–69.

321 Further, Hoffman . . . to diversify: Hoffman and Schrag, "Snowball Earth," 70, 74–75; Runnegar, Bruce, "Loophole for Snowball Earth," *Nature* 405 (May 25, 2000): 403–4.

321 There is, however . . . the fossil record": http://www.snowballearth.org/kick-start.html.

321 So what forced . . . ecosystem changes?: Robert Irion, "Slip-Sliding Away," *New Scientist* 171, no. 2304 (August 18, 2001): 35–37.

321 Paleontologist Kath Grey . . . old volcanic rock: V. A. Gostin, "The Acraman Asteroid Impact, South Australia," extended abstract for seminar at Macquarie University, November 1, 2002; Gostin, e-mail to author, November 29, 2007.

323 It was in 1979 . . . Australia's largest: George E. Williams, "Acraman, South Australia: Australia's Largest Meteorite Impact Structure," *Proceedings of the Royal Academy of Victoria* 106 (1994): 105–7.

324 Not long after . . . a puzzle: Vic Gostin, interview with author, November 21, 2003.

324 He had found . . . gross alteration": Vic Gostin, private notes shared with author.

324 Expecting the volcanic . . . 580 million years old: Gostin, "The Acraman Asteroid Impact."

324 Doubts nagged him . . . delirious": Vic Gostin, interview with author, November 19, 2003.

325 Doctoral student . . . iced up: Vic Gostin, e-mail to author, May 16, 2006.

325 Vic's ejecta blocks . . . Lake Acraman": Williams, "Acraman, South Australia," 106.

325 The next year . . . *Science:* Victor A. Gostin, Peter W. Haines, Richard J.F. Jenkins, William Compston, and Ian S. Williams, "Impact Ejecta Horizon Within Late Precambrian Shales, Adelaide Geosyncline, South Australia," *Science* 233 (July 11, 1986): 198–200. George E. Williams, "The Acraman Impact Structure: Source of Ejecta in Late Precambrian Shales, South Australia," *Science* 233 (July 11, 1986): 200–203.

325 Three years later . . . tag impact events: Victor A. Gostin, Reid R. Keays, and Malcolm W. Wallace, "Iridium Anomaly from the Acraman Impact Ejecta Horizon: Impacts Can Produce Sedimentary Iridium Peaks," *Nature* 340 (August 17, 1989): 542–44.

325 Acraman was . . . impact fragments: Gostin et al., "Impact Ejecta Horizon," 198.

326 This huge meteorite . . . just as deep: Gostin, "The Acraman Asteroid Impact."

326 Before I had . . . pall of dust: Williams, "Acraman, South Australia," 124.

327 The Cessna banks . . . crater: George E. Williams and Malcolm W. Wallace, "The Acraman Asteroid Impact, South Australia: Magnitude and Implications for the Late Vendian Environment," *Journal of the Geological Society* (London) 160 (2003): 545–54.

328 But in May . . . Western Australia: Gostin, May 16, 2006.

328 The Australian Centre . . . biostratigraphy: http://aca.mq.edu.au/News/Kath_Grey.htm.

328 For right after . . . became dominant: Kathleen Grey, Malcolm R. Walter, and Clive R. Calver, "Neoproterozoic Biotic Diversification: Snowball Earth or Aftermath of the Acraman Impact?," *Geology* 31, no. 5 (May 2003): 459–62.

328 Grey, Walter and Calver . . . diversifying species": Kath Grey, "A Baptism of Ice and Fire," *Australasian Science,* April 2005: 26–28.

328 The post-Acraman . . . ice and fire": Grey, "Baptism," 27–28.

329 The work behind . . . Grey says: Grey, "Baptism" 26.

329 When she began . . . her finds: Kath Grey, e-mail to author, 27 May 2003.

329 The rise of . . . tiny creatures: Grey, May 27, 2003.

329 Snowball Earther . . . wrong": Quoted in Rachel Nowak, *New Scientist* 178, no. 2393 (May 3, 2003): 17.

329 Grey replies . . . the K-T event": Kath Grey, e-mail to author, October 25, 2003.

330 Much more remains . . . further testing": Kath Grey, e-mail to author, May 9, 2006.

330 Moreover . . . spiny plankton: Grey, October 25, 2003.

330 Some four years . . . study is needed: Kath Grey, e-mails to author, July 18, 2007, and August 27, 2007.

330 In the time . . . the Acraman impact: Paul Hoffman, e-mail to author, August 9, 2007.

331 But Grey . . . she's found: Grey, May 27, 2003.

331 Her observations . . . far from over": Hoffman, August 9, 2007.

Chapter IX.2. Old Stones That Can Be Deciphered

333 Anaxagoras raised . . . type of matter: Anaxagoras, "On Nature," in *The Classical Greek Reader,* ed. Kenneth J. Atchity, (Oxford: Oxford University Press, 1996), 122–24.

334 As England's . . . not unscientific": Thomson quoted in Burke, 167.

334 Part of the impetus . . . a biological origin": Burke, 168.

334 And German naturalist . . . in our atmosphere": Hermann von Helmholtz, "Helmholtz on the Use and Abuse of the Deductive Method in Physical Science," *Nature,* January 14, 1875: 211–12.

334 Burke notes . . . death of the Earth: Burke, 169–70.

334 Otto Hahn . . . known from Earth: Burke, 171–72.

335 A new question . . . stir up the press: Burke, 312–13.

335 In the early 1960s . . . indigenous to the meteorite": George Claus and Bartholomew Nagy, "A Microbiological Examination of Some Carbonaceous Chondrites," *Nature* 192 no. 4803 (November 18, 1961): 594–96.

335 Two years after . . . nonbiological processes: Frank W. Fitch and Edward Anders, "Observations on the Nature of the 'Organized Elements' in Carbonaceous Chondrites," in *Annals of the New York Academy of Sciences* 108.Art 2 (June 29, 1963): 495–513, 505, 509–10.

335 Claus and Nagy disagreed: George Claus, Bartholomew Nagy, and Dominic L. Europa, "Further Observations on the Properties of the 'Organized Elements' in Carbonaceous Chondrites," in *Annals of the New York Academy of Sciences* 108.Art 2 (June 29, 1963): 580–605, 602.

335 In 1996 . . . Antarctica: David S. McKay, Everett K. Gibson Jr., Kathie L. Thomas-Keprta, Hojatollah Vali, Christopher S. Romanek, Simon J. Clemett, Xavier D. F. Chillier, Claude R. Maechling, and Richard N. Zare, "Search for Past Life on Mars: Possible Relict Biogenic Activity in Martian Meteorite ALH84001," *Science* 273, no. 5277 (August 16, 1996): 924–30.

335 The two major pieces . . . the universe it seems: Max P. Bernstein, Jason P. Dworkin, Scott A. Sandford, and Louis J. Allamandola, "Ultraviolet Irradiation of Naphthalene in H20 Ice: Implications for Meteorites and Biogenesis," *Meteoritics and Planetary Science* 36 (2001): 351–58, 351; "ISO Detects 'Ringed Molecule' Benzene Around Stars." http://www.spacedaily.com/news/extrasolar-01.html.

336 In ALH 84001 . . . contamination: Thomas Stephan, Elmar K. Jessberger, Christian H. Heiss, and Detlef Rost, "TOF-SIMS Analysis of Polycyclic Aromatic Hydrocarbons in Allan Hills 84001," *Meteoritics and Planetary Science* 38, no. 1 (2003): 109–16, 109.

336 As to the carbonate globules . . . terrestrial microfossils": Harry Y. McSween, Jr., "Evidence for Ancient Life in a Martian Meteorite (! or ?)," *Meteoritics and Planetary Science* 31 (1996): 691–92.

336 In fact . . . yet remain viable": J. Kelly Beatty, "Life at the Limit," *Sky & Telescope*, September 1999: 40–42, 41, 42.

336 ALH 84001 has yielded . . . unmelted: McSween, "Evidence for Ancient Life," 691.

336 The hydrogen inside . . . microbial processes: John M. Eiler, Nami Kitchen, Lauri Leshin, and Melissa Strausberg, "Hosts of Hydrogen in Allan Hills 84001: Evidence for Hydrous Martian Salts in the Oldest Martian Meteorite?," *Meteoritics and Planetary Science* 37 (2002): 395–405, 395.

336 And researchers also have . . . chemical process: J. Kelly Beatty, "Magnetite Chains Hint at Martian Microbe," *Sky & Telescope*, June 2001: 24. Kathie L. Thomas-Keprta, et al., "Elongated Prismatic Magnetite Crystals in ALH 84001 Carbonate Globules: Potential Martian Magnetofossils," abstract, *Geochimica et Cosmochimica Acta* 64, no. 23 (December 1, 2000). 4049–81; David J. Barber and Edward R.D. Scott, "Transmission Electron Microscopy of Minerals in the Martian Meteorite Allan Hills 84001," *Meteoritics & Planetary Science* 38, no. 6 (2003): 831–848, 831; Adrian J. Brearley. "Magnetite in ALH 84001: An Origin by Shock-Induced Thermal Decomposition of Iron Carbonate," *Meteoritics & Planetary Science* 38, no. 6 (2003): 849–70, 849.

337 However all . . . single-handedly": Richard Kerr, "New Signs of Ancient Life in Another Martian Meteorite?," *Science* 311 (March 31, 2006): 1858–59, 1858.

337 Meanwhile, panspermia . . . possible: "No Panspermia Between Stars," *Sky & Telescope*, May 2000: 23.

337 Microbes are tough . . . in outer space: Paul Davies, "Interplanetary Infestations," *Sky & Telescope* September 1999: 32–38, 35.

337 The universe itself . . . UV-saturated environments: David Warmflash and Benjamin Weiss, "Did Life Come from Another World?," *Scientific American*, November 2005: 64–71.

337 And we know . . . the hardy germs: http://science.nasa.gov/newhome/headlines/ast01sep98_1.htm.

337 Researchers have even . . . 250 million years ago: Ian Wright, review of *Astronomical Origins of Life: Steps Toward Panspermia*, in *Meteoritics and Planetary Science* 36 (2001): 321.

337 Meteoriticist Guy Consolmagno . . . spacelike conditions: Guy Consolmagno, e-mail to author. September 14, 2007.

338 Back in the 1960s . . . from the Moon: Burke, 313–16; "Panel Discussion," in *Annals of the New York Academy of Sciences* 108.Art 2 (June 29, 1963): 606–15.

338 So we know . . . pools here on Earth: "Two Extreme Asteroids," *Sky & Telescope*, December 1999: 26.

338 In fact, scientists . . . the early Earth": Victoria K. Pearson, Mark A. Sephton, Anton T. Kearsley, Philip A. Bland, Ian A. Franchi, and Iain Gilmour, "Clay Mineral-Organic Matter Relationships in the Early Solar System," *Meteoritics and Planetary Science* 37 (2002): 1829–33, 1829.

338 According to what . . . amino acids: G. Matrajt, S. Pizzarello, S. Taylor, and D. Brownlee, "Concentration and Variability of the AIB Amino Acid in Polar Micrometeorites: Implications for the Exogenous Delivery

of Amino Acids to the Primitive Earth," *Meteoritics & Planetary Science* 39, no. 11 (2004): 1849–58; Henner Busemann, Andrea F. Young, Conel M. O'D. Alexander, Peter Hoppe, Sujoy Mukhopadhyay, and Larry R. Nittler, "Intersteller Chemistry Recorded in Organic Matter from Primitive Meteorites," *Science* 312 (May 5, 2006): 727–30; G. Matrajt, D. Brownlee, M. Sadilek, and L. Kruse, "Survival of Organic Phases in Porous IDPs During Atmospheric Entry: A Pulse-Heating Study," *Meteoritics & Planetary Science* 41, no. 6 (2006): 903–11.

339 But just how . . . ultraviolet light: Robert Roy Britt, "Seeds of Life Are Everywhere, NASA Researchers Say," March 27, 2002. http://www.space.com/scienceastronomy/generalscience/amino_acids_020327.html.

339 There happens . . . organic materials: David F. Blake, and Peter Jenniskens, "The Ice of Life," *Scientific American* 265, no. 2 (August 2001): 44–51.

339 A few years ago . . . into a comet: "Close-Call Meteorite Yields Primordial Water," *Sky & Telescope*, January 2000: 18.

339 The halite . . . extraordinarily early: Larry O'Hanlon, "Meteor Hints at Earlier Origin of Life," June 12, 2000. www.discovery.com.

339 Researchers think . . . *on an asteroid:* Donald D. Bogard, Daniel H. Garrison, and Jozef Masarik, "The Monahans Chondrite and Halite: Argon-39/Argon-40 Age, Solar Gases, Cosmic-Ray Exposure Ages, and Parent Body Regolith Neutron Flux and Thickness," *Meteoritics and Planetary Science* 36 (2001): 107–22, 117–18.

339 One of my . . . so special: "Close-Call Meteorite."

339 One boy said . . . local police: Alan E. Rubin, Michael E. Zolensky, and Robert J. Bodnar, "The Halite-Bearing Zag and Monahans (1998) Meteorite Breccias: Shock Metamorphism, Thermal Metamorphism and Aqueous Alteration on the H-Chondrite Parent Body," *Meteoritics & Planetary Science* 37 (2002): 125–141, 136.

339 Because of the quick . . . Monahans and Zag: Rubin et al., "The Halite-Bearing Zag and Monahans," 140, 134–37.

340 Computer models . . . prevalent on Earth: J. Kelly Beatty, "Did Asteroids Supply Earth's Water?," *Sky & Telescope*, February 2001: 26–27.

340 Yet another . . . system's planets: Michael J. Drake, "Origin of Water in the Terrestrial Planets," *Meteoritics & Planetary Science* 40, no. 4 (2005): 519–27.

340 Meteorites have water . . . cometary material: "Meteorite Analysis Suggests Comets Delivered Life's Key Ingredients," February 27, 2001, http://www.spacedaily.com/news/life-01g.html; "Amino Acids Found in Orgueil," *Meteorite* (May 2001): 5–6; Matthieu Gounelle, Pavel Spurný, and Philip A. Bland, "The Orbit and Atmospheric Trajectory of the Orgueil Meteorite from Historical Records," *Meteoritics & Planetary Science* 41, no. 1 (2006): 135–50.

340 Bill Bryson . . . at a bar. Bill Bryson, *A Short History of Nearly Everything* (New York: Broadway Books, 2003), 288–90.

341 There's another . . . water: Bryson, 281.

341 Nearly fifty . . . on our planet: Kress, 26.

341 But of those . . . though not all: Kress, 26.

341 Meanwhile, scientists have . . . other worlds: Michael N. Mautner, "Meteorites, and the Origins and Future of Life," *Meteorite*, February 2002: 8–12.

342 But if such . . . "metallic, gunpowder notes": "Meteorite Perfume," *Sky & Telescope*, May 2000: 22.

342 The most exciting . . . the northern U.S. Alan R. Hildebrand, Phil J. A. McCausland, Peter G. Brown, Fred J. Longstaffe, Sam D. J. Russell, Edward Tagliaferri, John F. Wacker, and Michael J. Mazur, "The Fall and Recovery of the Tagish Lake Meteorite," *Meteoritics and Planetary Science* 41, no. 3 (2006): 407–31.

342 The quick collection . . . weakest meteorite known": P. G. Brown, A. R. Hildebrand, and M. E. Zolensky, "Tagish Lake," *Meteoritics & Planetary Science* 37, no. 5 (May 2002): 619–21, 620.

343 Outfitter and pilot . . . in a freezer: Dennis Urquhart, "Largest Meteorite Find in Canadian History," press release, University of Calgary, May 31, 2000; Hildebrand et al., "The Fall and Recovery," 407–31.

343 Peter Brown . . . in diameter: Peter Brown, e-mail to author, May 29, 2006.

343 Scientists from . . . meteorites, though: Urquhart, "The Fall and Recovery," 407–31.

343 Also distinguishing this . . . carbonaceous chondrite: "More Pieces of Rare Meteorite Recovered," *Sky & Telescope*, September 2000: 18–19.

343 The orbit . . . earth-crossers: Peter G. Brown et al., "The Fall, Recovery, Orbit, and Composition of the Tagish Lake Meteorite: A New Type of Carbonaceous Chondrite," *Science* 290 no. 5490 (October 13, 2000): 320–25, 321.

343 Tagish Lake . . . including L'Aigle: "Tagish Lake," 619.

343 This is due not . . . weird: Jeffrey N. Grossman, "A Meteorite Falls on Ice," *Science* 290, no. 5490 (October 13, 2000): 283–85, 283.

343 Everyone expected . . . Tagish Lake type materials": J. K. B., "Tagish Lake Meteorite Mystery Deepens," *Sky & Telescope* June 2001: 29. "Tagish Lake," 619–21. J. M. Friedrich, Ming-sheng Wang, and Michael E. Lipschutz, "Comparison of the Trace Element Composition of Tagish Lake with Other Primitive Carbonaceous Chondrites," *Meteoritics & Planetary Science* 37 (2002): 677–86, 685. There are several other relevant articles regarding Tagish Lake in this issue of *Meteoritics & Planetary Science*. Quote from M. E. Zolensky, K. Nakamura, M. Gounelle, T. Mikouchi, T. Kasama, O. Tachikawa, and E. Tonui, "Mineralogy of Tagish Lake: An Ungrouped Type 2 Carbonaceous Chondrite," *Meteoritics & Planetary Science* 37 (2002): 737–61, 759.

344 Later studies . . . meteorite collections": "The Fall and Recovery," 407–31.

344 Tagish Lake . . . home world: Keiko Nakamura-Messenger, Scott Messenger, Lindsay P. Keller, Simon J. Clemett, and Michael Zolensky, "Organic Globules in the Tagish Lake Meteorite: Remnants of the Protosolar Disk," *Science* 314 (December 1, 2006): 1439–42.

344 I never got to . . . a long time ago—Orgueil: Site visit, Paris, December 23, 2002.

346 Later, I'd learn . . . into Orgueil: Mark A. Sephton, Victoria K. Pearson, and Jon S. Watson, "The Sweet Smell of Orgueil," *Meteorite*, February 2003: 14–15.

Book X. Old Fire on Blue Ice

349 *For there is no escape anywhere*: Richard E. Byrd, *Alone* (New York: Grosset & Dunlap, 1938), 16.

Chapter X.1. Above the Clouds, Halfway Down

Site visit, Christchurch, November 29, 2003, to December 4, 2003. Site visit, Antarctica, December 4, 2003–January 20, 2004.

356 One picture . . . early 1990s: http://geology.cwru.edu/~ansmet/.

356 Cassidy was . . . before, anywhere: Ralph Harvey, "The Origin and Significance of Antarctic Meteorites," *Chemie der Erde/Geochemistry* 63, no. 2 (2003): 93–147, 94–95.

356 Now we know . . . Ralph writes: Harvey, "Origin and Significance," 115–16.

356 Some ice fields . . . Antarctic meteorites: Harvey, "Origin and Significance," 106.

356 The blue ice . . . they are: http://geology.cwru.edu/~ansmet/why_ant/index.html. Harvey, "Origin and Significance," 99–105.

356 Antarctica has become . . . search teams too: Harvey, "Origin and Significance," 95–96.

356 As Ralph puts . . . develop and succeed": http://www.cwru.edu/36720/affil/ansmet/faqs.html.

357 Antarctic meteorites have helped . . . ureilites come from Antarctica: Harvey, "Origin and Significance," 136–37.

357 So the simpler . . . them as such: http://www.cwru.edu/36720/affil/ansmet/faqs.html.

358 It looked like . . . glassy windshield": Harvey, "Origin and Significance," 141.

359 I was sitting . . . renunciation worth while": Wilfrid Noyce, *The Springs of Adventure* (Cleveland and New York: World, 1958), 33, 34.

360 So if I was . . . outlasts the doer: Noyce, 123.

361 During one of our . . . aided his survival: Multiple popular sources tell these stories.

361 Antarctica hadn't . . . cold and ice: Edward G. Atkins and Larry Engel, *Antarctica*, booklet (Arlington, VA: National Science Foundation, 2001), 24.

362 Of them, about 600 people have died there: *Antarctica: An Encyclopedia from Abbott Ice Shelf to Zooplankton* (Toronto: Firefly, 2002), "Deaths," 61.

362 Antarctica covers . . . a giant's sortilege: Multiple popular sources convey basic information about the continent.

365 The Ross Sea . . . we now stand: *Antarctica: An Encyclopedia,* "Ross Ice Shelf," 153.

366 McMurdo is . . . raison d'être: *McMurdo Station Guide,* United States Antarctic Program, National Science Foundation, booklet, no date, 7.

Chapter X.2. I Crap Through Disco Night at the South Pole

378 On its columns . . . 850 miles away: Elaine Hood, e-mail to author, September 21, 2004.

380 Because the pole's . . . about 100 millennia: *Antarctica: An Encyclopedia,* "South Pole," 174.

Chapter X.3. Bedlam

389 Years after . . . heaped up solitude": Quoted in *Antarctica: An Encyclopedia,* "Winds," 201.

402 One cloudy . . . experiment again: Marc Fries, e-mail to author, September 25, 2007.

Chapter X.4. Massif: An Unmaking

412 *The day we . . . more than 10,000 feet:* John Schutt, e-mail to author, May 22, 2005.

421 That morning . . . part of the strewnfield: William A. Cassidy, *Meteorites, Ice and Antarctica: A Personal Account* (Cambridge, MA: Cambridge University Press, 2003), 79. http://geology.cwru.edu/~amlamp/BDM/GRO/GRO63/GRO63text.html.

422 Monika eventually . . . very slow ice motion": M. E. Kress, P. S. Jandir, R. Carter, J. Schutt, R. P. Harvey, and G. K. Benedix, "An Unusual Strewn Field at the Otway Massif, Grosvenor Mountains, Antarctica," poster, Meteoritical Society Conference, Tucson, August 13–17, 2007.

432 A month later . . . never coming back: Oliver Botta, e-mail to author, February 15, 2004.

433 "I think virtually . . . for my tentmate": Ralph Harvey, e-mail to author, May 16, 2005.

434 In his forgotten . . . songs or stories: Thomas R. Henry, *The White Continent: The Story of Antarctica* (New York: William Sloane Associates, 1950), 83.

434 His rival . . . South Pole: Henry, 85.

434 Both were . . . beauty in action": Henry, 86.

434 "You are hemmed . . . off their fat": Byrd, 16–17.

434 "I find I . . . am undone": Byrd, 94.

434 "I don't think . . . phosphorus and calcium": Byrd, 129.

434 Perhaps more directly . . . he has fallen into a trap'": Jean Rivolier, Genevieve Cazes, and Ian McCormick, "The International Biomedical Expedition to the Antarctic: Psychological Evaluations of the Field Party," in *From Antarctica to Outer Space: Life in Isolation and Confinement,* ed. A. A. Harrison, Y. A. Clearwater, and C. P. McKay, (New York: Springer-Verlag, 1991), 283–90, 289.

435 Unlike my ANSMET . . . adolescent hostilities: Jack Stuster, *Bold Endeavors: Lessons from Polar and Space Exploration* (Annapolis: Naval Institute Press, 1996), 174–75.

435 I later read . . . varying degrees: Lawrence Palinkas and Peter Suedfeld, "Psychological Effects of Polar Expeditions," *The Lancet.* www.thelancet.com. Published online June 25, 2007: 1–11.

435 What Ralph . . . never be back: Harvey, May 16, 2005.

435 What the researchers . . . an absolute necessity": Rivolier et al., 290.

435 All told . . . field season: Ralph Harvey, e-mail to author, February 12, 2004.

Afterword

440 Not long after . . . *The Simpsons Movie:* Site visit, Tucson, August 12–14, 2007.

441 Tim also told me . . . about that age," he'd later tell me: Tim Swindle, e-mail to author, August 22, 2007.

441 Though Oliver Botta . . . another kind of bag: Daniel P. Glavin, et al, "Amino Acid Analyses of Antarctic CM2 Meteorites Using Liquid Chromatography-Time of Flight-Mass Spectrometry," *Meteoritics & Planetary Sciences* 41, no. 6 (2006): 889–902, 900.

442 This wasn't . . . get to Mars: Site visit, Houston, March 17–20, 2004.

445 Meteorites are even helping . . . and utterly random: John B. Barrow and John K. Webb, "Inconstant Constants," *Scientific American,* June 2005: 56–63.

446 A 2008 discovery . . . dark energy doesn't exist: Philip A. Bland, "Small-Scale Observations Tell a Cosmological Story," *Science* 320 (April 4, 2008): 61–62.

· *Acknowledgments, Permissions, and a Narrative Note* ·

I could not have written this book without the assistance and goodwill of many people. I thank all who have helped over the years, not only with the book but with the changes in my life that became a part of it. If I have inadvertently omitted the name of a source or someone who assisted me over the years, my apologies.

For my material on Ellis Hughes and Tomanowos, I thank Sandy McGuire, Patricia Kohnen, Brent Merrill, Dick Pugh, Gary Zarr, and Jeff Eden. I also thank Bob Kingston, Megan Friedel, Carolyn Hixson, and the Oregon Historical Society; staff at the American Museum of Natural History library, especially Mark Katzman; Kathryn Harrison; the Clackamas County Historical Society; and the Oregon State Archives reference staff.

For my chapter on Eliza Kimberly, F. W. Cragin, and the Kansas Meteorite Farm, I thank the University of Kansas Library, the Kansas State Historical Society, Kansas State University Library, the Mabee Library at Washburn University, the State Historical Society of Iowa, the Denver Public Library, and the University of Colorado Library. A special thanks to Jessy Randall, curator/archivist at the Special Collections and Archives department of Colorado College's Tutt Library, and the student assistants for finding material on and by Cragin.

My trip to Greenland could not have happened without the kindness of the Peregrine Fund and those associated with that stellar conservation group: I thank Bill Burnham, in memoriam, Kurt Burnham, and Regan Haswell. You were wonderful traveling companions in a beautiful place. At Thule, Jack Stephens could not have been more gracious. I also thank Jack Cafferty, for help with logistics and for answering panicky phone calls, as well as Torben Diklev, Henning Haack, and Holger Petersen for e-mail answers to my questions. I also thank Clif bar for providing a supply of my favorite flavor.

To Philippe Lherminier and Barbara Whiteman, both Kathe Lison and I offer deep gratitude for hosting us in the lovely Château du Fontenil. Thanks to Zelimir Gabelica for driving us to and fro. To the staff of the Regency Museum in Ensisheim, our thanks, especially for letting me touch the meteorite. Both Jean-Pierre Bruyére and Jean-Marie Blosser were kind hosts and patient sources of information. To Claude Perron of the National Museum of Natural History in Paris, thank you for showing me those historic meteorites and for answering my questions. Sarah Gordon kindly provided some translations of material that helped with historical detail.

I thank David Kring for reviewing my material on Meteor Crater, as well as Drew N. Barringer, Nancy Southgate, and Felicity Barringer for providing information on Daniel Moreau Barringer.

My mini-biography of Harvey Harlow Nininger could not have been written without the cooperation of his daughter Doris Banks and grandchildren Gary Huss and Peggy Schaller. I thank you for your time and answers and the unfettered access to Nininger's private papers. All three, as well as Sue Greiner, reviewed my chapters for accuracy while never constraining my handling of the material. I also thank Dorothy Richards and Barbara Richards Buskirk for their time, interviews, and material, which so richly rounded out my understanding of Nininger, Alex Richards, and the first Mexico expedition. My thanks to Drew N. Barringer for permission to quote from Brandon's letters. I also thank Harry Baldwin for sharing what papers he could from his father-in-law Lincoln LaPaz. I extend my appreciation to Glenda and Richard Hahn, Horace Collins, George Corner, Wolfgang Elston, Lora Varley, Horton Newsom, Cynthia Haynes of the *Norton Daily Telegram*, the Norton Public Library, Mike Mansur,

Kris Haglund, Jack Murphy, Cindy Gottsch, Linn Peterson, everyone at McPherson College, including Susan Taylor and Rowena Olsen, and everyone at the McPherson Museum, including David Flask, Mike Linblade, and Tashia Dare.

The impromptu expedition to the former home of the American Meteorite Laboratory was made possible by the enthusiasms of Gary Curtiss and Matt Morgan. Thank you both. Several dealers and collectors made time to talk to me in person or over e-mail. I thank them all, but in particular Blaine Reed, Marvin Killgore, Robert Haag, Steve Arnold, Darryl Pitt, Jeff Kuyken, Martin Horejsi, and Al Lang. Concerning the relationship between dealers and scientists, several of the latter also answered my questions, and I thank especially Jeff Grossman, Alan Rubin, and John Wasson. I thank Richard Norton for kindly showing me around my first dizzying Tucson Gem and Mineral Show. A very special thanks to Geoff Notkin for his sales, help in defining terms, and good cheer, especially as I neared completion of the book.

The trip to Germany would have been paltry had it not been for the enthusiastic guiding and assistance of Gisela Pösges (and her dog Joy). For answers to my questions, I thank Carolyn Shoemaker. Thanks also to Gerhard Klare for showing me Karl Reinmuth's "Plate of Detection."

I cannot thank Vic Gostin enough for his time, assistance, and friendship while I looked into the story of the Lake Acraman impact. To Vic, a toast. I thank also his wife Olga for her graciousness, Lorraine Edmunds for her naturalist's eye, George Williams for answering queries from afar, the staff at Wilpena Pound Resort for their help, Paul Hoffman for his time and not least of all, Kath Grey for patiently answering my questions regarding her work on post-Acraman plankton diversification. Vic and Kath also reviewed this portion of my manuscript for accuracy without constraining my handling of the content.

To the good people of the ice, you are esteemed: Guy Guthridge, Elaine Hood, everyone at McMurdo and Pole—with a special nod to Jennifer Armstrong, Traci Joan MacNamara, and Kristan Hutchison. Early on, Barry Lopez, Gretchen Legler, and Carl Safina offered advice on my application to the Antarctic Visiting Artists and Writers program at the National Science Foundation; to them, my thanks. I am ever grateful to the NSF for the grant. To the ANSMET team in 03–04—John Schutt, Bill McCormick, Andrew Dombard, Tim Swindle, Oliver

Botta, Gordon Osinski, Nancy Chabot, Gretchen Benedix, Barb Cohen, and Monika Kress—I will never forget you or our experiences on the polar plateau. I cannot thank you enough and I hope my words do some justice to our time there. Thank you for sharing your own words and impressions with me.

I single out Mongo, aka Ralph Harvey, who gave me a chance to collect meteorites in Antarctica, understood what happened to me there, and over the years has become more than a source for scientific information. He's become a friend. Ralph read the entire manuscript and offered "ex-cellent" feedback. Now go find another Martian, okay?

Several other scientists have answered questions, and to them, noted in the text, I offer my appreciation for taking time out of your busy lives to help me out.

Many friends have made it a point to send clippings and Web links to me that pertained to my research, as well as to offer words of encouragement and advice, including Robert Cook, Kelli Cargile Cook, Eric Lagergren (who led me away from a confusing title), Scott Minar, Ted Cable, C. A. Reynolds, John Calderazzo, Glen Chilton, Bob Johnson, and Julie Hensley.

To my writing friends in Logan, Utah—gathered as "Splinters," Kathe Lison, Charles Waugh, Michael Sowder, Maria Melendez, Jennifer Sinor, and the late (and missed) Ken Brewer—my deep thanks for your advice, love, and friendship. To the Wasatch Writers Alliance in Salt Lake City—especially Dorothee Kocks, Bill Kerig, Christy Karras, Melissa Bond, and Steve Trimble—thanks too for your advice and friendship.

I thank John Price, Jen Johnson, Jen Peebles, Kim Johnson, and Paisley Rekdal for their friendship. I also thank Kim Barnes, Chet Raymo, Bob Pyle, Terry Tempest Williams, Jen Sahn, David Gessner, and Patricia Hampl for good words at just the right time.

To all the fine folks of the Association for the Study of Literature and Environment, my appreciation for their fellowship and inspiration over the years. Yo, Mike Branch: The meteorite has hit neither my head nor my bone spurs.

At Utah State University, I thank the Department of English and the College of Humanities, Arts and the Social Sciences. The late Dean Gary Kiger was an enthusiastic supporter of my work. I thank department head Jeff Smitten for his support and patience over the years. I thank coworkers and friends Brian

McCuskey, Kris Miller, Marina Hall, Paul Crumbley, Phebe Jensen, and Leslie Brown. Leslie, my managing editor at *Isotope: A Journal of Literary Nature & Science Writing,* started that position just before I left for Antarctica. She has kept the magazine going strong over the years. I wish to gratefully acknowledge USU's support via a Gardner Fellowship and additional travel support for my research in New Zealand and Australia. Annie Nielsen patiently helped me with many forms.

To my two research assistants, Jen Henderson at Kansas State University and Darcy Minter at Utah State University, thanks for your help with some important if sometimes dull tasks.

To my students over the years, you have helped me pay attention to matters of craft and meaning through your own work and through our discussions. In particular, I want to thank Pete Gomben, Ben Quick, Sam Robinson, Kacy Lundstrom, Emily Burgon, Russ Beck, John Engler, Russ Winn, Darren Edwards, Lyra Hilliard, and Jacoba Mendelkow.

At Kansas State University several people offered cool heads and thoughtful words during a time of great turmoil. You know who you are, and I thank you.

To the Whiting Award committee, my still-stunned thanks. The prize made much of the travel for this book possible.

Ursula Marvin, meteoriticist and historian of all things meteoritical, read the manuscript and helped me to correct mistakes. To her, a debt of gratitude as big as Vesta.

Thanks again to Chet Raymo, for his wonderful words on behalf of this book, and to Richard Rhodes, who, after I contacted him out of the blue, also read an earlier version of the manuscript and offered his enthusiasms on its behalf. I'm grateful as well to Jonathan Rosen, for his endorsment of the book and his words elsewhere for my "bird book."

Any mistakes that remain are my own responsibility. Metric conversions were approximate. If readers note errors in such conversions or other mistakes in factual material, please contact me via this publisher so that I can make corrections to possible future editions.

To longtime friends Paul Sutton, Tim Ihssen, and Greg Rutzen, each a huzzah.

And to best friends Jen Henderson and Dane Webster—love and thanks for everything, including subtitle brainstorming along and in the Blacksmith Fork River.

I'd like to thank my first editor at Tarcher, Wendy Hubbert, who bought the proposal for this project and who started my career in nonfiction publishing. Senior Editor Mitch Horowitz guided this book initially and quoted me quoting *Patton* by way of showing how excited he was. Many thanks. And many thanks to Assistant Editor Gaby Moss, who thoughtfully edited the manuscript and fielded e-mails as numerous as dust grains in space while we worked to meet the deadline. Gaby, you've been terrific. Copy editor Toni Rachiele did a wonderful job polishing this book. I appreciate the support and hard work of the design, production, and copyediting departments at Tarcher in making this book possible.

Natasha Kern has been more than a fine agent over the years; she's a trusted friend.

I want to express my love and appreciation to my family, especially my late mother Marge Cokinos, my father George Cokinos, my stepmother Judie (yes, writing a book is like eating an elephant one bite at a time), my sister Vicki Wright, my niece Jessica, Aunt Bernice Lison, and sister-in-law Tiffany Lison, who always calls to check on me while Kathe's off on her own research travels.

To the three cats—Shackleton, Zinc, and Burchfield—yes, I'm almost done editing and will play Fling the String very soon.

Lastly, I thank Kathe Lison, my love and partner, without whom my life would be thwarted. Pum.

A NOTE ABOUT THE narrative: Because I conducted research and site visits on multiple subjects in sometimes a short span of time, then later might return to a particular subject or place, it became clear that a strictly chronological approach to how I presented the timing of my research would result in a confusing back-and-forth from one subject to another. Though the overall movement of both research and life events is from 2001 to the present, at times I folded in research conducted at different points into a single chapter; with memoir-event time pegs, at times I needed to make them less precise in order to avoid narrative sprawl. I also felt this was justified by the overall development of the historical narratives—a general movement from the early twentieth century to the present, with one chapter as a flashback to prescientific then scientific understanding of meteors and meteorites. The time pegs and tense shifts in each chapter indicate that some research visits and interviews were integrated from earlier or later times. The intent was to clarify the subject matter and not confuse the reader while at the same time preserving the integrity of the narrative spine by not falsifying shifts in time and place. The bulk of the site visits took place between early 2001 and early 2004. Dates are listed in the Notes.

THANKS TO DREW N. BARRINGER for permission to use letters quoted in the Harvey Nininger chapters from his family.

Thanks to Doris Banks, Gary Huss, Peggy Schaller, and Sue Greiner for permission to use letters, manuscripts, and other material written by Harvey Nininger and held by their family and Arizona State University.

I thank *Ecotone* for publishing a version of my Acraman chapter.

Earlier versions of material related to my Antarctica experience appeared in *Meteorite* and in the anthology *Antarctica: Life on the Ice*.

The preceding excludes those sources that are in the public domain and/or within the limits of fair use.

Credits and Permissions

The author acknowledges permission to quote from:

The Epic of Gilgamesh, translated with an introduction by N. K. Sandars (Penguin Classics 1960, third edition 1972). Copyright © N. K. Sandars, 1960, 1964, 1972. Reproduced by permission of Penguin Books Ltd.

"Dominion Road," by Donald Bain McGlashan. © Copyright Native Tongue Music Publishing. All print rights for Australia and New Zealand administered by Sasha Music Publishing, a division of All Music Publishing & Distribution Pty Ltd ACN 079 628 434, ABN 79 079 628 434. PO BOX 1031 Richmond North, Victoria 3121, Australia. www.ampd.com.au. All rights reserved. Used by permission. Unauthorised reproduction is illegal.

"Canyonero" from the Twentieth Century Fox Television Series *The Simpsons.* Music by Alf Clausen, lyrics by Donick Carey. Copyright © 1998, 1999 T C F Music Publishing, Inc. All rights reserved. Used by permission.

"Catch a Falling Star." Words and music by Paul Vance and Lee Pockriss. Copyright © 1957 (Renewed) by Music Sales Corporation and Emily Music Corp. International copyright secured. All rights reserved. Reprinted by permission.

"Catch a Falling Star." Words and music by Paul Vance and Lee Pockriss. © 1957 (Renewed) by Emily Music Corporation and Music Sales Corporation. All rights reserved. Used by permission of Alfred Publishing Co., Inc.

"The Mystery Meteors." Originally appeared in Eleanor Lerman's collection of poetry *The Mystery of Meteors* (Sarabande Books, 2001). Reprinted by permission.

"East Coker" in *Four Quartets,* copyright 1940 by T. S. Eliot and renewed 1968 by Esme Valerie Eliot, reprinted by permission of Houghton Mifflin Harcourt Publishing Company.

· *Index* ·

N

O

P

About the Author

Christopher Cokinos is the author of the critically acclaimed *Hope Is the Thing with Feathers: A Personal Chronicle of Vanished Birds* (Tarcher/Penguin) and the winner of several prizes, including a Whiting Award, the Glasgow Prize for an Emerging Writer in Nonfiction, and the Sigurd Olson Nature Writing Award. He has won fellowships and grants from the National Science Foundation, the American Antiquarian Society, and the Utah Arts Council. His nonfiction, reviews, and poems have appeared in the *Los Angeles Times, The American Scholar, Sky & Telescope, Orion, Shenandoah, Birder's World, Science,* and *Poetry,* among other publications. Cokinos teaches creative writing at Utah State University, where he has appointments in English and Natural Resources. He lives with his partner, Kathe Lison, along the Blacksmith Fork River in northern Utah's Cache Valley.